DIFFERENTIAL URBANIZATION

INTEGRATING SPATIAL MODELS

Edited by
H. S. Geyer
and
T. M. Kontuly

A member of the Hodder Headline Group
LONDON • NEW YORK • SYDNEY • AUCKLAND

First published in Great Britain in 1996 by
Arnold, a member of the Hodder Headline Group,
338 Euston Road, London NW1 3BH

Copublished in the US, Central and South America by
John Wiley & Sons, Inc.,
605 Third Avenue,
New York, NY10158-0012

©1996 Selection and editorial matter H. S. Geyer and T. M. Kontuly

All rights reserved. No part of this publication may be reproduced
or transmitted in any form or by any means, electronically or
mechanically, including photocopying, recording or any information
storage or retrieval system, without either prior permission in writing
from the publisher or a licence permitting restricted copying. In the
United Kingdom such licences are issued by the Copyright Licensing Agency:
90 Tottenham Court Road, London W1P 9HE.

British Library Cataloguing in Publication Data
A catalogue entry for this book is available from the British Library

Library of Congress Cataloging-in-Publication Data
Differential urbanization : integrating spatial models in developed and less
   developed countries / H.S. Geyer, T.M. Kontuly.
      p.   cm.
   Includes bibliographical references and index.
   ISBN 0-470-23635-3. — ISBN 0-470-23634-5 (pbk.)
   1. Migration, Internal.   2. Population geography.   I. Geyer, H.S.,
1951–   .   II. Kontuly, Thomas.
HB1952.D54   1996
304.8′09172′4—dc20   96-26231
                                                                     CIP

ISBN 0 340 66285 9 (Pb)
ISBN 0 470 23634 5 (Wiley)

ISBN 0 340 66286 7 (Hb)
ISBN 0 470 23635 3 (Wiley)

Composition by Phoenix Photosetting, Chatham, Kent
Printed and bound in Great Britain by J. W. Arrowsmith Ltd, Bristol

CONTENTS

Preface v
Acknowledgements vii

Introduction 1

PART ONE: MAJOR TRENDS IN MIGRATION IN THE DEVELOPED WORLD SINCE THE 1970s

Section One: Trends in the United States

1. B. J. L. Berry, 'The Counterurbanisation Process: Urban America since 1970' (1976) — 7
2. C. L. Beale, 'The Recent Shift of United States Population to Nonmetropolitan Areas, 1970–75' (1977) — 19
3. D. R. Vining Jr and A. Strauss, 'A Demonstration that the Current Deconcentration of Population in the United States Is a Clean Break with the Past' (1977) — 28
4. P. Gordon, 'Deconcentration without a "Clean Break"' (1979) — 37
5. K. Richter, 'Nonmetropolitan Growth in the Late 1970s: The End of the Turnaround?' (1985) — 47

Section Two: International Trends

6. D. R. Vining Jr and T. Kontuly, 'Population Dispersal from Major Metropolitan Regions: An International Comparison' (1978) — 67
7. S. G. Cochrane and D. R. Vining Jr, 'Recent Trends in Migration between Core and Peripheral Regions in Developed and Advanced Developing Countries' (1988) — 90
8. R. Koch, '"Counterurbanisation", also in Western Europe?' (1980) — 111
9. A. J. Fielding, 'Migration and Urbanisation in Western Europe since 1950' (1989) — 121
10. A. G. Champion, 'The Reversal of the Migration Turnaround: Resumption of Traditional Trends?' (1988) — 132

PART TWO: MAJOR TRENDS IN MIGRATION IN THE LESS DEVELOPED WORLD SINCE THE 1970s

Section One: International Trends

11 H. W. Richardson, 'Polarization Reversal in Developing Countries' (1980)   143

12 D. R. Vining Jr, 'Population Redistribution Towards Core Areas of Less Developed Countries, 1950–80' (1986)   161

Section Two: Polarization Reversal Trends

13 P. M. Townroe and D. Keen, 'Polarisation Reversal in the State of São Paulo, Brazil' (1984)   188

14 H. Lee, 'Growth Determinants in the Core–Periphery of Korea' (1989)   202

15 L. A. Brown and V. A. Lawson, 'Polarisation Reversal, Migration-related shifts in Human Resource Profiles, and Spatial Growth Policies: A Venezuelan Study' (1989)   216

16 H. S. Geyer, 'Implications of Differential Urbanisation on Deconcentration in the Pretoria–Witwatersrand–Vaal Triangle Metropolitan Area, South Africa' (1990)   238

PART THREE: LONG-TERM MIGRATION TRENDS IN DEVELOPED AND LESS DEVELOPED COUNTRIES

Section One: Migration Cycles

17 B. J. L. Berry, 'Migration Reversals in Perspective: The Long-Wave Evidence' (1988)   259

18 L. A. Brown and F. C. Stetzer, 'Development Aspects of Migration in Third World Settings: A Simulation, with Implications for Urbanisation' (1984)   264

19 H. S. Geyer and T. Kontuly, 'A Theoretical Foundation for the Concept of Differential Urbanization' (1993)   290

20 H. S. Geyer, 'Expanding the Theoretical Foundation of the Concept of Differential Urbanisation' (1996)   309

Conclusion   329

Subject Index   341

Index of Place Names   344

PREFACE

After a long history of population concentration in the developed and less developed world, no single issue aroused more interest among geographers, regional scientists, and planners than the sudden tendency towards population deconcentration experienced in North America and Western Europe during the early 1970s. But, barely a decade after counterurbanization was generally accepted in a large part of the developed world as a new long-term trend representing 'a clean break with past trends', many countries experiencing this 'rural renaissance' started to experience a reversal or a slowing down of these trends in the late 1970s and early 1980s. Albeit less obvious, equally interesting tendencies of 'polarization reversal' were observed in certain more advanced developing countries over the past two decades.

Considerable bodies of literature have developed on both phenomena, but the material has largely been compartmentalized, and very little effort has thus far been made to link the concepts of urbanization, polarization reversal, and counterurbanization across the development spectrum. Neither have all the matters of dispute in these related fields of research sufficiently been laid to rest. This is the *raison d'être* of this, the first book of what seems to be developing into a trilogy. Whereas an international testing of the concept of differential urbanization is attempted in the second book, and the planning implications of differential urbanization in developed and less developed countries are addressed in the third, this one attempts to eliminate the present threatening dichotomy in First/Third World migration theory by covering the much neglected 'middle ground' between observations in developed and less developed countries.

Teachers of demography, geography, regional science and planning who have attempted to transcend the gulf between migration literature of the First and Third World have long been aware of the need for an integration of the two bodies of research. The main objectives of the selection of works in this book are, first, to affect an integration of First World/Third World research in this field; second, to highlight overlapping theoretical premises which are often overlooked by researchers in developed and less developed environments; and third, to stimulate further research in this field. Consequently, the material included in the book should be of interest to urbanologists in the developed and less developed world, at both the entering and more advanced levels of research. It should be stressed, however, that the selection of works for this book does not in any way imply that other equally deserving works did not appear during the same period. Rather, articles were selected on the basis of their potential to contribute to the integration of the concepts of urbanization, polarization reversal, counterurbanization, and differential urbanization.

The papers presented in Part One combine the views of some leading scholars from various disciplines in a manner which portrays the evolution of thoughts in the field of population concentration versus deconcentration in the First World over the past 25 years. It portrays changing views on counterurbanization over the period by putting in perspective questions regarding the beginning and end of the turnaround. Emphasis is placed on the relevance of a cyclical approach to the explanation of the chronology in urbanization and counterurbanization tendencies.

Part Two brings together a corpus of papers debating the dominance of population concentration and deconcentration in the Third World. Part Three seeks to improve the present conceptual framework of the discipline and to stimulate further research, by highlighting overlapping views on population concentration and deconcentration in the First and Third Worlds. It enables the reader to determine the relevancy of related theoretical concepts of migration in both worlds, some of which have the potential to clarify controversial issues yet to be resolved.

It should be noted that certain changes in the original articles have been made to the papers that were included in this volume. In certain instances elaborate appendices referring to details in data in the original articles were omitted if they were not regarded as essential in the understanding of the contents of the article.

So many people have contributed to the compilation of this reader in one way or another that it is difficult to know where to begin. Many researchers in the fields of demography, regional planning, and regional science have contributed toward the uncovering of new aspects of the factors that are shaping our urban systems, but certain works have stood out as landmarks in the understanding of the fundamentals of the concept of differential urbanization. The articles that were included in this reader fall into this category. Borrowing heavily from the works of these authors, the concept of differential urbanization was born in South Africa at the University of Potchefstroom in 1988 and further developed in the United States at the University of Utah during the early 1990s.

We would like to thank all the authors of the papers that were included in this reader, as well as many other authors who published equally deserving research results in this field, for their contribution to the conceptualizing of the model of differential urbanization. This book serves as a monument for their dedication and hard work. We would also like to thank Cindy Pretorius, Laetitia Oosthuizen, Henk Swanevelder, Laurette Grobler, Ilze Oliver, Anton Erasmus and Ina van Rensburg, who assisted in the scanning and editing of the text and in the production of the figures. Finally, we would like to thank Laura McKelvie and Julie Delf of Arnold for their patience. It is Laura

McKelvie's vision of the potential academic value of the concept of differential urbanization which made the publication of this reader possible, and for that we are enormously grateful.

Manie Geyer, Potchefstroom, South Africa
Tom Kontuly, Salt Lake City, Utah

Acknowledgements

Pergamon Press plc for H.S. Geyer, 'Implications of differential urbanisation on deconcentration in the Pretoria-Witwatersrand-Vaal Triangle metropolitan area, South Africa', *Geoforum* 21 (1990), pp. 385–96; Pion Ltd. for D.R. Vining Jr and A. Strauss, 'A demonstration that the current deconcentration of population in the United States is a clean break with the past', *Environment and Planning A* 9 (1977), pp. 751–58, P. Gordon, 'Deconcentration without a "clean break"', *Environment and Planning A* 11 (1979), pp. 281–90, and L.A. Brown and F.C. Stetzer, 'Development aspects of migration in Third World settings: a simulation, with implications for urbanisation', *Environment and Planning A* 16 (1984), pp. 1583–603; Regional Studies Association for P.M. Townroe and D. Keen, 'Polarisation reversal in the state of São Paolo, Brazil', *Regional Studies* 18 (1984), pp. 45–54; The Regional Science Association International for H.W. Richardson, 'Polarization reversal in developing countries', *Papers of the Regional Science Association* 45 (1980), pp. 67–85; Population Association of America for K. Richter, 'Nonmetropolitain growth in the late 1970s: the end of the turnaround?', *Demography* 22 (1985), pp. 245–63; Royal Dutch Geographical Society for H.S. Geyer, 'Expanding the theoretical foundation for the concept of differential urbanisation', *Tijdschrift voor Economische en Sociale Geografie* 87, 1 (1996), pp. 44–59; Royal Geographical Society-Institute of British Geographers for A.J. Fielding, 'Migration and urbanisation in Western Europe since 1950', *The Geographical Journal* 155 (1989), pp. 60–69; Sage Publications, Inc. for B.J.L. Berry, 'The counterurbanization process: urban America since 1970', *Urban Affairs Annual Review* 11 (1976), pp. 17–30; West Virginia University for Calvin Beale, 'The recent shift of United States population to nonmetropolitan areas, 1970–75', *International Regional Science Review* 2 (1977), pp. 113–22, D.R. Vining Jr and T.M. Kontuly, 'Population dispersal from major metropolitan regions: an international comparison', *International Regional Science Review* 3 (1978), pp. 49–73, S.G. Cochrane and D.R. Vining Jr, 'Recent trends in migration between core and peripheral regions in developed and advanced developing countries', *International Regional Science Review* 11 (1988), pp. 215–43, D.R. Vining Jr, 'Population redistribution towards core areas of less developed countries 1950–80', *International Regional Science Review* 10 (1986), pp. 1–45, H. Lee, 'Growth determinants in the core-periphery of Korea', *International Regional Science Review* 12 (1989), pp. 147–63, L.A. Brown and V.A. Lawson, 'Polarisation reversal, migration-related shifts in human resource profiles, and spatial growth policies: a Venezuelan study', *International Regional Science Review* 12 (1989), pp. 165–88, B.J.L. Berry, 'Migration reversals in perspective: the long-wave evidence', *International Regional Science Review* 11 (1988), pp. 245–51, H.S. Geyer abd T.M. Kontuly, 'A theoretical foundation for the concept of differential urbanization', *International Regional Science Review* 17 (1993), pp. 157–77, and A.G. Champion, 'The reversal of the migration turnaround: resumption of traditional trends?', *International Regional Science Review* 11 (1988), pp. 253–60.

INTRODUCTION

Various periods are known in the history of human civilization in which changes in the distribution of population occurred in particular regions at a sufficient scale to be of global significance. Certain changes were due to natural disasters of some kind, such as the Black Plague of the fourteenth century when between a third and a half of the population of Western Europe was wiped out. Other changes were economically or politically driven, notably the organized waves of colonization of parts of Europe by the Greeks during the seventh to third centuries BC and the Romans from the second century BC to the second century AD, and the occupation of large parts of the globe by Western European communities between the fifteenth and the nineteenth centuries. Also, spontaneous migration patterns seem to have occurred in waves throughout history. One of the most outstanding examples in modern times is the unprecedented scale of urbanization which occurred during the 'Industrial Revolution'.

Over the past fifty years the First World has experienced two significant spontaneous changes in the redistribution of population which are at least equal in significance to the changes in migration that were recorded in Western Europe during the Industrial Revolution. The first is the rate of urbanization and especially of primate city growth in Western Europe and North America after the Second World War. In an effect similar to that of mechanization on industrial development and population movement during the Industrial Revolution, markets expanded sufficiently during the second half of the twentieth century for multinational companies to become a common urban economic feature globally. During this period the advantage of economies of scale coupled with the ever-increasing size of markets – factors which in large part were responsible for the waves of industrialization and of urbanization during the Industrial Revolution – was intensified to such an extent that markets soon expanded to encompass the entire world. This was the period when certain cities grew at such an unprecedented rate that they were regarded as 'world cities'. Certain schools of thought believed that many of these world cities would continue to grow until they linked up with one another to form 'ecumenopolises', i.e. urban agglomerations that eventually would span continents.

2 Differential Urbanization

Although the growth rate of certain large megalopolitan areas started slowing down towards the end of the 1960s, few scholars in the fields of economic, urban, and demographic development in the First World would have believed that a reversal of the urbanization trend of the past decades was imminent and that they would soon experience one of the most dramatic and unexpected spontaneous changes in the direction of mainstream migration in the recorded history of population redistribution. In fact, so sudden was the change that it was still generally believed at the end of the 1960s that the urbanization process would continue indefinitely.

If the beginning of the 'turnaround' was met with surprise, its unexpected slowdown in certain countries by the turn of the 1980s and its apparent end in others were met with an even larger measure of surprise. Whether the slowdown or end of counterurbanization will be lasting is still a debatable issue. Certain scholars take a short-term view and regard the turnaround period as merely a phase during which the urban structure readjusted itself to a restructuring of the market worldwide, or the reaction of people to normal economic cyclical trends. They regard the phenomenon as a short-term issue. Others take a more fundamental position. Their argument is based on the view that there is a limit to the degree of urbanization in any society. It is argued that the degree of urbanization in certain developed countries has reached (or is approaching) its upper limit and that it is unlikely that the share of the urbanized population will increase much further. Whether the process of counterurbanization is accompanied by periodic interruptions of concentration or whether it proceeds continuously is a matter that still needs to be determined. This is one of the issues which is put forward for debate in this book. Closely linked to this issue is the question whether the turnaround was a continuation of the urbanization wave or whether it was a clean break with past trends.

Two other noteworthy matters are also raised in this book. One is the explanations given for the occurrence of counterurbanization in the First World. The other, which is linked to the question whether counterurbanization as a phenomenon will continue in the future or whether it could be expected to re-emerge in the future in those countries where it has subsequently slowed down or come to an end, deals with the criteria that have been used to determine the occurrence of counterurbanization. These criteria include metropolitan versus non-metropolitan growth differentials, population deconcentration at different regional scales, core–peripheral net migration differentials, and the relationship between net migration rates and settlement density.

Whether counterurbanization could safely be regarded as a complete turnaround in migration trends in the First World, as was generally accepted in the 1980s, or whether it was merely a continuation of the wave-theory largely depends on the spatial criteria which were used to measure migration trends. Indications are that the position of the boundary between what was regarded as the core and the periphery in each case study could play a decisive role in the determination whether the turnaround really represented a break with past trends or whether it was nothing more than an advanced stage of the migration wave-theory. In this book the merits of the deliberate overbounding of core regions by certain scholars to prove the validity of the counterurbanization concept in the 1970s as a unique phenomenon are discussed. If the concept of counterurbanization is to be assessed in its proper context, answers to at least these major issues need to be found.

Within the less developed economic environment two opposing views in demography developed over the past two decades. One maintains that there has been an uninterrupted flow of people from rural to urban areas throughout the Third World. Although the urbanization process has never been constant in the Third World it is likely, according to this view, that urbanization will remain the dominant migration process in these areas for the foreseeable future.

Other scholars maintain that a turning point has been observed in the concentration tendencies in certain advanced developing countries over the period. However, unlike in the First World early indications of a turnaround in the Third World have mainly been observed in economic terms rather than in migration trends.

The body of research that was sparked off by these observations of the 1970s in the First and Third World brought forth a set of fresh and seemingly contradicting issues which previously either did not exist, or slipped by without attracting much interest. Consequently a considerable and growing dichotomy has developed in the literature dealing with population migration in developed and less developed countries. Despite the extensive body of literature on the subjects, not enough effort has thus far been made to explicitly integrate the work on migration trends in the developed and less developed worlds. Some of the issues were resolved, others are still controversial and deserve further attention.

One of the main aims of this volume is to highlight similarities and differences in the fundamentals of demographic trends in the First and Third Worlds to stimulate further research in this field. As an introduction to issues yet to be resolved in developed and less developed countries and to give the reader an overview of the issues at stake at present, the book highlights issues revolving around concentration and deconcentration in a number of selected First and Third World countries. Some may argue that the question whether the onset of counterurbanization in certain developed countries implied a clean break with past trends, or whether it was merely an advanced phase of urbanization, was sufficiently resolved during the late 1970s. However, when one weighs the central line of reasoning in the works of proponents of the 'clean break' proposition against that of sceptics, four interrelated issues remain of interest. These issues are: reasons for the first and second turnarounds; urban/non-urban versus regionally oriented migration flows; mainstream versus substream migration differentials; and short-term perspectives versus long-term explanations of the migration processes. These are the main controversies addressed in the book, none of which has sufficiently been laid to rest thus far.

The book attempts to eliminate the present dichotomy in First/Third World migration theory by covering the much-neglected 'middle ground' between observations in developed and less developed countries. It highlights those issues in migration in the Third World which could potentially assist in the explanation of specific trends in migration in a First World setting, and vice versa. Because the concept of polarization reversal was introduced in, and is almost exclusively associated with spatial development in Third World settings, its demonstrated relevance in First World migration revives certain analytical and explanatory options in migration theory which have remained latent until now. On the basis of an analysis of long-term migration trends in several developed and less developed countries, the book gives a prognosis of expected future migration trends. Emphasis is placed on the relevance of a cyclical approach to the explanation of chronology in migration patterns. The book specifically deals with the concepts of counterurbanization, polarization reversal, and differential urbanization, and demonstrates the potential usefulness of the latter in the explanation of migration cycles in the First World.

The concept of differential urbanization, which links the notions of urbanization, polarization reversal, and counterurbanization across the development spectrum, is put in perspective in the book. The theory underlying the differential urbanization model covers the presently neglected 'middle ground' between observations on inter-urban net migration in developed and less developed countries.

PART ONE

Major Trends in Migration in the Developing World Since the 1970s

SECTION ONE
TRENDS IN THE UNITED STATES

1 B. J. L. Berry,
'The Counterurbanisation Process: Urban America since 1970'

From: *Urban Affairs Annual Review* 11, 17–30 (1976)

A turning point has been reached in the American urban experience. Counterurbanisation has replaced urbanisation as the dominant force shaping the nation's settlement patterns. A similar tendency has been noted in other Western nations (Alexandersson and Falk, 1974). This paper lays out the facts of the change, and speculates about the nature of the process.

To those who wrote about nineteenth- and early twentieth-century industrial urbanisation, the essence was size, density, and heterogeneity. 'Urbanisation is a process of population concentration,' wrote Hope Tisdale in 1942. 'It implies a movement from a state of less concentration to a state of more concentration.' But since 1970 American metropolitan regions have grown less rapidly than the nation and have actually lost population to nonmetropolitan territory – 1.8 million persons between March 1970 and March 1974 according to the estimates of the US Bureau of the Census. Because migration has been selective of particular social and economic groups, very specific subgroups have been left behind.

The process of counterurbanisation therefore has as its essence decreasing size, decreasing density, and decreasing heterogeneity. To mimic Tisdale: *counterurbanisation is a process of population deconcentration; it implies a movement from a state of more concentration to a state of less concentration.*

Recent population changes: the facts

Many of the facts of the recent changes have been spelled out in an excellent report by Forstall (1975), from whose materials Table 1.1 and the following summary have been assemble.

Table 1.1 Population, change, and components of change for various groups of metropolitan and nonmetropolitan counties: 1960–70 and 1970–73

Residence category	Population			Population change				Natural increase				Net migration			
	1 July, 1973 (provisional)	1 April, 1970 (census)[d]	1 April, 1960 (census)	1970–73 No.	%	1960–70 No.	%	1970–73 No.	%	1960–70 No.	%	1970–73 No.	%	1960–70 No.	%
Total United States	209 851	203 300	179 323	6 551	3.2	23 977	13.4	4 917	2.4	20 841	11.6	1 634	0.8	3 135	1.7
Inside SMSAs[a]	153 350	149 093	127 348	4 258	2.9	21 744	17.1	3 768	2.5	15 637	12.3	489	0.3	6 107	4.8
Outside SMSAs	56 501	54 207	51 975	2 293	4.2	2 232	4.3	1 149	2.1	5 204	10.0	1 144	2.1	−2 972	−5.7
Metropolitan areas over 3 000 000[b]															
New York area	56 189	55 635	47 763	554	1.0	7 872	16.5	1 218	2.2	5 464	11.4	−664	−1.2	2 408	5.0
Los Angeles area	16 657	16 701	15 126	−45	−0.3	1 576	10.4					−305	−1.8	218	1.4
Chicago area	10 131	9 983	7 752	147	1.5	2 231	28.8					−119	−1.2	1 164	15.0
Philadelphia area	7 689	7 611	6 794	78	1.0	817	12.0					−124	−1.6	−17	−0.2
Detroit area	5 653	5 628	5 024	25	0.4	604	12.0					−75	−1.3	91	1.8
San Francisco area	4 691	4 669	4 122	22	0.5	547	13.3					−114	−2.4	9	0.2
Boston area	4 544	4 423	3 492	121	2.7	932	26.7					23	0.5	485	13.9
Washington SMSA	3 783	3 710	3 358	73	2.0	352	10.5					15	0.4	32	0.9
	3 042	2 910	2 097	132	4.5	813	38.8					34	1.2	426	20.3
Metropolitan areas of 1–3 000 000 by region															
Northeast	35 705	34 448	28 497	1 257	3.6	5 951	20.9	861	2.5	3 510	12.3	396	1.1	2 441	8.6
Pittsburgh SMSA	3 720	3 751	3 712	−30	−0.8	38	1.0	42	1.1	289	7.8	−73	−1.9	−251	−6.8
Buffalo SMSA	2 367	2 401	2 405	−35	−1.4	−4	−0.2					−56	−2.3	−167	−6.9
North Central	1 353	1 349	1 307	4	0.3	42	3.2					−16	−1.2	−84	−6.4
St Louis SMSA	12 427	12 381	10 868	46	0.4	1 513	13.9	318	2.6	1 369	12.6	−272	−2.2	144	1.3
Cleveland SMSA	2 388	2 411	2 144	−23	−0.9	266	12.4					−78	−3.2	21	1.0
Minneapolis–St. Paul SMSA	1 997	2 064	1 909	−67	−3.2	154	8.1					−109	−5.3	−45	−2.4
Milwaukee SMSA	1 994	1 965	1 598	28	1.4	368	23.0					−30	−1.5	119	7.4
Cincinnati SMSA (part)[c]	1 432	1 404	1 279	28	2.0	125	9.8					−4	−0.3	−38	−3.0
Kansas City SMSA	1 126	1 134	1 039	−8	−0.7	95	9.2					−36	−3.2	−31	−3.0
Indianapolis SMSA	1 295	1 274	1 109	21	1.6	165	14.9					−14	−1.1	30	2.7
Columbus SMSA	1 139	1 111	944	28	2.5	167	17.7					−8	−0.7	37	3.9
Florida	1 057	1 018	845	39	3.8	173	20.4					6	0.6	52	6.1
Miami Area[b]	3 376	2 976	2 078	400	13.4	898	43.2	21	0.7	135	6.5	379	12.7	764	36.7
Tampa–St Petersburg SMSA	2 106	1 888	1 269	218	11.5	619	48.8					196	10.4	511	40.2
Other South Atlantic	1 271	1 089	809	182	16.7	279	34.5	102	2.8	411	13.8	182	16.7	253	31.3
Baltimore SMSA	3 845	3 667	2 973	178	4.9	694	23.3					76	2.1	282	9.5
Atlanta SMSA	2 117	2 117	2 071	45	2.2	267	14.8					5	0.3	53	2.9
	1 728	1 596	1 169	133	8.3	426	36.5					71	4.4	230	19.7

Area															
South Central[e]	5 930	5 675	4 305	255	4.5	1 370	31.8	213	3.8	693	16.1	42	0.7	677	15.7
Dallas–Fort Worth SMSA	2 442	2 378	1 738	63	2.7	640	36.8					−24	−1.0	362	20.8
Houston SMSA	2 138	1 999	1 430	139	7.0	569	39.8					52	2.6	311	21.8
Cincinnati SMSA (part)[c]	257	251	229	6	2.4	22	9.4	163	2.7	613	13.4	(z)	0.1	−5	−2.1
New Orleans SMSA	1 093	1 046	907	47	4.4	139	15.4					14	1.4	8	0.9
West	6 406	5 998	4 561	408	6.8	1 438	31.5					244	4.1	825	18.1
San Diego SMSA	1 470	1 358	1 033	112	8.2	325	31.4					72	5.3	169	16.4
Seattle SMSA	1 385	1 425	1 107	−40	−2.8	317	28.7					−68	−4.8	187	16.9
Denver SMSA	1 366	1 239	935	127	10.2	305	32.6					86	7.0	163	17.4
Phoenix SMSA	1 119	969	664	150	15.5	306	46.1					114	11.7	190	28.6
Portland SMSA	1 066	1 007	822	59	5.9	185	22.5					40	4.0	117	14.2
Other SMSA territory by region	61 456	59 009	51 088	2 447	4.1	7 921	15.5	1 690	2.9	6 662	13.0	757	1.3	1 259	2.5
Northeast	13 517	13 225	11 828	292	2.2	1 397	11.8					60	0.5	312	2.6
North Central	14 751	14 447	12 820	313	2.2	1 627	12.7					−88	−0.6	6	(z)
Florida	3 072	2 735	2 015	338	12.3	720	35.7					271	9.9	443	22.0
Other South Atlantic	7 556	7 317	6 285	238	3.3	1 032	16.4					6	0.1	126	2.0
South Central[e]	14 458	13 753	12 178	705	5.1	1 575	12.9					218	1.6	−248	−2.0
West	8 093	7 532	5 962	561	7.4	1 570	26.3					290	3.8	620	10.4
Counties with 20% or more commuters to SMSAs	4 099	3 848	3 474	251	6.5	373	10.7	74	1.9	315	9.1	177	4.6	58	1.7
Northeast (11 counties)	1 047	970	794	77	8.0	176	22.2					63	6.4	111	14.0
North Central (51 counties)	1 212	1 154	1 077	58	5.0	78	7.2					37	3.2	−5	−0.5
Florida (6 counties)	78	67	48	11	16.7	19	39.6					11	16.7	17	34.9
Other South Atlantic (41 counties)	730	697	659	33	4.7	38	5.8					17	2.4	−37	−5.6
South Central[e] (57 counties)	963	898	848	65	7.2	50	5.9					45	5.0	−36	−4.2
West (5 counties)	58	61	49	6	10.6	12	25.4					5	7.8	8	16.5
Counties with 10–19% commuters to SMSAs	9 633	9 269	8 636	414	4.5	633	7.3	182	2.0	792	9.2	232	2.5	−159	−1.8
North Central (27 counties)	1 933	1 843	1 703	90	4.9	140	8.2					64	3.5	23	1.3
Florida (8 counties)	3 327	3 228	3 019	99	3.1	209	6.9					38	1.2	−51	−1.7
Other South Atlantic (61 counties)	293	246	193	37	15.2	53	27.3					35	14.3	37	19.2
	1 559	1 511	1 468	58	3.8	43	2.9					20	1.3	−124	−8.4

Table 1.1 continued

	Population			Population change				Natural increase				Net migration			
	1 July, 1973	1 April, 1970	1 April, 1960	1970–73		1960–70		1970–73		1960–70		1970–73		1960–70	
Residence category	(provisional)	(census)[d]	(census)	No.	%	No.	%	No.	%	No.	%	No.	%	No.	%
South Central[e] (96 counties)	2 174	2 083	1 952	91	4.4	131	6.7					46	2.2	−67	−3.4
West (16 counties)	396	357	300	39	10.8	57	19.0					30	8.3	22	7.4
Peripheral counties by region															
Northeast	42 719	41 091	39 865	1 628	4.0	1 226	3.1	893	2.2	4 097	10.3	735	1.8	−2 871	−7.2
North Central	3 823	3 673	3 490	150	4.1	183	5.2					84	2.3	−119	−3.4
	13 493	13 101	12 919	392	3.0	182	1.4					201	1.5	−823	−6.4
Florida	868	767	617	100	13.1	150	24.4					80	10.5	67	10.9
Other South Atlantic	7 585	7 347	7 183	239	3.2	164	2.3					44	0.6	−694	−9.7
South Central[e]	10 021	9 723	9 718	298	3.1	5	0.1					76	0.8	−1 061	−10.9
West	6 929	6 481	5 938	449	6.9	542	9.1					250	3.9	−243	−4.1

Sources: Richard L. Forstall, *Trends in metropolitan and nonmetropolitan population growth since 1970* (Washington, DC: Population Division, US Bureau of the Census, Rev. 20 May 1975), summarising the following: 1960 population and 1960–70 natural increase from US Bureau of the Census, *Current Population Reports*, Series P-25, no. 461, 'Components of population change by county: 1960 to 1970'; and 1970 *Census of Population and Housing*, PHC(2)-1, 'General demographic trends for metropolitan areas, 1960 to 1970'. 1970 and 1973 populations and 1970–73 natural increase and net migration from US Bureau of the Census, *Current Population Reports*, Series P-25, nos. 527, 530–532, and 535, 'Estimates of the population of [state] counties and metropolitan areas: July 1, 1972 and 1973'; for New York, Maryland, Alaska, California, and Texas, respectively; *Current Population Reports*, Series P-26, Nos. 49–93, 'Estimates of the population of [state] counties and metropolitan areas: July 1, 1972 and 1973', for the other 45 states. 1960–70 population change computed from 1960 and 1970 populations; 1960–70 net migration computed by subtracting 1960–70 natural increase from 1960–70 population change; these data may differ from those in Series P-25, No. 461 and Series PHC(2), no. 1 as a result of reflecting corrections in 1970 local and national totals

Note: Numbers are in thousands. Minus sign (−) denotes decrease

(z) Less than 500 or 0.05%

[a] SMSAs (Standard Metropolitan Statistical Areas) as defined by OMB as of 31 December 1974, except in New England, where definitions in terms of entire counties have been substituted
[b] Reflects certain combinations of SMSAs, as specified below. Population size groups are as of 1973
[c] Boone, Campbell, and Kenton Counties, Ky., are in the South Central divisions; the remainder of the Cincinnati SMSA is in the North Central Region
[d] Includes corrections in local and national totals determined after 1970 census complete-count tabulations were made
[e] Comprises East South Central and West South Central divisions

SMSAs combinations are as follows: *New York area* comprises New York SMSA, Jersey City SMSA, Long Branch–Asbury Park SMSA, Nassau–Suffolk SMSA, New Brunswick–Perth Amboy–Sayreville SMSA, Newark SMSA, and Paterson–Clifton–Passaic SMSA; *Philadelphia area* comprises Philadelphia SMSA, Trenton SMSA, and Wilmington SMSA; *Boston area* comprises Essex, Middlesex, Norfolk, Plymouth, and Suffolk Counties, Mass.; *Chicago area* comprises Chicago and Gary–Hammond–East Chicago SMSAs; *Detroit area* comprises Detroit and Ann Arbor SMSAs; *Miami area* comprises Miami and Fort Lauderdale–Hollywood SMSAs; *Los Angeles area* comprises Los Angeles–Long Beach, Anaheim–Santa Ana–Garden Grove, Oxnard–Simi Valley–Ventura, and Riverside–San Bernardino–Ontario SMSAs; *San Francisco area* comprises San Francisco–Oakland, San Jose, and Vallejo–Napa SMSAs

1. Since 1970, US metropolitan areas have grown more slowly than the nation as a whole, and substantially less rapidly than nonmetropolitan America, a development that stands in contrast to all preceding decades back to the early nineteenth century.
2. On a net basis, metropolitan areas are now losing migrants to nonmetropolitan territory, although they still show a slight total increase in immigration because of recent immigrants from abroad.
3. The decline in metropolitan growth is largely accounted for by the largest metropolitan areas, particularly those located in the Northeast and North Central regions. The eight metropolitan areas exceeding 3 million population have lost two-thirds of a million net migrants since 1970, and their central counties have declined in population absolutely by more than a quarter of a million. Altogether the central cities of the nation's SMSAs [Standard Metropolitan Statistical Areas] grew at an average annual rate of 0.6 per cent between 1960 and 1970, but declined at an average annual rate of −0.4 per cent after 1970 (annexations excluded). Much of the decline is attributable to a post-1970 decline of central city white populations at a rate of 1 per cent per annum (1960–70 the white population remained stable in the aggregate). Meanwhile Black and other minority populations have continued to decline in nonmetropolitan America since 1970, and the farm population has stabilised at approximately 9.5 million persons.
4. Rapid growth has taken place in smaller metropolitan areas, particularly in Florida, the South, and the West; in exurban counties located outside SMSAs as currently defined, but with substantial daily commuting to metropolitan areas; and in peripheral counties not tied into metropolitan labour markets.
5. Particularly impressive are the reversals in migration trends in the largest metropolitan areas and in the furthermost peripheral counties: the metropolitan regions with populations exceeding 3 million gained migrants between 1960 and 1970 but have lost since 1970; the nation's peripheral nonmetropolitan counties lost migrants between 1960 and 1970 but have gained since 1970. The balance of migration flows has been reversed.

Some sense of the accompanying restructuring of older metropolitan regions can be obtained by examining Figures 1.1 and 1.2, drawn from a nation-wide study by the author (Berry and Gillard, 1976). The data relate to the 1960–70 decade. Figure 1.1 shows the commuting field of the city of Cleveland in 1970; the longest one-way daily journey to work in the central city exceeded 70 miles in that year, revealing that many workers in Cleveland's factories, offices, and shops selected places of residence not simply outside the central city or Cuyahoga County, but beyond the Cleveland SMSA as currently defined. Figure 1.2 shows the decadal changes in Cleveland's daily urban system, as indexed by changes in the percentages of workers commuting to the central city. Crosshatched zones are those areas into which Cleveland's workers extended their places of residence. The thrust was outwards, in association with newly constructed expressways. Dashed areas to the east and south of the city reveal zones in which there were dramatic decreases in the volume of

12 Differential Urbanization

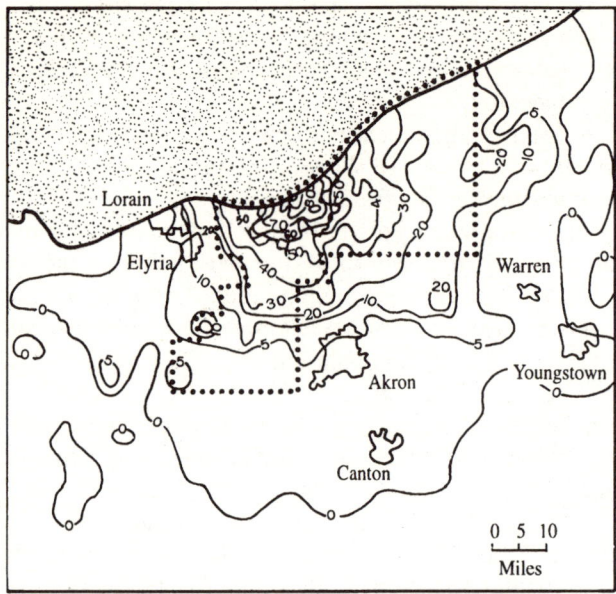

Fig. 1.1 Commuting field of Cleveland, Ohio, in 1970. The contours show the pecentage of workers commuting to the central city of Cleveland in 1970, based on census tract data. The outer limit is some 70 miles from the city centre, and the daily urban system is revealed to extend far beyond the limits of the Cleveland SMSA, shown by the dotted lines

daily commuting to the city. These were the zones of active suburban and exurban development, where new residential and employment complexes enabled people to seek out new life-styles and to cut their ties to the older central city.

Similar illustrations for Akron reveal that city's commuting field to be substantially the same as Cleveland's, changing during the decade in the same ways. Some of Akron's workers moved much further afield, stretching their choice of place of residence outwards along new interstate highways. But the dependence of many areas on Akron's jobs declined drastically, particularly in that interurban belt between Akron and Cleveland that had also freed itself of dependency on Cleveland jobs during the decade. Thus, what had previously been the intermetropolitan periphery was now displaying newly found independence as one of the region's new growth centres – not in the form of a traditional concentrated industrial-urban node, but rather in a low-slung and far-flung form, like the new metropolitan regions of the South and West (Holleb, 1975). Northeastern Ohio had always been a multicentred urban region. During the 1960s it became more thoroughly dispersed as the older central cities declined, decentralisation proceeded apace, metropolitan regions were restructured internally, and more amenity-rich outlying areas were brought into daily interaction with other parts of metropolitan America by expressway-related accessibility changes (Lamb, 1975; Berry and Gillard, 1976).

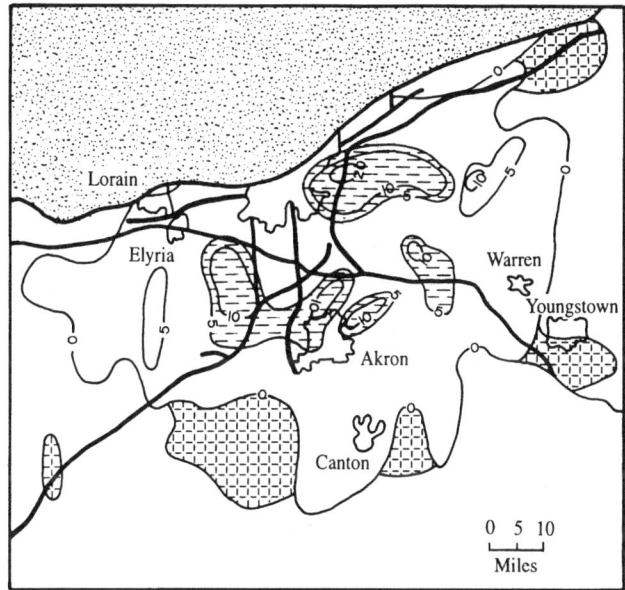

Fig. 1.2 Changes in the Cleveland commuting field, 1960–70. The contours show the change in percentage of workers commuting to the central city of Cleveland between 1960 and 1970. Dashed shades pick out the zones in which the commuting percentage dropped by 5 points or more – east and south of Cleveland within the SMSA, and outside the SMSA north of Akron. These were the zones of rapid suburban residential and industrial growth in the decade. Cross-hatched areas pick out the zones into which the commuting field extended during the decade, generally associated with expressway extensions (solid black lines)

Temporary perturbation, long-term trend, or cultural predisposition?

To some, the census changes summarised in Table 1.1 are but a temporary perturbation, an anomaly caused by the recession that will vanish when the health of the economy improves. But such an attitude is hardly credible; twentieth-century trends have all pointed in the same direction – creation of nothing less than 'an urban civilisation without cities', at least in the classical sense (Kristol, 1972). As early as 1902, H. G. Wells wrote that the 'railway-begotten giant cities' he knew were

> in all probability ... destined to such a process of dissection and diffusion as to amount almost to obliteration ... within a measurable further space of years. These coming cities ... will present a new and entirely different phase of human distribution. ... The city will diffuse itself until it has taken up considerable areas and many of the characteristics of what is now country. ... The country will take itself many of the qualities of the city. The old antithesis ... will cease, the boundary lines will altogether disappear.

Similarly, Adna Weber suggested in his remarkable 1899 study that

> the most encouraging feature of the whole situation is the tendency ... towards the development of suburban towns [which] denotes a diminution in the *intensity* of concentration. ... The rise of the suburbs it is, which furnishes the solid basis of hope that the evils of city life, so far as they result from overcrowding, may in large part be removed. If concentration of population seems desired to continue, it will be a modified concentration which offers the advantages of both city and country life.

Later Frank Lloyd Wright argued that 'Broadacre City' was the most desirable settlement pattern for mankind, and Lewis Mumford called for a new reintegration of men and nature in dispersed urban regions, to cite but a few cases.

Throughout the twentieth century all trends have pointed in the directions suggested by these writers (Berry, 1970; Lamb, 1975). Every public opinion survey has indicated that popular preferences are for smaller places and lower densities, with richer environmental amenities (Sundquist, 1975). The trend has been one leading unremittingly toward the reversal of the processes of population concentration unleashed by technologies of the Industrial Revolution, a reversal finally achieved after 1970.

Viewed more generally, though, what has finally been achieved in the 1970s is not something new, but something old – the reassertion of fundamental predispositions of the American culture that, because they are antithetical to urban concentration, have resulted in many of the contradictions and conflicts of recent decades.

We should go back two hundred years, to Hector de Crèvecoeur's *Letters from an American farmer*. 'Who, then, is this new man, the American?' he asked, and his answer was a description of basic American culture traits. Foremost among these was *a love of newness*. Second was the overwhelming desire to be *near to nature*. *Freedom to move* was essential if goals were to be realised, and *individualism* was basic to the self-made man's pursuit of his goals, yet *violence* was the accompaniment if not the condition of success – the competitive urge, the struggle to succeed, the fight to win. Finally, Crèvecoeur (though he did not use the later, now-familiar phrases) perceived a great *melting pot* of peoples and a sense of *manifest destiny* (Watson, 1970).

The love of newness

There has been no more evocative description of the consequences of the love of newness for American metropolitan structure than Homer Hoyt's discussion of *The structure and growth of residential neighborhoods in American cities*, published in 1939. Hoyt said:

> The erection of new dwellings on the periphery ... sets in motion forces tending to draw population from older houses and to cause all groups to move up a step, leaving the oldest and cheapest houses to be occupied by the poorest families or to be vacated. The constant competition of new areas is itself a cause of neighborhood shifts. Every building boom, with its crop of structures equipped with the latest modern devices, pushes all existing structures a notch down in the scale of desirability.... The high grade areas tend to preempt the most desirable residential land.

> ... Intermediate rental groups tend to occupy the sectors in each city that are adjacent to the high rent area. ... Occupants of houses in the low rent categories tend to move out in bands from the center of the city by filtering up. ... There is a constant outward movement of neighborhoods because as neighborhoods become older they tend to be less desirable. A neighborhood composed of new houses in the latest modern style ... is at its apex. ... Physical deterioration of structures and the aging of families ... constantly lessen the vital powers of the neighborhood. ... The steady process of deterioration is hastened by obsolescence; a new and more modern type of structure relegates all existing structures to lower ranks of desirability.

Hoyt's perceptions cut right to the core of much of that which has transpired, for the accompaniment of the process of counterurbanisation is urban decay and the abandonment of the nonachieving social underclass (Berry, 1975); ghetto growth is a product of the white exodus (Long, 1975).

Near to nature

The love of newness joins with the desire to be near nature. H. G. Wells's 1902 forecasts should be recalled:

> Many of our railway-begotten giant cities are destined to such a process of dissection and diffusion as to amount almost to obliteration ... within a measurable further space of years. ... These coming cities ... will present a new and entirely different phase of human distribution. ... The social history of the middle and later thirds of the nineteenth century ... all over the civilised world has been the history of a gigantic rush of population into the magic radius of – for most people – four miles, to suffer there physical and moral disaster ... far more appalling than any famine or pestilence that ever swept the world. ... But new forces ... bring with them ... the distinct promise of a centrifugal application that may finally be equal to the complete reduction of all our present congestions. ... What will be the forces acting upon the prosperous household? The passion for nature ... and that craving for a little private *imperium* are the chief centrifugal inducements. ... The city will diffuse itself until it has taken upon considerable areas and many of the characteristics of what is now country. ... We may call ... these coming town provinces 'urban regions.'

Almost as an echo of Wells comes that sociological essay of the 1950s that proclaimed in its title that 'the suburbs are the frontier' and Lamb's (1975) analyses of the role of amenities in peripheral expansion during the 1960s. But again antithetically, the greater the numbers trying to get near to nature, the more that which is sought is degraded.

Freedom to move

To occupy this new frontier, close to nature, and to keep on adjusting to succeeding waves of growth have demanded freedom to move. Americans are the world's most mobile people. Forty million Americans change residence each year; Americans change residence an average of 14 times in a lifetime. As Peter Morrison has remarked recently (1974):

> The typical American's life might be characterised as a prolonged odyssey. Marriage, childbearing, military service, higher education, changes from one employer to

another or shifts from one plant or office location to another with the same employer, divorce, retirement – all may bring a change in residence and locale, not to speak of upward social mobility which may impel people to move for other reasons as well. ... Now as in the past Americans continue to migrate for reasons that are connected to the working of national economic and social systems. ... The quick exploitation of new resources or knowledge requires the abandonment of old enterprises along with the development of the new ... and migration is also an *assortive* mechanism, filtering and sifting the population as its members undergo social mobility.

Yet again, there is an antiphonal note. Filtering in housing markets, for example, is a process that has positive welfare consequences if new construction exceeds the rate necessary to normal growth of housing and produces an excess housing supply at the point where the filtering originates; if such new construction exerts a downward pressure on the rents and prices of existing housing, permitting lower-income families to obtain better housing bargains relative to their existing housing quarters; if the upward mobility is apart from any changes caused by rising incomes and/or declining rent–income ratios; and if a decline in quality is not necessarily forced by reductions in maintenance and repair to the extent that rents and prices are forced down; and finally if a mechanism exists to remove the worst housing from the market without adversely affecting rents and prices of housing at the lowest level. Part of the reason for urban decay is that the last two conditions have not been met: deterioration has accelerated in many older neighbourhoods, and abandonment has become contagious, frequently adversely affecting access by low-income residents to the better-quality housing available locally.

Individualism

Contrary to the views of most radicals, however, urban expansion and urban decay are not caused by a single-minded conspiracy among large-scale institutions and investors. They result instead from myriad decisions made individually, within a tradition of privatism. This tradition has been called by Sam Bass Warner (1968)

> the most important element of American culture for understanding the development of cities. It has meant that the cities of the United States depended for their wages, employment, and general prosperity on the aggregate successes and failures of thousands of individual enterprises, not upon community action. It has also meant that the physical forms of American cities, their lots, houses, factories and streets have been the outcome of a real estate market of profit-seeking builders, land speculators, and large investors. And it has meant that the local politics of American cities have depended for their actors, and for a good deal of their subject matter, on the changing focus of men's private economic activities.

Privatism has prevailed throughout America's history, and a consequence is a preference for governmental fragmentation and for interest-group politics under presumed conditions of democratic pluralism. Antithetically, it has also meant that American city planning has been curative rather than future-oriented, reactive rather than going somewhere.

Violence

Although achievement in the mainstream has involved an individual fight to succeed, violence also is a pervasive underpinning of American life if only because fights have to have more that one participant. Acrimonious confrontations mark the fights to control turf within cities, while, for the underclass abandoned in deteriorating ghettos, crime and violence are a way of life. President Johnson's Commission on Crimes of Violence reported that if present trends continue,

> we can expect further social fragmentation of the urban environment, greater segregation of different racial groups and economic classes ... and the polarization of attitudes on a variety of issues. It is logical to expect the establishment of the 'defensive city' consisting of an economically declining central business district in the inner city protected by people shopping or working in buildings during daylight hours and 'sealed off' by police during night-time hours. High-rise apartments and residential 'compounds' will be fortified cells for upper-, middle-, and high-income populations living in prime locations.... Suburban neighborhoods, geographically removed from the central city, will be 'safe areas,' protected mainly by racial and economic homogeneity.

The melting pot(?)

In the expanding frontiers of suburban America, upwardly mobile individuals from a variety of backgrounds have been readily integrated into the achievement-oriented mainstream of society. When the heterogeneity of American cities was caused primarily by the influx of successive immigrant waves, the policy of encouraging such assimilation was taken for granted ideologically. But even in the suburbs, what poured out of the melting pot rapidly crystallised into a complex mosaic of sharply differentiated communities of achievers, counterposed against those who have been unable or unwilling to move out of the cities. Thus, the national ideal of integration remains inaccessible for many – in particular, for the unassimilable blacks, browns, and reds, for whom segregation within the central cities remains the rule as battle lines are drawn along neighbourhood boundaries and at the gates of the schools. For others, integration is perceived as undesirable: unassimilated ethnics regard it as destructive of self-identity; members of new communities avoid it as they seek a return to simpler ways, sometimes in rural communes. Why, then, we might ask, all the surprise at the discovery that forced racial integration of the schools accelerates white emigration from the central cities? It's the American way – the way in which the individuality of the homogeneous subgroup is maintained.

Manifest destiny

Yet alongside individualised withdrawal to the periphery, there remains the continuing feeling that Americans should no longer be willing simply to *react* to problems, or to *act* in a privatised mode, but should – collectively – *go somewhere*, to achieve goals, to 'win' the 'wars' on poverty, underprivilege, and urban decay – to perfect America and Americans. One version is a favourite of planning professionals. How many recall the call of the National Resources Committee in 1937, repeated many times since?:

> If the city fails, America fails. The Nation cannot flourish without its urban-industrial centres. ... City planning, county planning, rural planning, State planning, regional planning, must be linked together in the strategy of American national planning and policy, to the end that our national and local resources may best be conserved and developed for our human use. ... The Committee is of the opinion that the realistic answer to the question of a desirable urban environment lies ... in the judicious reshaping of the urban community and region by systematic development and redevelopment in accordance with forward looking and intelligent plans.

And how many public programs pointed in such directions have been failures? For the other version *is* the collective decision of individual Americans. Planners' predilections notwithstanding, the American mainstream is one of counterurbanisation processes that are being stemmed by planning – federal, state, and local – as effectively as King Canute stopped the tide.

References

Alexandersson, G. and Falk, T. 1974: Changes in the urban pattern of Sweden, 1960–1970: the beginning of a return to small urban places? *Geoforum* 8, 87–92.

Berry, B. J. L. 1970: The geography of the United States in the year 2000. *Transactions of the Institute of British Geographers* 51, 21–53.

Berry, B. J. L. 1975: The decline of the aging metropolis: Cultural bases and social process. Paper prepared for a conference at the Center for Urban Policy Research, Rutgers University.

Berry, B. J. L. and Gillard, Q. 1976: *The changing shape of metropolitan America: Commuting patterns, urban fields and decentralization processes, 1960–1970.* Cambridge, Mass.: Ballinger.

Crèvecoeur, J. H. St John de. 1782: *Letters from an American farmer.* London: Thomas Davies.

Forstall, R. L. 1975: *Trends in metropolitan and nonmetropolitan population growth since 1970.* Washington, DC: Population Division, US Bureau of the Census.

Holleb, D. B. 1975: *Moving towards megacity: Urbanization and population trends.* Chicago: Centre for Urban Studies, University of Chicago.

Hoyt, H. 1939: *The structure and growth of residential neighborhoods in American cities.* Washington, DC: Federal Housing Administration.

Kristol, I. 1972: *An urban civilization without cities.* Washington Post Outlook, Sunday, 3 December, p. B1.

Lamb, R. 1975: *Metropolitan impacts on rural America.* Department of Geography Research Paper no. 162, University of Chicago.

Long, L. H. 1975: How the racial composition of cities changes. *Land Economics* 51, 258–67.

Morrison, P. 1974: *Toward a policy planner's view of urban settlement systems.* Unpublished manuscript. Santa Monica, Calif.: Rand Corporation.

Sundquist, J. L. 1975: *Dispersing population: What America can learn from Europe.* Washington, DC: Brookings Institution.

Tisdale, H. 1942: The process of urbanisation. *Social Forces* 20, 311–16.

Warner, S. B. Jr 1968: *The private city*. Philadelphia: University of Pennsylvania Press.

Watson, J. W. 1970: Image geography: The myth of America in the American scene. *Advancement of Science* 27, 1–9.

Weber, A. F. 1899: *The growth of cities in the nineteenth century*. New York: Macmillan.

Wells, H. G. 1902: Anticipations: *The reaction of mechanical and scientific progress on human life and thought*. London: Harper.

2 C. L. Beale,
'The Recent Shift of United States Population to Nonmetropolitan Areas, 1970–75'

From: *International Regional Science Review* 2, 113–22 (1977)

In the last several years it has become evident that the historic position of rural and small-town areas of the United States as net exporters of people to metropolitan centres has changed (Beale, 1974, 1975; Morrison and Wheeler, 1976). It is the purpose of this paper to document the extent of this change from 1970 to 1975 and to review the relative involvement of different types of counties in the trend, thereby enhancing our understanding of it. The study focuses on counties that remained nonmetropolitan after the adoption in 1971 of new metropolitan criteria, which resulted in substantial enlargement of the boundaries of Standard Metropolitan Statistical Areas. In general, nonmetropolitan counties have no urban nucleus of 50 000 or more people and do not meet certain criteria of metropolitan character and worker commuting (Office of Management and Budget, 1975).

From April 1970 to July 1975, the nonmetropolitan population increased by an annual average of 1.2 per cent compared with a metropolitan average of 0.8 per cent (Table 2.1). By contrast, in the 1960–70 decade population grew by just 0.4 per cent per year in the nonmetropolitan counties compared with 1.6 percent in the metropolitan areas. As a result, 37 per cent of national population increase occurred in nonmetropolitan territory from 1970 to 1975, whereas only 10 per cent went into the same areas in the 1960s. Nonmetropolitan areas have experienced a net annual inflow of about 0.35 million persons from 1970 to 1975, which more than reverses the annual net outflow of 0.3 million observed in the previous decade. The nonmetropolitan growth rate exceeded that in metropolitan areas solely because of migration. Natural increase – the excess of births over deaths – continues to be somewhat higher in the metropolitan areas because of age composition.

In the decades following World War II, many rural counties adjoining metropolitan areas grew rapidly as the metropolitan population burgeoned and

Table 2.1 Population change by metropolitan status and selected county characteristics

	Number of counties	Population					Net migration			
		Number (millions)			Annual change[a]		Number (millions)	Annual rate[b]	Number (millions)	Annual rate[b]
		1975	1970	1960	1970-75	1960-70	1970-75	1970-75	1960-70	1960-70
Total United States	3 097	213.1	203.3	179.3	0.9	1.3	2.5	0.2	3.0	0.2
Metropolitan status[c]										
Metropolitan counties	628	155.0	148.9	127.2	0.8	1.6	0.6	0.1	6.0	0.5
Nonmetropolitan counties	2 469	58.0	54.4	52.1	1.2	0.4	1.8	0.6	-3.0	-0.6
Adjacent counties[d]	969	30.1	28.0	26.1	1.3	0.7	1.1	0.8	-0.7	-0.3
Nonadjacent counties	1 500	27.9	26.4	26.0	1.1	0.1	0.7	0.5	-2.3	-0.9
Characteristics of nonmetropolitan counties[e]										
Counties with:										
10% or more net immigration at retirement age[f]	360	8.8	7.6	6.3	3.1	1.8	1.1	2.7	0.6	1.0
A senior State college	187	9.1	8.4	7.5	1.5	1.2	0.4	0.8	0.1	0.1
30% or more employed in manufacturing	638	20.3	19.3	18.2	1.0	0.6	0.4	0.4	-0.7	-0.4
30% or more employed in agriculture	331	2.1	2.1	2.3	[g]	-1.1	[h]	-0.3	-0.4	-1.8
40% or more black population	189	3.5	3.5	3.7	0.3	-0.6	-0.1	-0.5	-0.7	-2.0
10% or more military population	29	1.2	1.2	1.0	0.2	5.6	-0.1	-1.4	[h]	0.3
Size of largest city:										
25 000 or more persons	141	10.9	10.2	9.1	1.2	1.1	0.2	0.4	-0.1	-0.1
10 000-24 999	389	18.4	17.2	15.9	1.3	0.8	0.6	0.7	-0.5	-0.3
2500-9999	1 076	21.7	20.5	20.3	1.1	0.1	0.7	0.6	-1.7	-0.8
Less than 2500	863	7.1	6.6	6.8	1.3	-0.3	0.3	1.0	-0.7	-1.1
Population density per square mile										
150 or more persons	56	5.5	5.1	4.5	1.3	1.3	0.2	0.8	0.1[h]	0.3[g]
100-149	104	6.8	6.4	5.7	1.2	1.1	0.2	0.5	-0.2	-0.4
75-99	130	6.3	6.0	5.6	1.1	0.7	0.1	0.3	-0.5	-0.6
50-74	248	8.9	8.4	8.0	1.1	0.4	0.2	0.5	-1.0	-0.7
25-49	703	16.2	15.2	14.8	1.2	0.3	0.5	0.7	-0.8	-0.9
10-24	624	9.3	8.7	8.8	1.2	-0.1	0.4	0.8	-0.8	-0.9
Less than 10	604	5.0	4.6	4.7	1.6	-0.1	0.2	0.9	-0.6	-1.3

Median family income in 1969									
$10 000 and over	61	3.0	2.8	2.4	1.3	1.7	0.1	0.5	0.1	0.6
$9000–$9999	187	8.4	7.8	6.9	1.3	1.2	0.3	0.7	0.2	0.2
$8000–$8999	454	13.7	12.8	11.9	1.3	0.7	0.4	0.6	-0.4	-0.3
$7000–$7999	533	12.6	11.9	11.5	1.1	0.3	0.3	0.6	-0.8	-0.7
$6000–$6999	548	10.3	9.6	9.4	1.3	0.2	0.4	0.8	-0.6	-0.7
$5000–$5999	426	6.5	6.1	6.3	1.2	-0.2	0.2	0.7	-0.8	-1.2
$4000–$4999	198	2.8	2.7	3.0	0.8	-0.8	^{h}	0.2	-0.6	-1.9
Less than $4000	62	0.7	0.6	0.7	0.7	-1.1	^{h}	-0.1	-0.2	-2.3

Sources: *US Census of Population: 1970* and *1960*, and *Current Population Reports*, US Bureau of the Census; and Bowles, Gladys K. and others. *Net Migration of the Population, 1960–70, by Age, Sex, and Color*. US Department of Agriculture, University of Georgia, and National Science Foundation, cooperating 1975

^{a} Population change expressed as an average annual percentage rate of change
^{b} Net migration expressed as an annualized percentage of the population at beginning of specified period
^{c} Metropolitan status as of 1974
^{d} Nonmetropolitan counties adjacent to Standard Metropolitan Statistical Areas
^{e} Characteristic as of 1970 unless otherwise stated
^{f} Counties with specified 1960–70 net migration rate for white persons 60 years old and over, 1970
^{g} Between ±0.05 per cent change
^{h} Between ± 50 000 migrants

spread out. They became metropolitan in character and economic dependence, and scores of them have been reclassified as officially metropolitan. Thus, given the typical process of metropolitan growth at the periphery, it is not surprising that nonmetropolitan counties adjacent to metropolitan areas have had somewhat higher average growth rates since 1970 than have those not adjacent. The adjacent ones have been gaining people at an annual rate about twice as high as they experienced in the 1960s, but the difference in post-1970 annual growth of the adjacent and nonadjacent groups (1.3 per cent vs. 1.1 per cent) is not great, although rather regular and pervasive in most parts of the country. The more impressive fact would seem to be the convergence of growth rates of these two classes of counties. In the 1960s, the adjacent group increased seven times as rapidly as the counties more distant from metropolitan influence compared with a margin of just under one-fifth higher growth since then. Thus, the revived growth pattern is not merely one of accentuated metropolitan sprawl. It is both close-in growth of quasi-metropolitan nature and more remote growth not stimulated by metropolitan proximity.

Of the classes of counties for which data are shown in Table 2.1, five stand out as areas of far above average nonmetropolitan population gain in the 1960s, when slow growth or outright decline was the pattern for most nonmetropolitan counties. These are counties that had either high income, high population density, a four-year State college or university, or a relatively large military population, or were retirement areas. It is instructive to see what happened in such counties after 1970.

A strong direct relationship in the 1960s between county income and population change is shown in Table 2.1.[1] Generally, each increment of $1000 in average family income was associated with 3 to 5 per cent higher population growth rates with high-income counties growing rapidly and low-income counties declining rapidly. In the 1970s thus far, this relationship has almost ended. In the first six of the eight income intervals, county population increase ranged from 1.1 to 1.3 per cent annually from 1970 to 1975. Only at the lowest two intervals was it below 1.0 per cent, and even in these cases a strong turnaround from earlier loss to current gain was evident. The low growth at the lowest income levels is accounted for by counties in the Deep South with large percentages of blacks. On the average, people are simply not moving to or between nonmetropolitan counties in a manner associated with the income levels of areas, suggesting that many of them are not moving for monetary motivations.

Density of population was also related to nonmetropolitan growth in a positive manner in the 1960s, although not quite so much as income. Here the previous association has disappeared, and net inmigration now occurs at all density levels. In an abrupt reversal, the highest recent population growth and migration rates have been at the lowest density class, i.e. less than 10 persons per square mile. There is no better illustration of the penetration of the current redistribution trend into the most remote and previously least settled areas of the nation. The growth of the lowest density classes stems heavily from events in the Western region, and occurs despite the partly offsetting continued loss of people from sparsely inhabited areas of the Great Plains.

Counties with senior State colleges acquired about 42 per cent of net nonmetropolitan population growth in the 1960s, a period of rapidly rising college enrolments and upgrading of the functions and size of many former teachers colleges. In the 1970s, these counties have continued to attract population at an above average pace and have somewhat further increased their annual growth rate. However, their relative role in nonmetropolitan increase has been considerably diminished, because so many other counties have shifted from loss to gain.

Counties with a dominant military influence in their demography are not numerous, but they contributed to nonmetropolitan growth quite out of proportion to their numbers during the 1960s, when the armed forces were being expanded and when the presence of wives and families around military bases became more common than in the past. Since 1970 and the end of the Vietnamese war, the military counties have become the most notable exception to the general pattern of increased rates of nonmetropolitan population growth. In fact, these counties shifted to outmigration during the period, although their high rate of natural increase (resulting from a high proportion of young adults) has prevented overall loss.

Excluding the military counties, areas with high inmigration of people of retirement age were the most rapidly growing class of nonmetropolitan counties in the 1960s. The rate of growth in nonmetropolitan retirement areas has accelerated in the 1970s, despite the higher base of growth from which any acceleration had to occur. Counties identified as retirement destinations on the basis of their 1960-70 experience have increased their annual rate of population growth by slightly more than two-thirds since 1970. With diminished natural increase, this has involved more than a tripling of inmigration. These counties received an annual average of 0.06 million net inmigration of all ages in the 1960s and 0.21 million annually since 1970. Estimates of population over age 65 by county for 1975 (derived from Medicare statistics) indicate that the number of nonmetropolitan retirement counties is increasing and spreading further geographically since 1970, when 360 of them were identified by the procedure used here. Even so, the retirement counties, like the college counties, accounted for a smaller percentage of total nonmetropolitan growth from 1970 to 1975 than from 1960 to 1970, because so many other counties, formerly stationary or declining, have begun to grow.[2]

The types of counties mentioned thus far have all had a diminished *relative* role in the current decade (even though some have higher absolute growth rates). What other kinds of counties are responsible for the changes in nonmetropolitan growth? It is difficult to answer in terms of economic functions, but the most marked changes have come in the most rural counties (lacking any town of 10 000 people), sparsely settled counties (those of less than 25 persons per square mile), counties with less than $6000 of median family income in 1969, and counties located in the Upper Great Lakes, the Southern Appalachian coalfields, the Rio Grande area, and the Rocky Mountains–Utah Valleys and Columbia Basin subregion. These settlement, income, and regional categories are somewhat interrelated.

Two classes of counties that experienced heavy outmigration and population loss in the 1960s are those primarily dependent on farm employment or that have a high proportion of black residents. (The two groups are almost entirely mutually exclusive.) Only occasionally was a county with 30 per cent or more of its labour force in agriculture able to more than offset declines in farm employment with gains in other sources of employment. In heavily black areas net outmigration of blacks occurred almost regardless of the degree of dependence on farming or on the pattern of movement of the white residents. Since 1970, the aggregate level of population in both classes of counties has stabilised, although some net outmigration still goes on. The demographic situation in these counties is often still negative in regard to population retention, but the trend is clearly positive in relation to the recent past.

In effect, some major patterns of the past have been broken, and certain indicators or causal attributes that were formerly reliable as positive predictors of county population growth no longer function. Growth is occurring in hundreds of counties that conventional analysis in the 1960s would have consigned to continued demographic stagnation and decline. Any cut of the nonmetropolitan counties will show some degree of renewed population retention, except for the military areas and the counties of the Lower Great Lakes Industrial Subregion. However, the change in trend would be comparatively prosaic and unremarkable were it not for the intensity of the reversal in those areas that seemed least likely to attract population because of their smallness, rurality, remoteness, and low incomes. In some of these counties, the change is clearly linked to new economic opportunity in the form of mineral developments; for example, the oilfields of Duchesne County, Utah, the natural gas fields of Sutton County, Texas, or the new coal mines of Rosebud County, Montana. However, in broader terms, the change strongly implies an attraction, exceeding mere pecuniary considerations, that has prompted people to stay in or move to areas that will continue to be comparatively small, remote, rural, and low in income even after the migration has occurred.

The effect of some of the characteristics discussed may be linked with the simultaneous actions of others, or may be particular to specific areas of the country. Therefore, multiple regression has been used to provide a more concise appraisal of the associations between population change and migration and 10 socio-economic variables and 6 regional location dummy variables. The 16 independent variables, which were all premised to influence county population change, yielded coefficients of determination (R^2) of 0.34 and 0.40 with respect to 1970–75 and 1960–70 growth, respectively (Table 2.2). Thus, there is some overall decline in the explanatory power of these factors. However, in both periods they accounted for less than half the variation in population change.[3]

The only bivariate correlations of more than 0.20 found in both the 1970–75 and 1960–70 period between independent and dependent variables were those between population growth and retirement county status (0.40 in 1970–75 and 0.39 in 1960–70) and between population growth and percent employed in agriculture (–0.27 in 1970–75 and –0.39 in 1960–70). In the 1960–70 period, bivariate correlations with population change of more than

Table 2.2 Multiple regression analysis of population change, 1970–75 and 1960–70, and selected variables, nonmetropolitan counties

	1970–75	1960–70
Regression coefficients[a]		
Population density[b]	−0.001	−0.022**
Median family income, 1969	**[c]	0.003**
% employed in manufacturing[b]	−0.113**	0.053**
% employed in agriculture[b]	−0.251**	−0.094**
% population black[b]	−0.128**	−0.069**
Size of largest city, 1970	−1.212**	1.355**
Presence of a senior State college	0.662	8.748**
% military population, 1970	−4.256**	9.614**
Retirement status, 1970	6.311**	11.371**
Metropolitan adjacency status, 1970	1.757**	2.903**
Dummy location variables		
Northeast	2.020**	1.794
East North Central	0.786	1.673**
South Atlantic	6.484**	7.571**
East South Central	4.740**	9.991**
West South Central	0.264	3.860**
West	7.926**	3.330**
Intercept	8.82	−31.20
Coefficient of multiple determination (R^2)	0.34	0.40
Number of counties	2469	2469

[a] Regression coefficients are expressed in unstandardized form. The symbol ** indicates statistically significant at the 0.01 level and the symbol * indicates statistically significant at the 0.05 level
[b] Computed using 1970 figures for the 1970–75 regressions and 1960 figures for the 1960–70 regressions
[c] Less than 0.0005

0.20 (positive or negative) were also found for median income, size of largest city, and percentage engaged in manufacturing, but they dropped below this level of association in the 1970s as the pattern of nonmetropolitan distribution shifted.

Without the regional location variables, the R^2 in 1970–75 is reduced from 0.34 to 0.27, whereas deletion of regional location in 1960–70 reduces the R^2 only from 0.40 to 0.38. Thus, in the current decade, the regional variable has become more influential as an independent explanatory factor than it was earlier. An examination of the more detailed data shows this effect to arise primarily from the enhanced recent association of growth with western location.

Regressions of 10 variables were run within regions, using census regions in the Northeast and West, but divisions in the South and North Central states where the number of counties is largest.[4] This procedure showed the variables to have different degrees of association with population change by region, ranging in 1970–75 from an R^2 of 0.56 in the Northeast, where the association with manufacturing employment, military presence, and size of largest city become stronger (all in a negative way), to 0.23 in the West, where the pattern of relationships is basically similar to that of the United States as a whole, but more muted, particularly in the case of retirement. Retirement is the strongest explanatory factor in most of the areas in terms of its contribution to R^2.

Table 2.3 Multiple regression analysis of population change, 1970–75 and 1960–70, and selected variables for geographic areas, nonmetropolitan counties

	Northeast		East North Central		West North Central		South Atlantic		East South Central		West South Central		West	
	1970–75	1960–70	1970–75	1960–70	1970–75	1960–70	1970–75	1960–70	1970–75	1960–70	1970–75	1960–70	1970–75	1960–70
Regression coefficients[a]														
Population density[b]	0.047**	0.020	0.004	0.002	−0.026	−0.012	−0.024**	−0.107**	−0.010	−0.067**	0.050**	0.076*	0.009	−0.035
Median family income, 1969	0.001**	0.008**	−0.000	0.005**	−0.001**	0.001**	0.002**	0.007**	−0.001**	0.004**	−0.002**	0.001**	0.002**	0.004**
% employed in manufacturing[b]	−0.279**	−0.145	0.038	0.232**	0.062*	0.263**	−0.193**	−0.086**	−0.032	0.161**	0.007	0.305**	−0.257**	−0.314**
% employed in agriculture[b]	−0.154	0.325**	−0.218**	−0.022	−0.138**	−0.059*	0.114*	0.104*	−0.262**	0.036	−0.287**	−0.032	−0.350**	−0.380**
% population black[b]	0.070	−0.396	−0.225**	−0.170	−0.218**	−0.647**	−0.157**	−0.148**	−0.174**	−0.153**	−0.267**	−0.033	0.181	−0.617
Size of largest city, 1970	−3.792**	−1.130	−1.953**	−1.389**	0.212	2.925**	0.781	4.070**	1.135**	2.701**	−1.064*	0.327	−4.020**	−1.421
Presence of a senior state college	1.981	4.827**	4.001**	17.813**	0.837	10.948**	−0.016	5.121**	0.202	10.117**	2.690**	11.395**	1.521	9.007**
% military population, 1970	−30.620**	−3.331	−2.609	8.244**	0.301	12.557**	−4.935**	2.215	−9.027**	10.485**	−2.303	30.440**	−6.079**	4.563
Retirement status, 1970	4.286**	12.359**	7.211**	9.904**	5.453**	7.210**	6.272**	10.907**	0.424	6.926**	4.379**	8.728**	6.307**	15.353**
Metropolitan adjacency status, 1970	−1.648	2.074	−1.350**	−1.085	1.781**	3.266**	2.778**	1.897*	2.890**	1.713**	1.992**	0.278	3.888	9.295**
Intercept	12.44	−64.22	10.33	−38.18	8.99	−16.80	1.88	−42.51	13.57	−27.84	20.31	−17.21	11.80	−22.86
Coefficient of multiple determination (R^2)	0.56	0.65	0.47	0.61	0.26	0.46	0.39	0.50	0.51	0.64	0.42	0.36	0.23	0.37
Number of counties	116	116	306	306	568	568	421	421	303	303	382	382	373	373

[a] Regression coefficients are expressed in unstandardized form with the symbols ** and * having the same meaning as in Table 2.2
[b] Computed using 1970 figures for the 1970–75 regressions and 1960 figures for the 1960–70 regressions

The most notable exception is the East South Central states, where current population change is very negatively associated with the percentage of blacks in a county ($r = -0.54$) and where few intensely developed retirement areas have arisen. Even though some of the factors have insignificant effects from 1970 to 1975 at the national level, all show significant association in at least one region. See Table 2.3 for complete results of this analysis.

Additional variables, in terms of baseline characteristics of the counties, that offer major additional insights were sought unsuccessfully. Counties with state capitals or large Indian populations are positively associated with nonmetropolitan growth, but are not very numerous. Interstate highway location is positively associated and could logically be in the analysis, but it is known to be strongly related to pre-interstate 1950–60 growth patterns that antedate the construction of most of these highways (Fuguitt and Beale, 1976). A good identifier of recreational areas would contribute further explanation, but many such counties are already subsumed in the retirement group.

In sum, the revived population growth in rural and small-town areas, here approximated by use of nonmetropolitan counties, has continued through 1975. Indeed, reduced population growth is so highly associated with the metropolitan areas of the North Atlantic Coast, the Lower Great Lakes, and the Pacific Coast that the rest of the nation, metropolitan and nonmetropolitan combined, has been increasing in population at a faster rate in the 1970s than in the 1960s, despite the reduction of the birth rate to subreplacement levels. Basically, the United States has been having renewed growth or diminished loss of nonmetropolitan population throughout most of the country and increased regional growth of small and medium-sized metropolitan areas in the South and the West. How far the forces producing it will take the trend is unclear, whether one considers those forces attracting people to live in the smaller areas or those impelling people to leave or avoid the larger metropolitan areas. However, it is too late to view the trend as a potentially insignificant or ephemeral phenomenon, as might have been the temptation when it was first noted. The underpinning of the movement is complex and not readily altered overnight. The net movement of 1.8 million people into nonmetropolitan areas in just five years is a substantial one. It has already had far-reaching effects on many receiving communities, and is subtly changing the demographic outlook of many of the largest metropolitan areas.

Notes

1 For each characteristic to be discussed, counties are grouped by the value of that characteristic in the 1970 Census. Logically it might be superior to use the characteristic values for 1960 for tabulations of 1960–70 population change. In practice, the 1960 and 1970 values are so highly correlated that no meaningful difference in the patterns of population change would result from such usage. For convenience, therefore, only the 1970 characteristic values were used.

2 All of the data in Table 2.1 that are discussed in the preceding paragraphs were run by adjacency status of the counties. For any given class of counties those adjacent to metropolitan areas tended to average somewhat higher population growth (or less population loss) than those not adjacent. However, the pattern of relationships

between population change and a given variable is generally very similar for adjacent and nonadjacent counties for both time periods measured.
3 Because of the wide disparity among counties in population size, the data were also run weighted for population size to reflect the greater impact of changes in large counties. This procedure – in which the characteristic values for each county were weighted by the number of people in the county – yields somewhat higher and presumably more valid relationships. For 1970–75 the weighted R^2 was 0.42, against the unweighted value of 0.34 and for 1960–70 weighting increased the R^2 from 0.40 to 0.47. The direction and magnitude of the coefficients of individual variables changed only modestly; the structure of the relationships between the variable and population change is consistent between the weighted and unweighted equations. The weighted regression results are available from the author.
4 Weighted regressions were not run within regions. R^2 values would be affected somewhat by doing so, but the use of regional data is in itself a partial control for variation in size of counties.

References

Beale, C. L. 1975: *The revival of population growth in nonmetropolitan America.* ERS-605. US Department of Agriculture.
Beale, C. L. 1974: Rural development: population and settlement prospects. *Journal of Soil and Water Conservation* 29, 23–27.
Fuguitt, G. V. and Beale, C. L. 1976: *Population change in nonmetropolitan cities and towns.* Agricultural Economic Report no. 323. US Department of Agriculture.
Fuguitt, G. V. and Beale, C. L. 1975: *Population trends of nonmetropolitan cities and villages in subregions of the United States.* Working Paper 75-30. Center for Demography and Ecology, University of Wisconsin.
Morrison, P. A. and Wheeler, J. P. 1976: Rural renaissance in America?: The revival of population growth in remote areas. *Population Bulletin* 31, 1–26.
Office of Management and Budget, Executive Office of the President. 1975: *Standard Metropolitan Statistical Areas.*

---

3   D. R. Vining Jr and A. Strauss,
'A Demonstration that the Current Deconcentration of Population in the United States is a Clean Break with the Past'

From: *Environment and Planning A* 9, 751–8 (1977)

Two schools of thought have emerged from among students of America's population geography in reaction to recent Bureau of the Census data showing a higher rate of population growth in the nonmetropolitan counties than in the metropolitan counties. One school interprets this development as simply a continuation of past trends, representing 'primarily an accelerated overspill of metropolitan areas into their exurban counties' (Regional Plan Association,

1975, p. 54); the other as a clean and wholly unprecedented break with past trends: 'the decentralization trend is not confined to metropolitan sprawl. It affects non-metropolitan counties well removed from metropolitan influence' (Beale, 1975, p. 7).

If the debate is confined to the comparison of average rates of growth among various classes of nonmetropolitan and metropolitan counties, the second school has clearly won its case. Nonmetropolitan counties well removed from the commuting range of our 250 or so SMSAs [Standard Metropolitan Statistical Areas] are growing at a significantly higher rate than these SMSAs themselves, though at a somewhat lower rate than the nonmetropolitan counties adjacent to these SMSAs (Morrison, 1975, p. 10). This fact represents a clear and unmistakable break with past trends of a long duration.

It seems to us that now is the proper time, the second school having won its case for these averages, to approach this question from a different angle, so to speak; to employ a different kind of statistic to see if this 'break in trend' is likewise indicated.

In our opinion, the heart of the debate is over the American population's tendency to concentrate itself and whether or not there has been an abatement in or exhaustion of this tendency in recent years. Certainly, we all 'feel' somehow that over the last fifty or so years the population has been arranging itself into a more and more uneven, or concentrated, distribution. The challenge is to quantify this tendency, given the data available, and to ask if the behaviour of this quantity between 1970 and the present is significantly different from its behaviour between 1900 and 1970, say, the period of rapid urbanisation and increasing population concentration in the USA. Comparisons of average rates of growth among different classes of counties do not directly answer this question.

Perhaps the most easily understood statistic of concentration is the so-called Hoover index, or index of concentration, named after its originator, Edgar Hoover (Hoover, 1941). Let p_{it} be the fraction of a nation's population in sub-area i in year t, a_i be its fraction of the nation's land area, and k be the total number of subareas. Then, the Hoover index, H_t, is given by

$$H_t = \tfrac{1}{2} \sum_{i=1}^{k} |p_{it} - a_i| 100$$

If $p_{it} = a_i$ for all i, then the population is perfectly uniformly distributed and H_t will equal 0; all subareas have the same population density. If all of the population is located in one subarea, and if the subareas are of negligible size relative to the total size of the nation, then H_t will approach 100 as a limit; there is perfect concentration. A value of H_t between 0 and 100 tells us the percentage of the population that would have to be resettled in order to have a uniform density of population across the k subareas, an intuitively appealing way of quantifying the degree of population concentration across the k subareas.

There have been numerous objections raised against this measure (Duncan, 1957, pp. 31–2). The most damaging, in the minds of many, is that its movement over time (showing increasing, decreasing, or constant concentration) can be different for different ways of subdividing the nation into subareas. For example, the index might move in one direction if the 50 states were the unit

30 Differential Urbanization

of subarea used in its computation, and in an entirely opposite direction if the 3000 or so counties were used. In short, a time series of H_t will not necessarily give an unambiguous answer to the question posed above: is the unevenness of population distribution increasing or decreasing over time and has the trend shifted direction in recent years? The answer may depend on how one disaggregates the nation into subareas.

This ambiguity turns out not just to be a mathematical possibility, demonstrated for a few artificial examples, but actually the case for the empirical data on real populations. In the USA, for example, H_t computed on the basis of the 48 conterminous states declined between 1910 and 1940; when computed on the basis of the 3000-odd counties, it increased over this same time period. Figure 3.1, a reproduction of Duncan et al.'s (1961) graph of H_t for five different ways of subdividing the nation into subareas, for the period 1900–50, demonstrates this fact.[1]

Our attitude towards this lack of 'invariance' in the Hoover index with respect to the areal units chosen as the basis for its computation is that rather than being a defect, it is actually a resource that may be exploited rather nicely to answer the question posed at the start of this paper: has the trend in population concentration in the USA been broken or not?[2]

Duncan et al. (1961) give the following interpretation to Figure 3.1:

> (i) The decrease in concentration shown for 'large' units like States and divisions may reflect a gross smoothing out of population distribution over the continental area, a continuation of the east – west pattern of settlement. (ii) The decrease in concentration for the decade 1900–1910 shown by indexes based on subregions, SEAs, and counties ... probably means that this gross redistribution overshadowed any local patterns of redistribution that would be reflected in the indexes.

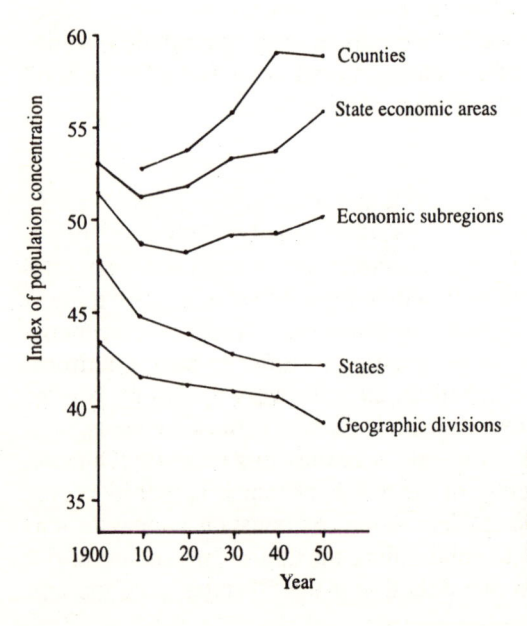

Fig. 3.1 Indexes of population concentration for various systems of areal subdivision of the United States, 1900–50. (Source: Duncan et al., 1961)

(iii) The increase in the county indexes, 1910–1940, the SEA indexes, 1910–1950, and the subregion indexes, 1920–1950, doubtlessly represent urban and metropolitan concentration occurring concomitantly with the broad regional deconcentration evidenced in the State and division indexes. (iv) The slight decrease in the county index, 1940–1950, probably represents deconcentration within metropolitan areas occurring concomitantly with the disproportionate growth of entire metropolitan units, as reflected in the increase, 1940–1950, in the index based on SEAs.

(Duncan *et al.*, 1961, pp. 84–7)

In short, the American population dispersed itself from the densely settled East to the sparsely settled West over the first half of this century and, more recently, from the densely settled city centres to the sparsely settled suburbs. Concentration occurred with the movement from rural regions to urban regions, broadly defined. The 'clean break' school would clearly predict a continuation in the first two patterns of dispersal, or deconcentration, between 1950 and the present, and a reversal in the third pattern of concentration sometime in this latter period, most likely in the 1970s. That is, deconcentration should be an overall characteristic of the American population, an occurrence at all levels of regional disaggregation.

Figure 3.2 extends Figure 3.1 to the years 1960, 1970, and 1980 (the latter being projected on the basis of the 1974 estimates of county populations). The 'clean break' school is clearly vindicated. Deconcentration appears to be a universal phenomenon, as it shows up in 1970 for the first time at all five levels of regional disaggregation. An equally interesting pattern revealed in this second

Table 3.1 Indexes of population concentration for various systems of areal subdivision of the United States: 1900–74

Year	Geographic divisions	States	Economic subregions	State economic areas	Counties
1980[a]	37.5	41.6	51.2	57.9	62.0
1974	38.2	42.4	51.8	58.7	62.7
1970	38.7	42.9	52.2	59.2	63.2
1960	38.5	42.8	51.3	57.9	61.5
1950	39.2	42.2	50.1	55.8	58.9
1940	40.5	42.2	49.2	53.7	59.1
1930	40.8	42.8	49.1	53.4	55.9
1920	41.2	43.0	48.3	51.8	53.8
1910	41.6	44.9	48.7	51.3	52.8
1900	43.6	48.0	51.6	53.3	—[b]

Sources: 1900–50, Duncan *et al.* (1961, table 2, p. 85); 1960–74, US Bureau of the Census (1963, 1967, 1972, 1973, 1976). Land-area statistics for geographical divisions, states, and counties were taken from US Bureau of the Census (1973); for economic subregions and state economic areas, from the original worksheets used by Duncan *et al* (1961). 1970 land-area figures were used for geographical divisions, counties, and states. 1950 definitions of state economic areas and economic subregions (*see* Bogue and Beale, 1953) were followed in this study to render it compatible with the Duncan study. The independent cities of Virginia as well as Baltimore City in Maryland were included in the counties with which they have the longest common boundary.

[a] Projected on the basis of the 1974 index
[b] Not computed

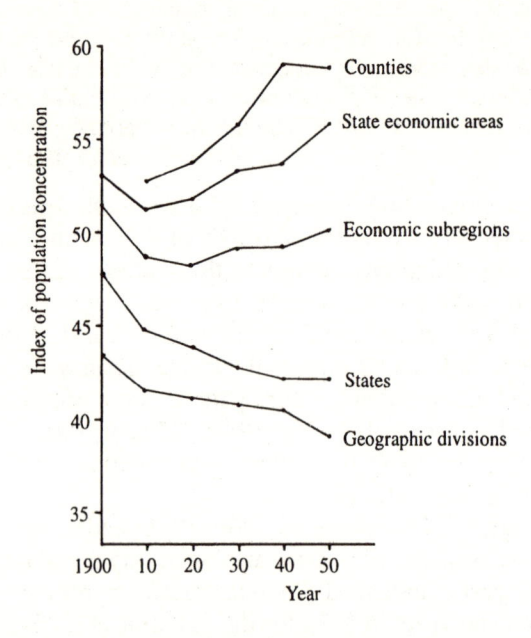

Fig. 3.2 Indexes of population concentration for various systems of areal subdivision of the United States, 1900–74. (Sources: 1900–50, Duncan et al., 1961; 1960–74, US Bureau of the Census, 1963, 1967, 1972, 1973, 1976; for further details see legend to Table 3.1)

figure is the nearly universal (the sole exception being at the division level between 1950 and 1960) rise in concentration between 1950 and 1970. Population concentration accelerated in the post-war period, achieving its maximum rate of increase and greatest universality in the 1960s [this explosion of concentration in the 1960s has also been observed for most countries of Western Europe as well as Japan (Vining and Kontuly, 1977b)]. The abruptness of its decline in the early 1970s is for that reason all the more remarkable.

There remains a second question – namely how general is this deconcentration from region to region in the USA. Since the Hoover index is an average measure of concentration, declines in this average may not reflect an overall deconcentration in all areas of the country but rather a large decline in concentration in some regions of the country, coupled with a general dispersal from densely populated to sparsely populated regions that overcompensates for a slight or moderate increase in concentration *within* other regions. In particular, since the Japanese data demonstrate what one author has called 'dispersed concentration' (Tachi, 1971, p. 19), that is, a movement from urban to rural regions but within the latter a continuing concentration in certain local cities at the expense of the peripheral areas in these rural regions, it is of interest to ask if a similar dichotomy exists in the USA, namely, a dispersal within the urban regions and from urban to rural regions but a continued increase in concentration *within* the latter.

To answer this question, we computed Hoover indexes (using SEAs and counties as our subareas) for each of the 48 conterminous states and then compared their movement between 1960 and 1970 and between 1970 and 1974.

Figures 3.3 and 3.4 summarise the behaviour of the individual state Hoover indexes. Both figures reveal a similar pattern, of a type consistent with the Japanese experience, namely, a continued rise in concentration within many of the rural states of the USA in the period 1970–74. At the same time we find an almost equally large number of states that show declines in concentration between 1970 and 1974 after experiencing increases in concentration between 1960 and 1970. Though this latter group is, on the average, more urban in character than the states experiencing continuous increases in concentration, there is a surprisingly large number of rural states in this group. Finally, there is a small group of states, consisting in the main of the highly urbanised states of the Northeast, which have experienced continuously declining concentration over the last 15 years. There is at least a suggestion, then, of a pattern whereby regions move toward a more dispersed state as they become more urbanised.

Dispersal within rural regions, some incipient signs of which can be observed in the USA, is probably rather rare in the industrialised nations, though the literature does suggest something similar in West Germany (George, 1972, p. 535). West Germany, however, is unusual among the countries of Europe in that it lacks large expanses of rural (city-less) territory:

> Although almost one-half of West Germany's population live in 10 major metropolitan regions, the country still enjoys a highly favourable distribution of population. Other European countries are often dominated by a single metropolis or have only a few metropolitan regions that constantly grow at the further expense of the less populous or less prosperous districts. This rather even population spread is one beneficial dividend of Germany's traditional regionalism, a factor that has militated against the dominance of any city or district.
> (Kirby, 1974, p. 55)

From this passage, we would speculate that most of West Germany's rural areas are in the hinterlands of these ten metropolitan regions and that, therefore, dispersal to these areas is more in the nature of an extended suburbanisation than an actual abandonment of the metropolitan regions of the kind that we are observing some signs of in the United States. Japan probably has the more common pattern: in none of its rural prefectures do we find much growth in districts (*gun*) other than those containing the capital cities of these prefectures (Japan, 1975; Ogasawara, 1975; Schöller, 1973). The predominantly rural districts of these rural prefectures are still losing population, though more and more to their regional capitals than to the urban prefectures of the Tokkaido megalopolis.

The pattern of population redistribution in the USA is indeed, in Duncan's word, 'complex'. But even so crude a measure as the Hoover index of concentration reveals certain broad regularities in this pattern. Although there has been a sharp break in concentration tendency at the national level, a number of the predominantly rural states in the peripheral areas of the country

Fig. 3.3

LEGEND

- ▨ Increasing concentration, 1960–1974
- ▤ Increasing concentration, 1960–1970
  Decreasing concentration, 1970–1974
- ⋯ Decreasing concentration, 1960–1974
- ☐ Anomalous states

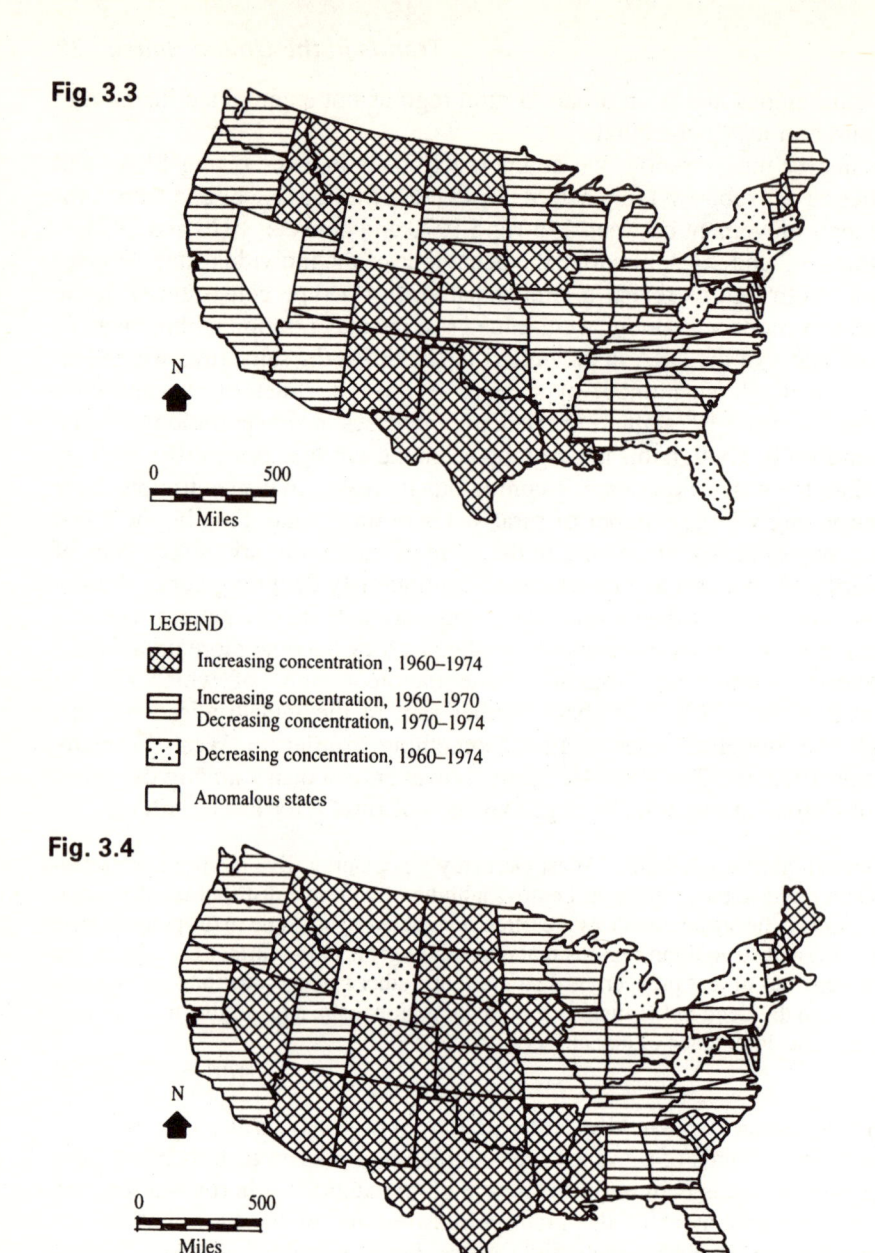

Fig. 3.4

Fig. 3.3 and 3.4 Show changes in indexes of population concentration, 1960–74. Indexes were computed individually for states on the basis of their constituent SEAs (Fig 3.3) and their constituent counties (Fig 3.4) (those lying outside of its traditional 'industrial heartland') continue to experience increasing concentration. Thus, deconcentration at the national level, from densely populated to sparsely populated regions, may mask in many instances a continued concentration within the latter. However, we also find that a large number of the sparsely populated states have experienced declines in concentration since 1970 and that, furthermore, for most of those with increases, the Hoover index has increased at a reduced rate between 1970 and 1974.

Since the USA is widely accepted to be the country most advanced along the course of modernisation and industrialisation, we might anticipate other countries as they move through the stages of development that the USA has experienced over the last 70 years to follow a similar spatial development: first, decentralisation within the urban regions (which is occurring in most of the highly modernised countries of Europe as well as in Japan); second, decentralisation from urban to rural regions (which appears to be imminent in Western Europe and Japan (Vining and Kontuly, 1977a, 1977b)); and third, decentralisation within the rural regions (which has only been observed in the USA and possibly West Germany). Though this simple paradigm seems to suggest an equilibrium of sorts, with the completion of decentralisation in the rural regions, we are bound to repeat Henry Adams's maxim, 'If one physical law exists more absolute than another, it is the law that stable equilibrium is death' (Adams, 1919, p. 79). Doubtless, equally surprising developments in the industrial world's population geography are in store for us.

Acknowledgements

This study was made possible by a grant from the National Science Foundation (SOC 76-04821). We also wish to thank Professor Otis Duncan of Arizona University for providing us with the original worksheets that were used in his study of population concentration in the United States, and Kenneth Bieri of the Regional Science Research Institute for his assistance in the preparation of the maps and charts in this paper. This paper originally appeared as discussion paper no. 90 of that institute. Some minor errors in that earlier version have been corrected here.

Notes

1 The following systems of areal units were used by Duncan et al. in their study: '(a) counties, which number about 3000 though they vary in number from census to census; (b) State Economic Areas, which number 443 and are combinations of counties (the SEA system employed here distinguishes metropolitan from non-metropolitan areas); (c) economic subregions, which number 119 and are combinations of SEAs (economic subregions may include SEAs lying in different States, and most metropolitan SEAs do not constitute separate subregions); (d) the 48 States and the District of Columbia; and (e) the 9 geographic divisions delimited by the Bureau of the Census, which are combinations of States' (Duncan et al., 1961, p. 83). In our extension of Duncan et al.'s analysis, to be described below, we have followed the 1950 definitions of SEAs and economic subregions that they used; we have ignored boundary changes in the states, counties, and geographic divisions.

2 Initially, our choice of the Hoover index of concentration over other indexes, such as the centrographic measures developed by Bachi (Duncan et al., 1961, pp. 90–3) and entropy measures, was motivated by purely practical considerations; historical series exist only for the Hoover index and we did not have sufficient resources to develop such series for the other indexes. In fact, we were able to find very few applications of these alternative measures to the data of any country, historical or current. This is but a specific instance of a general distaste among developers of new techniques in our field to go much beyond simple illustrative examples. However, Bachi's forthcoming book, *Geostatistical analysis of territories and populations* (manuscript in preparation), may remedy this situation in the particular case here of measures of spatial concentration. But there are also substantive reasons for doubting if concentration measures like those advocated by Bachi that are 'independent of the manner of subdivision into districts' (Duncan et al., 1961, p. 90) are really desirable after all. As Duncan argues elsewhere (1957, p. 32), the changes in population distribution may follow a complex pattern which would not be detected by a measure that was designed to be invariant with the regionalization scheme used in its computation. 'The contrary results obtained with alternative indexes [by 'alternative', Duncan means 'computed for different areal subdivisions'] may ... reflect a basic ambiguity inherent in any concept of concentration that does not specify the system of areal units to which it refers, rather than a defect in the operational definition of the measure of concentration.'

References

The data used to compute the Hoover indexes in this study were taken from references marked *.

Adams, H. 1919: *The degradation of the democratic dogma.* New York: Macmillan.

Beale, C. 1975: *The revival of population growth in nonmetropolitan America.* ERS-605. Economic Research Service, United States Department of Agriculture, Washington, DC.

Bogue, D. and Beale, C. 1953: *Economic subregions of the United States.* Series Census BAE, no. 19. US Government Printing Office, Washington, DC, Table A, 5–21.

Duncan, O. 1957: The measurement of population distribution. *Population Studies* 11, 27–45.

Duncan, O., Cuzzort, R. and Duncan, B. 1961: *Statistical geography.* Glencoe, Ill.: The Free Press.

George, P. 1972: Questions de géographie de la population en République Féderal Allemande. *Annales de Géographie* 81, 525–37.

Hoover, E. 1941: Interstate redistribution of population, 1850–1940. *Journal of Economic History* 1, 199–205.

Japan. 1975: *1975 Population census of Japan, preliminary count of population.* Tokyo: Bureau of Statistics, Office of the Prime Minister.

Kirby, G. 1974: Germany, Federal Republic of (in part) *Encyclopaedia Britannica, Macropaedia,* 15th edition, vol. 8, 45–56.

Morrison, P. 1975: *The current demographic context of national growth and development.* P-5514. Santa Monica, Calif.: Rand Corporation.

Ogasawara, S. 1975: *Population growth of several cities in Iwate prefecture.* Science Reports of the Tohoku University, Seventh Series (Geography), vol. 25, 75–9.

Regional Plan Association. 1975: *Growth and settlement in the US; Past trends and future issues.* RPA Bulletin 124. New York: Regional Plan Association.

Schöller, P. 1973: Wanderungszentralität und Wanderungsfolgen in Japan. *Erdkunde* 27, 290–8.
Tachi, M. 1971: The inter-regional movement of population as revealed by the 1970 census. *Area Development in Japan* 4, 3–24.
* US Bureau of the Census. 1963: *US Census of Population: 1960*. Selected Area Reports, State Economic Areas, Final Report PC (3)-1A. US Government Printing Office, Washington, DC, Table 1, 1–46.
* US Bureau of the Census 1967: *County and City Data Book, 1967*. (A Statistical Abstract Supplement.) US Government Printing Office, Washington, DC, Table 2, 12–432.
* US Bureau of the Census 1972: *Census of Population: 1970*. State Economic Areas, Final Report PC(2)-1OB. US Government Printing Office, Washington, DC, Table 1, 1–34.
* US Bureau of the Census 1973: *County and City Data Book, 1972*. (A Statistical Abstract Supplement.) US Government Printing Office, Washington, DC, Table 2, 29–545.
* US Bureau of the Census 1976: *Current Population Reports*. Series P-25, no. 620. Estimates of the Population of Counties: July 1, 1973 and 1974. US Government Printing Office, Washington, DC, Table 1, 3–65.
Vinning, D. R. Jr. and Kontuly, T. 1977a: Increasing returns to city size in the face of an impending decline in the sizes of large cities: Which is the bogus fact? *Environment and Planning A* 9, 59–62.
Vining, D. R. Jr and Kontuly, T. 1977b: Population dispersal from large metropolitan regions: an international comparison. Discussion Paper, Regional Science Research Institute, Philadelphia, Pa. (forthcoming) (*see* this volume, Chapter 6).

4 P. Gordon
'Deconcentration without a "Clean Break"'

From: *Environment and Planning A* 11, 281–90 (1979)

Introduction

A number of recent papers have argued that settlement patterns in the USA may be characterised by a clear 'reversal' of past trends, by 'significant changes', by a 'rural renaissance', or by a 'clean break with the past'. Much less has been written about human settlement changes in the other developed countries. This paper reviews some of the recent literature on counterurbanisation in the USA. We also look at a new data file to describe recent settlement trends in Europe and Japan. Both the literature review and the analysis of the new data file cause us to register some scepticism of the 'clean break' thesis. Rather, a continued 'wave' of urban decentralisation as well as renewed rural growth seem to be in progress.

Background

Although scholars interpret the US evidence with varying certitude, most conclude that we are witnessing fundamentally new phenomena and that the 'shift' occurred either in the later 1960s or the early 1970s. Berry and Dahmann note that

> for the first time the growth rate of metropolitan areas has dropped below that of non-metropolitan areas. More significantly, the long term inflow of persons from non-metropolitan areas has been reversed; as recently as the 1960s there was a net flow of migrants from non-metropolitan areas. Since then, however, these areas have added residents largely as the result of increased out-migration from metropolitan areas.... While the total population increased 13.3 percent during the 1960s, the number of individuals residing in metropolitan areas increased 16.6 percent, a rate of metropolitan increase that was 25 times the rate for non-metropolitan areas. Since 1970, however, a reversal has occurred; nation-wide statistics for the first half of the 1970s indicate that population has increased 6.3 percent in non-metropolitan areas and only 3.6 percent in metropolitan areas.
>
> (Berry and Dahmann, 1977, p. 444)[1]

Vining and Kontuly (1977) have suggested that the 'new' patterns of settlement can also be detected in other economically advanced countries. In documenting declining in-migration to core areas, spatial units as large as 20 per cent to 30 per cent of each nation's territory were chosen. This was done in order to contain most of the spread effects of the populations from central cities. Yet even this approach cannot detect whether *intra*metropolitan relocations are of increasing length and ever more exurban, as a 'wave theory' of development might predict.

The fact that there are bound to be major measurement problems is significant. It suggests that the issue is not really resolved. Zelinsky admits that 'what is abundantly clear is that our attempts to understand the turnaround phenomenon have been straining our factual and theoretical resources to their limits' (Zelinsky, 1978, p. 15).

The data which we present in this paper contain evidence which supports the wave theory as an alternative hypothesis to the clean break. The wave theory has been around for some time and it suggests that we might be observing more of some very traditional trends: growth takes place at the centres of smaller cities and is ever more removed from the centre as the city gets larger. The diseconomies of agglomeration are not simply to be associated with bigness but can be *located* in older central cities.

We are not the first to suggest that the US data, which most often underline clean-break reports, are unable really to test the hypothesis of a reversal against the idea of continued spillover growth (Wardwell, 1977). Yet it is the ambiguity of the US results which underlines our interest in the new data file. We shall argue that since the US data cannot defeat the wave hypothesis and since the new data file does support it, the notion will have to stand for a while longer.

Rural-to-urban population shifts are a trend of long standing through most of the world. Thus it would certainly be intriguing to find that this process has suddenly been reversed. Yet it should be obvious that metropolitan-to-nonmetropolitan movements, using the US Census Bureau definitions, (1) do not necessarily imply urban to rural movements, and (2) can just as readily reflect a continuation of outward growth. We need only imagine that the large metropolitan areas are continuing their long-established outward growth and that this growth has now extended beyond the formally defined current boundaries of the Standard Metropolitan Statistical Areas (SMSAs). It thus shows up as nonmetropolitan growth. We need further imagine that urban development continues in the smaller cities and within their metropolitan boundaries. Of course the attractiveness of rural areas may be increasing at the same time.

It must be mentioned that clean-break advocates have entertained the possibility of a continued wave effect but have rejected it by noting that the most dramatic net migration changes have taken place in those US counties that are nonadjacent to the metropolitan areas (Morrison, 1977). However, arranging the US data in terms of a locational breakdown of nonmetropolitan growth (Table 4.1) reveals that in the most recent years annual growth is greatest in those nonmetropolitan counties which are most linked to the metropolitan centres. Annual net in-migration rates *diminish regularly* as we move away from SMSAs (*see also* Tucker, 1976). Thus, the US data do not rule out the wave theory and statements such as 'clearly the migration reversal cannot be explained away as just more metropolitan sprawl or spillover because it is affecting distinctly remote and totally rural non-metropolitan areas, *as well as those adjacent to metropolitan centres*' (Morrison, 1977, p. 6) are not really conclusive. In fact, the most compelling position is probably that of Wardwell, who underlines the complexity of recent trends as well as our inability to interpret them unequivocally. Wardwell cites the fact that 63 per cent of in-migration to nonmetropolitan counties takes place in those nonmetropolitan counties that are adjacent to metropolitan counties and says that 'this suggests that the spillover effect of continued deconcentration of metropolitan centres is a substantial force in producing the observed patterns of nonmetropolitan county growth'. He also reports that the growth rate of counties classified as nonmetropolitan in 1970 but reclassified to metropolitan in 1974 'is substantially greater (10 per cent) during this period than that of counties which retained their nonmetropolitan classification' (Wardwell, 1977, p. 159). Beale (1977, p. 116) counters that 'The more impressive fact would seem to be the convergence of growth rates of these two classes of counties.' Berry and Dahmann report that

> In the South ... the central cities of metropolitan areas with less than one million residents have gained population. ... In the West the largest gains have been occurring in central cities of metropolitan areas with less than one million residents.
> (Berry and Dahmann, 1977, p. 450)

All of these observations are consistent with the version of the wave theory outlined above.

Obviously, there is something going on in the nonadjacent counties which demands attention. Wardwell suggests that this growth can be explained by new propensities to retire and recreate *and* that these new phenomena can be analysed on top of the wave effect rather than in its place.

40 Differential Urbanization

Table 4.1 Locational breakdown of US population growth

Population category	Provisional 1975 population (x10³)	Annual population change rate		Annual natural increase rate		Annual net migration rate[a]	
		1960–70	1970–75	1960–70	1970–75	1960–70	1970–75
United States total	213 051	1.3	0.9	1.1	0.7	0.2	0.2
Metropolitan							
Total, all SMSAs[b]	156 098	1.6	0.8	1.1	0.7	0.5	0.1
>1.0 million	94 537	1.6	0.5	1.1	0.6	0.6	–0.2
0.5–1.0 million	23 782	1.5	1.0	1.2	0.8	0.4	0.3
0.25–0.5 million	19 554	1.4	1.3	1.2	0.8	0.2	0.5
<0.25 million	18 225	1.4	1.5	1.2	0.8	0.2	0.7
Nonmetropolitan							
Total, all nonmetropolitan counties	56 954	0.4	1.2	0.9	0.6	–0.5	0.6
In counties from which:							
>20% commute to SMSAs	4 407	0.9	1.8	0.8	0.5	0.1	1.3
10–19% commute to SMSAs	10 011	0.7	1.3	0.8	0.5	–0.1	0.8
3–9% commute to SMSAs	14 338	0.5	1.2	0.9	0.6	–0.4	0.6
<3% commute to SMSAs	28 197	0.2	1.1	1.0	0.6	–0.8	0.5
Entirely rural counties[c] not adjacent to an SMSA	4 661	–0.4	1.3	0.8	0.4	–1.2	0.9

Source: unpublished preliminary statistics furnished by R.L. Forstall, Population Division, US Bureau of the Census; and C.L. Beale, Economic Research Service, US Department of Agriculture

[a] Includes net immigration from abroad, which contributes newcomers to the USA as a whole and to the metropolitan sector, thereby producing positive net migration rates for both
[b] Population inside Standard Metropolitan Statistical Areas (SMSAs) or, where defined, Standard Consolidated Statistical Areas (SCSAs). In New England, New England County Metropolitan Areas (NECMAs) are used
[c] 'Entirely rural' means the counties contain no town of 2500 or more inhabitants (reproduced from Morrison, 1977)

The most stirring of the reversal reports is the one by Vining and Strauss, in which it is stated that

> Nonmetropolitan counties well-removed from the commuting range of our 250 or so SMSAs are growing at a significantly higher rate than the SMSAs themselves, *though at a somewhat lower rate than the nonmetropolitan counties adjacent to these SMSAs* (Morrison, 1975, p. 10, italics added). This fact represents a clear and unmistakable break with past trends of long duration.
>
> (Vining and Strauss, 1977, p. 751)

We have added the italics to emphasise a possible *non sequitur*. Vining and Strauss go on to look for evidence from a source other than the migration data; they process population stock data through the well-known Hoover index of population dispersion.[2] Interpreting trends in the index in a novel way, the authors conclude that a wave effect can be rejected and that a clean break is, in fact, observed.

In describing the pre-1970 US settlement changes, the authors note that the Hoover index, calculated for various levels of spatial aggregation, moves in opposing directions. They view this quirk in the index as a 'resource'. Previously, for example, the index would turn up when the spatial units were US counties, indicating urbanisation. At the same time, the index would turn down when the units were states, indicating a movement of the population to the less-populated Midwest and West. Thus, a clean break is announced when the index, computed for *all* levels of aggregation, turns down, as it does for the most recent years. However, computations of the Hoover index for small spatial units can show a downturn and still be consistent with the wave effect. Table 4.1 underlines this view: the small or lightly populated nonmetropolitan counties and the smaller SMSAs are the major gainers; looking at *where* the major nonmetropolitan growth is taking place, we are back to spillover effects. In other words, if we were to compute the Hoover index for US spatial units which combine metropolitan areas with only the adjacent counties, Table 4.1 suggests that we may not get a downturn after all.

That test would consider a subset of spatial units and would be distinct from Vining and Strauss's Hoover index computations for the US state economic areas (SEAs). Their SEA test deals with a collectively exhaustive set of spatial units and could be marred by the effect which Vining and Strauss describe in their paper: the Hoover index computed over states may show decentralisation even while substantial urbanisation is taking place because of the movement to urban centres (from rural areas) which happen to be in the less populous regions of the country. In any event, a different application of the Hoover index was developed for the FUR data file (*see below*) which avoids some of these difficulties.

Most of the evidence that has been cited up to this point has been from the works of the clean-break advocates. Clearly neither side has proven its case. The problem lies with the way in which the data are reported. The US Census Bureau divides the country into two population concentrations: metropolitan and nonmetropolitan areas. The former are made up of a central city and a suburban area. Any additional large cities within the metropolitan areas are included as part of the central city. Nonmetropolitan areas include all the area outside metropolitan areas. Unfortunately this way of reporting data is not 'functional'. Since SMSA boundaries tend to be county boundaries, the exact or near-exact limit of the commuting field is usually not adequately approximated. The same applies to temporal change in the labour-market area. Thus as the wave of development spreads outward and spills over SMSA lines, a 'reversal' is perceived though none may have occurred.

Cliff and Robson report that since most reporting units

> are defined as distinct physical nucleations rather than in functional terms, then in studying changes over time, the researcher is caught on the horns of two dilemmas: whether to use an unchanging areal definition of each town or to alter the definition so as to match most closely the changing form of the town at successive dates, *and* whether to use a fixed or fluctuating number of towns throughout the period.
> (Cliff and Robson, 1978, p. 163)

The ambiguity of the US data arises precisely because of these two dilemmas. Yet we do not want to continue to plumb the US data, having maintained that they cannot hold the answer. Rather, we want to look at a new data file for some indication of what happened in the recent experience of Europe and Japan.

The data file and definitions

In an effort to initiate wide-ranging comparative investigations of patterns of urban growth and decline as well as to test the effects on urban growth of

42 Differential Urbanization

various national policies, a network of scholars from the International Institute for Applied Systems Analysis and from collaborating institutions in a number of countries have joined to define comparable sets of urban areas for seventeen nations in Western Europe and Eastern Europe and for Japan. To date, population, employment, and area data have been stored for these countries for the years 1950, 1960, and 1970, with post-1970 data available for five countries. The actual delineations have emphasised urban core areas, their hinterlands, and the residual rural areas. The core areas and their associated hinterlands make up 'functional urban regions' (FURs). These are defined so that commuting across FUR boundaries is minimal. In that sense, they are similar to the US Bureau of Economic Analysis (BEA) regions and represent functional labour markets.

The most useful aspect of this data file is consistency and comparability between the various nations. Enough data are available to compute a variety of Hoover indices for many regional sets and subsets. For this we adopt the following notation:

$H_i(t)$ is the Hoover index computed for some nation over the set of regions i, for year t;
$H_{ij}(t)$ is the Hoover index computed for a nation over the union of the set of regions i and j, for year t.

It should be noted that the index will be computed for sets of regions which are exhaustive as well as for subsets of regions. Vining and Strauss looked at Hoover indices for a variety of regional delineations for the USA; yet all of these were exhaustive delineations. If the set of regions for which we compute the index is exhaustive then the proportions of population and area are defined with the national totals as denominators. If, however, the set is some subset, such as the set of all *urban* areas, then the denominators used in computing percentages refer to total *urban area* and population. The reason for this convention is that we wish to observe trends in H_i which are not affected by trends in other subsets of regions. We hope to show that this modified version of the Hoover index renders it a more powerful tool.

We denote:

u as the set of all urban core areas;
h as the set of all hinterland areas;
r as the set of all rural areas;
s as the set of all functional urban areas, each of which is $u + h$;
uhr, sr are exhaustive unions of regional subsets.

A compact way of representing Hoover index trends for 18 developed countries over the 20-year span 1950 to 1970 is the array of index changes, or concentration changes, as shown in Table 4.2. Post-1970 performance is shown in Table 4.3 for some countries. Overall population concentration is measured by looking at the behaviour of the first two indices which are defined over exhaustive sets of areas. We note that three groupings are possible. Since far more

Table 4.2 Population concentration trends indicated by direction of Hoover index changes

Country	Period	H_{uhr}	H_{sr}	H_u	H_h	H_s
Group A						
Spain	1950–70	+	+	+	+	+
Japan[a]	1960–70	+	+	+	+	+
Finland	1955–70	+	+	+	+	+
Italy	1950–70	+	+	+	+	+
Group B						
Norway	1950–70	+	+	–	+	+
Sweden	1950–70	+	+	–	+	+
Denmark	1950–70	+	+	–	+	+
Portugal	1950–70	+	+	–	+	+
France	1950–70	+	+	–	+	+
Ireland	1950–70	+	+	–	+	+
Hungary[a]	1960–70	+	+	–	+	+
German Federal Republic[a]	1950–70	+	+	–	+	+
Group C						
Great Britain	1950–70	–	–	–	+	–
Netherlands	1950–70	–	–	–	+	–
Switzerland	1950–70	–	–	–	+	–
Belgium	1950–70	+	–	–	+	NC
Austria[a]	1950–70	NC	–	–	–	–
Poland[a]	1950–70	–	–	–	–	–

Key: NC = no change
[a] Delineated in terms of urban cores and hinterlands only; there are no nonhinterland rural areas

Table 4.3 Post-1970 population concentration trends for those countries for which 'functional urban region' (FUR) data are available

Country	Period	H_{uhr}	H_{sr}	H_u	H_h	H_s
Poland	1970–73	+(R)[a]	NC	+ (R)	+ (R)	NC
Japan	1970–75	+	+	+	+	+
Hungary	1970–75	+	NC	–	+	NC
Finland	1970–74	+	+	NC	+	+
Denmark	1970–75	– (R)	– (R)	–	+	– (R)

Key: NC = no change
[a] (R) indicates a reversal from the pre-1970 trends shown in Table 4.2

data are available for the years up to and including 1970, those results are examined first. An obvious grouping of nations can be seen. The countries in group A show increasing concentrations of their populations for all spatial levels of aggregation; most growth took place in the most populous spatial units. (Actually, the index only allows change towards more or less dense settlements to be detected. Yet the strong correlation between size and density allows us to use the more useful size characterisations.)

The countries in group B are of interest because they show increasing concentration of the population *except* with respect to urban cores. The straight

44 Differential Urbanization

column of minuses for H_u, group B, shows that the smaller urban cores are getting more of the growth than the larger urban cores. This should be linked with the pluses in the next column. In fact, across groupings and for as many as 16 of the 18 countries, the larger hinterlands grew faster than the smaller ones. If we recall that large urban cores are associated with the larger hinterlands, then spillover growth is suggested. In fact, for the 12 countries which have negative changes in H_u *along with* positive changes in H_h, it seems that the diminishing importance of the largest urban core areas and the concurrent increasing importance of the large hinterland areas is strong evidence of a wave effect and reinforces scepticism as to the clean break. The countries in group C show deconcentration in light of the signs on Hoover index changes computed for exhaustive sets of areas. In other words, the overall figures are *heavily weighted* by the effect noted for the urban cores. (A similar table was computed for the interval 1960–70. Surprisingly, very little changed. Poland and Switzerland exchanged places between groups B and C. The number of countries for which we have a negative change for H_u along with a positive change for H_h remained at 12.)

Of course post-1970 data are more interesting because the alleged reversals are a recent phenomenon. Unfortunately, those data are limited to five countries. Table 4.3 shows that Japan continued to concentrate at all levels of aggregation. Yet the actual numbers show that the rate of increase in Hoover index values falls for each year between 1970 and 1975. Perhaps Japan will soon be in group B. Denmark is the clearest example of transition from group B to group C, suggesting that there may be a natural evolutionary sequence.

The case of Poland is the most difficult to decipher. The raw data suggest that there is a decline in the relative importance of the large cities yet, within that set, growth is skewed towards the larger urban cores.

Problems of inference

As mentioned, the delineations on which our data file is based for functional urban areas are defined by commuting patterns for 1970. The hinterland is usually defined as areas from which at least 15 per cent of commuting is to the central city. Obviously areas which were functional spatial units in 1970 might not have been so for 1950 or for 1960. Thus a bias similar to that which we have discussed with respect to the US data is built into our sample. The crucial difference is that the definitional units differ. Hinterlands are much more spatially extensive than the US metropolitan suburban areas. Thus our computed Hoover index over the set of hinterland areas would be akin to looking at some suburban and some adjacent as well as some nonadjacent but linked counties for the US sample. Also, our use of the Hoover index is somewhat novel in that we have been able to look at hinterland and core areas separately. It is these crucial differences which cause us to believe that our approach permits analysis that is surely possible with the US data but is not often practised because of the convenient availability of standard reporting units (metropolitan versus nonmetropolitan).

We should also consider the extent to which the fixed boundaries of our sample have biased our own computations. Looking at the definition of the Hoover index, we find that all the P_{it} certainly changed over time while constant values for a_i were used. Yet a_i should also have a 't' subscript because the boundaries of the functional areas certainly advance with population growth. In fact, if the FUR boundaries advanced such that areal proportions kept exact pace with population changes, then the Hoover index would remain constant. This could not occur in a situation such as the Vining–Strauss investigation because of their use of fixed administrative boundaries, but it is very much a problem when the regional definitions are supposedly functional and encompass a subset of regions. We recognise this problem of possible bias in our fixed-area regions and counter by asserting that, over the relatively short time span considered, it is likely that population changes were much greater than areal changes. As such the calculated indices should certainly change, although the rate of change may be overstated in our results.

Cliff and Robson (1978) suggest the obvious: any sort of functional regions which are studied over time must be made up of constituent units for which data are available so that recalculations can be made for alternative areal units. Zelinsky does precisely this in his study of Pennsylvania settlement systems. Of course this procedure brings on new problems of *how* to reclassify the smaller spatial units. In spite of this, Zelinsky gets closer to events than many of the other cited studies and comes out on the side of a wave effect. Writing about the period 1950–70, he concludes that what is observed in the USA is 'a reconcentration of people within distances of some 25 to 35 miles of the metropolitan centre' (Zelinsky, 1978, p. 37).

Conclusion

Our survey of some of the evidence presented for US settlement patterns suggests that there is cause for scepticism of a clean break. For the 18 countries of our sample, we have been able to look at developments beyond the metropolitan areas and these suggest that a continuing wave effect is taking place rather than a clean break. Since there is no reason to expect that settlement patterns in Europe, Japan, and the USA evolve in opposite fashion, the findings from the FUR file lend some support for the wave-effect conclusion in the USA. In that event, we side with Wardwell's (1977) judgement that the US record alone is too complex to denote a clean break with the past.

Yet the Hoover index values that have been computed are perhaps also suitable for the testing of some demoeconomic hypotheses. Human settlement patterns, it has been suggested, change in response to new technologies, a new age structure of the population, and new social arrangements, especially with regard to pensioning and retirement practices. McCarthy and Morrison (1977) sustain similar hypotheses for the US case. An attempt to develop similar tests for the FUR data file was less successful. Collinearity hampered the proper test specifications and generated ambiguous results.

The standard urban economic models of Alonso and Mills suggest that rising incomes and declining travel costs explain flatter bid–rent curves and eventual expansion of the metropolis. Other urban and regional economic theories, of various degrees of formality, are available in support of the wave effect. A preference for small-town life has long been used to explain suburbanisation. The data seem to suggest that this trend is as strong as ever and that it is taking place at ever greater distances from central cities, especially if these central cities are large. Wardwell concludes that people are showing 'a clear desire for living in smaller sized places within commuting radius of the metropolitan centre, *and* for smaller sized places beyond that radius in preference to living within the centre itself' (Wardwell, 1977, p. 176, italics added).

None of this is really new. Commuting radii are growing as usual. Central-city decline, as W. Thompson suggests, is a cause as well as an effect. For example, if we detect central-city growth in the smaller cities and peripheral growth in the larger cities we may hypothesise that agglomeration diseconomies emerge *in central locations* when the metropolis is mature. Wardwell quotes Thompson's detailing of this hypothesis: Thompson suggests that large urban areas are the natural incubus to new industry formation and innovation only as long as their industries are centrally located. As soon as plants begin to decentralise, as they inevitably do at their stage of maturity, the centres of the larger cities lose this important function and begin to decline.

This is related to the Vernon hypothesis (Vernon, 1960). The latter suggests that central cities are hospitable to innovation and new industrial processes because they are the scene of external economies. Yet, as plants grow, they seek scale economies rather than external economies and therefore seek cheap lands in the peripheral areas. Thus they leave the centre and add to its decline in two ways: by not being there and by no longer providing external economies to newcomers.

The theory that is available on behalf of a reversal thesis (*see*, for example, Friedmann, 1973) is much slimmer.

Obviously, more theory building and more testing are required. Working across an international cross section with the aid of a small sample does not guarantee definitive results. Yet policy issues such as whether planning ought to be done at metropolitan levels or not depend, in part, on whether metropolitan areas are expanding or whether they are becoming ever less important.

Notes

1 It is important to note that Berry and Dahmann (1977, p. 488) do *not* restrict their analysis to 1970 and beyond. They assert that 'Signs of a shift away from the long term trend of metropolitan growth exceeding that of non-metropolitan areas in the United States first appeared during the 1960s'.
2 The Hoover index is given by $H_t = \frac{1}{2} \sum_{i=1}^{k} |p_{it} - a_i| 100$, where p_{it} refers to the proportion of a country's population residing in area *i* at time *t*; a_i refers to the proportion of that nation's area taken up by subarea *i*. The index varies from 0 to 100, or from a reading of perfectly uniformly distributed population to perfect concentration.

References

Beale, C. L. 1977: The recent shift of United States population to nonmetropolitan areas, 1970–75. *International Regional Science Review* 2, 113–22 (*see* this volume Chapter 2).

Berry, B. J. L. and Dahmann, D.C. 1977: Population redistribution in the United States in the 1970s. *Population Development Review* 3, 443–71.

Cliff, A. D. and Robson, B. T. 1978: Changes in the size distribution of settlements in England and Wales 1801–1968. *Environment and Planning A* 10, 163–71.

Friedmann, J. 1973: *Urbanisation, planning, and national development*. Beverly Hills and London: Sage Publications.

McCarthy, K. F. and Morrison, P. A. 1977: The changing demographic and economic structure of nonmetropolitan areas in the United States. *International Regional Science Review* 2, 123–42.

Morrison, P. A. 1975: *The current demographic context of national growth and development*. P-5514. Santa Monica, Calif.: Rand Corporation.

Morrison, P.A. 1977: *Current demographic change in regions of the United States*. P-6000. Santa Monica, Calif.: Rand Corporation.

Tucker, C. J. 1976: Changing patterns of migration between metropolitan and nonmetropolitan areas in the United States: recent evidence. *Demography* 13, 435–43.

Vernon, R. 1960: *Metropolis 1985*. Cambridge, Mass.: Harvard University Press.

Vining, D. R. Jr and Kontuly, T. 1977: Population dispersal from major metropolitan regions: an international comparison. RSRI-100. Philadelphia, Pa.: Regional Science Research Institute (*see* this volume, Chapter 6).

Vining, D. R. Jr and Strauss, A. 1977: A demonstration that the current deconcentration of population in the United States is a clean break with the past. *Environment and Planning A* 9, 751–8 (*see* this volume, Chapter 3).

Wardwell, J. M. 1977: Equilibrium and change in non-metropolitan growth. *Rural Sociology* 42, 156–79.

Zelinsky, W. 1978: Is nonmetropolitan America being repopulated? The evidence from Pennsylvania's minor civil divisions. *Demography* 15 (1), 13–39.

5 K. Richter
'Nonmetropolitan Growth in the Late 1970s: The End of the Turnaround?'

From: *Demography* 22, (2), 245–63 (1985)

Introduction

The 'discovery' of the turnaround in growth patterns between metropolitan and nonmetropolitan areas in the early 1970s generated a large body of both theoretical and descriptive literature. Early research documented that the switch from negative to positive net migration into nonmetro areas was more than just a continuation of urban sprawl and that real growth was occurring in remote rural areas. Explanations for the trend included both the economic (deconcentration of manufacturing, expanding energy extraction, growth in the government and service sectors) and the non-economic (preference for rural living, retirement migration, and the modernisation of nonmetro areas, including greater accessibility of urban centres) (Beale, 1977; Beale and Fuguitt, 1978a; Dillman, 1979; McCarthy and Morrison, 1979; Heaton *et al.*, 1981). Though the shift took many researchers by surprise, it was soon incorporated into theories of urban evolution and economic differentiation. Many voiced concern, however, that the new trend would level off or 'bottom out', for several reasons. If nonmetropolitan growth rates reflected a preference for rural living, growth and continued development in such areas could counteract these forces. The growth in new sectors of the economy such as recreational and service-related industries and the expansion of government employment and higher education were also seen as trends that would not continue indefinitely. The most frequently voiced concern was that rising energy costs, which accelerated after the oil embargo of 1973–74, would hamper the accessibility of remote regions to urban areas and thus lessen their appeal (Beale, 1976; Phillips and Brunn, 1978; Voss and Fuguitt, 1979). Long and DeAre (1980) specifically addressed these issues in their examination of turnaround trends as of 1978; they found that differential growth between nonmetropolitan and metropolitan counties was actually larger in 1974–78 than in the 1970–74 period. They concluded that the momentum of nonmetro growth would continue owing to higher rates of natural increase, decentralisation of employment, and convergence of nonmetropolitan and metropolitan income levels. Tucker (1982) examines Current Population Survey (CPS) estimates of net migration in the 1970–75 and 1975–80 periods and concurs that the propensity of metropolitan residents to leave SMSAs and for nonmetro residents to remain in nonmetro areas remained higher throughout the 1970s than in the 1965–70 period.

Besides Tucker's and Long and DeAre's studies, there has been relatively little analysis of nonmetropolitan growth within the 1970–80 period. Most of the turnaround literature compares the post-1970 period with the 1960s and previous decades. The objective of this paper is to examine possible changes in the extent of the turnaround throughout the full decade of the 1970s. Comparisons of trends in nonmetropolitan growth are made across three time periods: 1970–74, 1974–77 and 1977–80. These intervals were chosen to conform to previous research on the turnaround which examined the 1970–74 period, and to divide the remainder of the decade into intervals of similar length. Counties are grouped by size of place to determine if the inverse pattern of the turnaround, with the most remote and least dense counties showing the greatest increase in growth rates, continued throughout the decade. How the components of growth, net migration and natural increase interacted over the course of the 1970s is compared for metro and nonmetro counties. Explanations of trends in net migration are explored by grouping counties by region and economic characteristics. Regression analysis is used to examine which factors continue to explain these trends throughout the decade and which diminish in importance.

Data

The data used in this analysis are from a special file of intercensal county estimates prepared by the Census Bureau from the Federal–State Cooperative Series (Current Population Reports Series P-25 and P-26). The annual estimates were prepared using a combination of Administrative Records (using federal tax returns), multiple regression, and vital components techniques. While the methods used to develop the estimates were not completely consistent for all years (the Administrative Records technique was not used in 1971 and 1972), inaccuracies due to this inconsistency should not be significant. The estimates were adjusted by the Census Bureau for differences between the 1 April 1980 estimate for each county and the 1980 census count (defined as the error of closure due to estimation error). This process utilised a curvilinear procedure which took into account both the length of time from the previous census and the size of the estimation error (see Appendix, p. 66). Comparison of these intercensal estimates with net migration figures from the Current Population Survey for 1970–75 and 1975–80 show similar results (see Tucker, 1982).

The analysis reported here is based upon 3097 county units and their equivalents (election districts in Alaska, independent cities combined with adjacent counties in Virginia, and SMSA equivalents in New England). Annual births and deaths for these counties and county equivalents were obtained from the National Center for Health Statistics in order to examine natural increase and net migration in the decade.

A problem in examining trends in metro and nonmetro growth is that the official definition of which counties are metropolitan shifted throughout the decade. A nonmetro county which has been reclassified as metropolitan by the end of the decade is more accurately viewed as undergoing metropolitan growth or expansion rather than receiving turnaround migration. Long and DeAre (1980) compared growth rates using a 1970, 1974 and 1980 definition and found that the same general trends prevailed regardless of the definition used. In this paper, both the 1974 (based on 1970 commuting data) and 1980 definitions are used to examine migration and growth trends. The 1974 definition gives a picture of what occurred in counties classified as nonmetro at the outset of the decade, and is used in a detailed comparison of population change and migration for different types of metropolitan and nonmetropolitan counties. On the other hand, the 1980 definition is the more restrictive classification of which counties remained nonmetropolitan throughout the 1970s, and is used in the subsequent examination of nonmetro trends.

Nonmetropolitan growth trends

A first look at the data for the late 1970s reveals that nonmetropolitan growth appears to have slowed. Table 5.1 shows annual growth rates over the three time periods for various metropolitan categories. These rates are the average percentage increase in a 1-year period and facilitate comparisons between the three varying-length time periods (of 4.25, 3.00 and 2.75 years). While

50 Differential Urbanization

Table 5.1 Annual growth rates[a] by metropolitan status

Area	1960–70	1970–74[b]	1974–77[b]	1977–80[b]
US total	1.25	1.14.	1.00	1.11
1980 metropolitan definition				
Metropolitan	1.57	1.04	0.85	1.07
Nonmetropolitan	0.30	1.46	1.47	1.23
Adjacent	0.52	1.61	1.49	1.43
Nonadjacent	–0.04	1.23	1.43	0.90
Ratio nonmetropolitan/metropolitan	0.19	1.40	1.73	1.15
1974 metropolitan definition				
Metropolitan	1.57	1.00	0.80	1.04
SMSA >500 000				
Core	1.20	0.39	0.19	0.53
Fringe	2.76	1.69	1.51	1.64
SMSA 100 000–500 000				
Core	1.55	1.67	1.36	1.42
Fringe	1.41	2.37	2.04	2.16
SMSA <100 000 (core)	1.02	1.26	1.79	2.02
Nonmetropolitan[c]	0.42	1.53	1.51	1.30
Adjacent	0.70	1.70	1.52	1.52
SLP 2500+	0.75	1.68	1.46	1.48
SLP <2500	0.22	1.95	2.10	1.92
Nonadjacent	0.14	1.34	1.50	1.06
SLP 10 000+	0.62	1.47	1.42	1.14
SLP 2500-10 000	–0.17	1.14	1.49	0.96
SLP <2500	–0.34	1.44	1.72	1.06
Ratio nonmetropolitan/metropolitan	0.27	1.53	1.89	1.25

[a] Growth rate is computed by $\frac{p_2 - p_1}{0.5k(p_2 + p_1)} \times 100$, where k is the length of the time period in years
[b] Does not include Washington DC
[c] Nonmetropolitan counties are classified by size of largest place (SLP) as of 1970

nonmetropolitan growth rates continued to exceed both the national rate and that of metropolitan areas, by the late 1970s the differential narrowed. Using the 1980 metropolitan definition (Table 5.1) the ratio of nonmetro to metro growth fell to 1.15 in 1977–80 after a 1.73 ratio in 1974–77. The drop in the growth rates for nonmetro areas in the late 1970s occurred almost entirely in counties that were not adjacent to metropolitan areas (0.90 in 1977–80 versus 1.43 in 1974–77). Thus the 'turnaround', or higher growth in nonmetro than metro counties, was found only in counties adjacent to an SMSA after 1977. Nonadjacent counties had lower growth rates than metro areas (0.90 versus 1.07) while adjacent counties continued to grow at about the same rate as in the 1974–77 period (1.43 versus 1.49). Meanwhile metropolitan counties actually experienced higher growth in the 1977–80 period than at any other point in the decade.

Table 5.1 also shows growth rates using the more detailed metropolitan classification as of 1974. Throughout the 1970s, growth in the larger metropolitan areas (those over 500 000) occurred mainly in fringe counties outside the central city, continuing the pattern of the 1960s. But in a reversal of the 1960s pattern, fringe counties of smaller SMSAs were growing faster than those of the larger SMSAs in the 1970s. Growth in core areas of the largest metro areas (over 500 000) was much lower in the 1970s than in the 1960s, though the 1977–80 period showed an upturn in growth for those areas. The growth rate for the smallest SMSAs (those under 100 000) increased consistently over the course of the decade, with a rate almost twice that of the nation as a whole by 1977–80 (2.02 per cent annually versus 1.11 per cent nationally). The non-metropolitan, nonadjacent counties growing fastest at mid-decade were those with the smallest population centres, a group that had experienced negative growth in the 1960s. By the 1977–80 period, however, growth in the remotest regions appears to have slowed. The decline in growth rates for nonadjacent areas was greatest in completely rural counties (those with no centre of 2500), where growth dropped to 1.06 after a 1974–77 rate of 1.72. In nonmetro counties adjacent to an SMSA, growth rates remained fairly constant throughout the decade, with a small decline for counties having cities of over 2500 population. Thus while nonmetro counties continued to grow in the 1977–80 period, several of the patterns emphasised in the turnaround literature of the early 1970s appear to have shifted. In particular, the growth of the 1977–80 period was characterised by the expansion of metropolitan areas, with adjacent counties having the highest growth rates, and by a revival in SMSA growth for the smaller SMSAs.

Migration trends

While nonmetropolitan counties growing at a faster rate than SMSAs in the early 1970s represented a marked change from previous patterns, an even more remarkable finding was that these counties had shifted from negative to positive net migration. Areas which had experienced an outflow of population for two decades became the recipients of new migrants while the traditional drawing power of metropolitan centres diminished. It is thus important to examine migration rates over the course of the decade to see if the momentum of this turnaround has continued. In addition, growth rates may mask trends in migration as shifts in the age structure of metro and nonmetro areas lead to changes in rates of natural increase.

Table 5.2 shows annual net migration rates for the 1960s and 1970s. The metropolitanization of the 1960s is shown by the negative growth rates in all non-metro areas and in SMSAs under 100 000. The turnaround of the 1970–74 period was thus even more dramatic than revealed by growth rates, with nonmetro net migration nearly three times that of metropolitan counties (0.88 per cent annually versus 0.32 per cent in metro areas). This differential increased in the 1974–77 period (0.92 per cent versus 0.24 per cent) and the highest migration rates were found in counties with the smallest population centres (1.65 per cent in adjacent counties and 1.26 per cent in nonadjacent counties with centres of less than 2500). But the flow of migration in the 1977–80 period into non-metro and particularly nonadjacent counties slowed even more than would be

52 Differential Urbanization

Table 5.2 Annual net migration rates[a] by metropolitan status

Area	1960–70	1970–74[b]	1974–77[b]	1977–80[b]
1980 metropolitan definition				
Metropolitan	0.43	0.32	0.24	0.40
Nonmetropolitan	−0.64	0.88	0.92	0.60
Adjacent	−0.42	1.02	0.96	0.82
Nonadjacent	−0.98	0.65	0.86	0.23
Ratio nonmetropolitan/metropolitan		2.75	3.83	1.50
1974 metropolitan definition				
Metropolitan	0.44	0.28	0.20	0.37
SMSA >500 000				
Core	0.13	−0.25	−0.34	−0.07
Fringe	1.57	0.93	0.88	0.96
SMSA 100 000–500 000				
Core	0.29	0.84	0.64	0.64
Fringe	−0.33	1.59	1.36	1.45
SMSA <100 000 (core)	−0.33	0.35	1.00	1.13
Nonmetropolitan[c]	−0.54	0.92	0.95	0.66
Adjacent	−0.24	1.10	0.99	0.92
SLP 2500+	−0.20	1.06	0.92	0.87
SLP <2500	−0.62	1.46	1.65	1.37
Nonadjacent	−0.85	0.73	0.91	0.38
SLP 10 000+	−0.54	0.72	0.73	0.37
SLP 2500–10 000	−1.05	0.62	0.96	0.34
SLP <2500	−1.16	1.01	1.26	0.49
Ratio nonmetropolitan/metropolitan		3.29	4.75	1.78

[a] Net migration rate is computed by $\dfrac{p_2 - p_1 - \text{Natural increase}}{0.5k(p_1+p_2)}$ where k is the length of the time period in years.
[b] Does not include Washington DC
[c] Nonmetropolitan counties are classified by size of largest place (SLP) as of 1970

indicated by growth rates; nonadjacent counties grew only 0.23 per cent annually from net migration in this period, less than a third of the 1974–77 rate. As seen in Table 5.2, migration into metropolitan areas picked up in this period, mainly because of greatly reduced levels of outmigration in the core counties of the largest SMSAs and small increases in fringe counties and small SMSAs. While nonmetropolitan migration was still 1.5 times that of metropolitan counties, much of this migration was to areas adjacent to a metropolitan centre rather than to the most rural counties. And for both adjacent and nonadjacent nonmetro counties, the biggest drops in net migration rates between the 1974–77 and 1977–80 period were in areas with smaller population centres.

Changes in natural increase in the 1970s

The change in the differential between metro and nonmetro growth rates is further explained by changes in natural increase over the period. The ageing of the nonmetropolitan population in the 1960s, as young people of childbearing ages moved out, helped to contribute to the low growth rates in these areas in the past. The turnaround in migration patterns in the 1970s means that imbalances in the age structure of nonmetropolitan areas may have begun to dissipate, causing increases in the rates of natural increase in nonmetro areas. But these age structure differentials would continue if retirement migration is a major explanation for the turnaround, as has been suggested by much of the literature (Wardwell, 1977; Beale and Fuguitt, 1978a; Lichter et al., 1979).

Figure 5.1 shows the dramatic drop in crude natural increase rates (births minus deaths per 1000) in the early 1970s, particularly for metropolitan areas. While the gap between metro and nonmetro counties was large in the early 1970s, it narrowed by the mid-1970s, and nearly disappeared for nonadjacent counties by 1977. If migration into nonadjacent counties was mainly by young families, increasing birth rates and declining death rates in these areas may explain the closing gap. These two components of natural increase are examined in Figure 5.2. Crude death rates declined throughout the decade for both metro and nonmetro areas, and the gap narrowed as the nonmetropolitan rates declined more steeply. Crude birth rates dropped sharply from 1970 to 1973, then levelled off before increasing again in 1977. Nonmetropolitan birth rates were lower than metropolitan rates in 1970 but they were higher throughout the rest of the decade, and the gap widened particularly for nonadjacent counties. Thus the narrowed gap in natural increase between nonmetro and metro areas by the end of the decade was due to the fact that the birth rates in nonmetro areas became high enough to offset the higher nonmetro death rates, which in addition declined relative to metropolitan death rates over the course of the decade.

The interaction between natural increase and net migration in producing the growth rates for nonmetro and metro areas is summarised in Table 5.3.

Table 5.3 Annual growth, birth, death, and net migration rates by 1980 metropolitan status

Interval	Growth	=	Birth	−	Death	+	Net migration[a]
Total							
1970–74	1.14	=	1.62	−	0.93	+	0.45
1974–77	1.00	=	1.47	−	0.88	+	0.41
1977–80	1.11	=	1.51	−	0.85	+	0.45
Metropolitan							
1970–74	1.04	=	1.60	−	0.89	+	0.32
1974–77	0.85	=	1.45	−	0.84	+	0.24
1977–80	1.07	=	1.49	−	0.82	+	0.40
Nonmetropolitan							
1970–74	1.46	=	1.65	−	1.06	+	0.88
1974–77	1.47	=	1.54	−	0.99	+	0.92
1977–80	1.23	=	1.57	−	0.94	+	0.60
Adjacent							
1970–74	1.61	=	1.64	−	1.05	+	1.02
1974–77	1.49	=	1.51	−	0.98	+	0.96
1977–80	1.43	=	1.54	−	0.93	+	0.82
Nonadjacent							
1970–74	1.23	=	1.67	−	1.09	+	0.65
1974–77	1.43	=	1.59	−	1.02	+	0.86
1977–80	0.90	=	1.62	−	0.96	+	0.23

[a] US total figures include international migration but also reflect differences in census coverage over the period

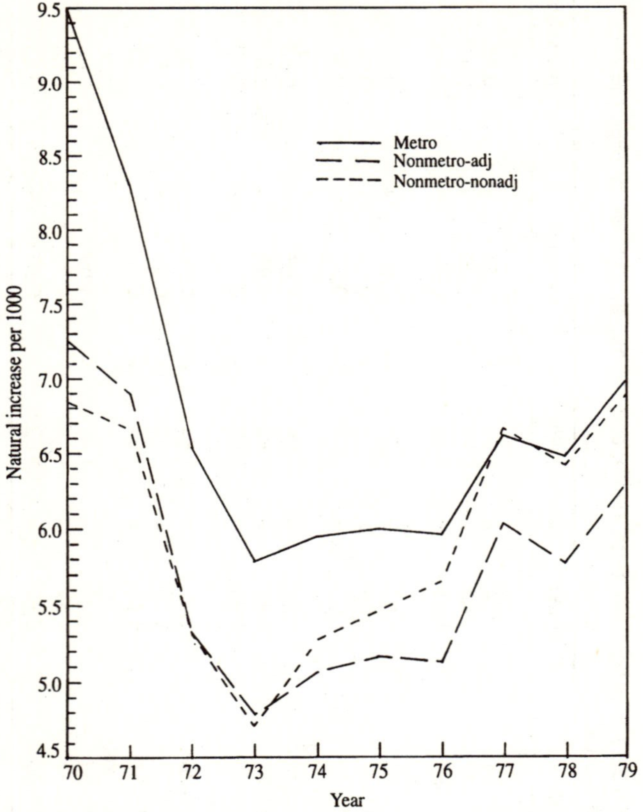

Fig 5.1 Natural increase by 1980 metropolitan status

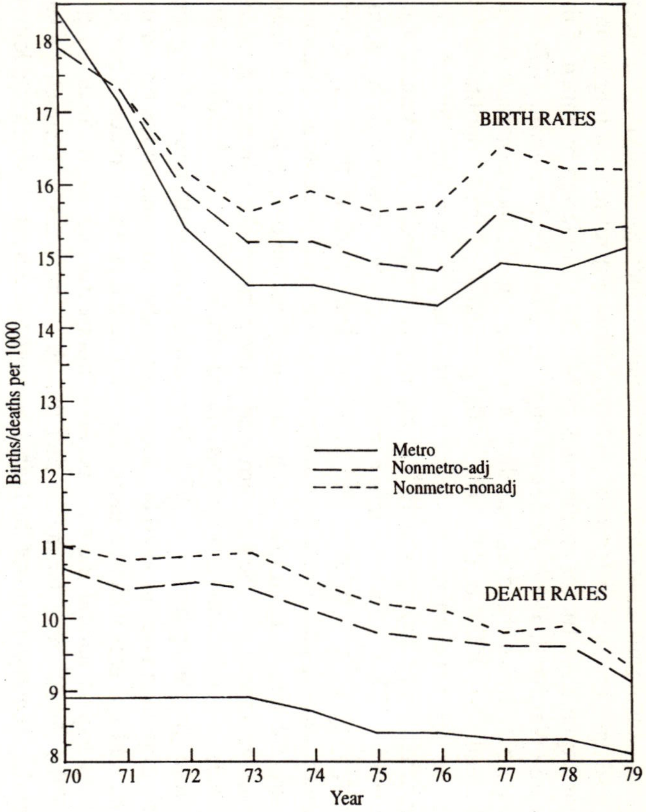

Fig 5.2 Crude birth and death rates by 1980 metropolitan status

Here birth and death rates are annualised over the three periods in the same way as net migration rates, so that the contribution of each to growth (births minus deaths plus net migration) is shown explicitly. These components are shown graphically in Figure 5.3. Much of the growth of metro areas in the 1970–74 period was due to natural increase, and the decline in growth by 1974–77 was due mainly to a drop in this component of growth. Both components increased in the 1977–80 period, but it is net migration that contributed most to the upturn in metro growth. For nonmetro areas the drop in death rates along with continued relatively high birth rates created a momentum of growth by the late 1970s that helped to offset the decline in net migration. In nonadjacent counties this relationship was most extreme; a comparison of the last period with the first shows that while annual crude death rates dropped 0.13 points, birth rates dropped only 0.05 and hence while migration dropped 0.42 points annual growth dropped only 0.33. Figure 5.3 clearly shows, however, that net migration continued to be the largest component of growth for counties adjacent to a metropolitan area throughout the 1970s, although for nonmetro counties overall natural increase contributed more to growth rates by 1977.

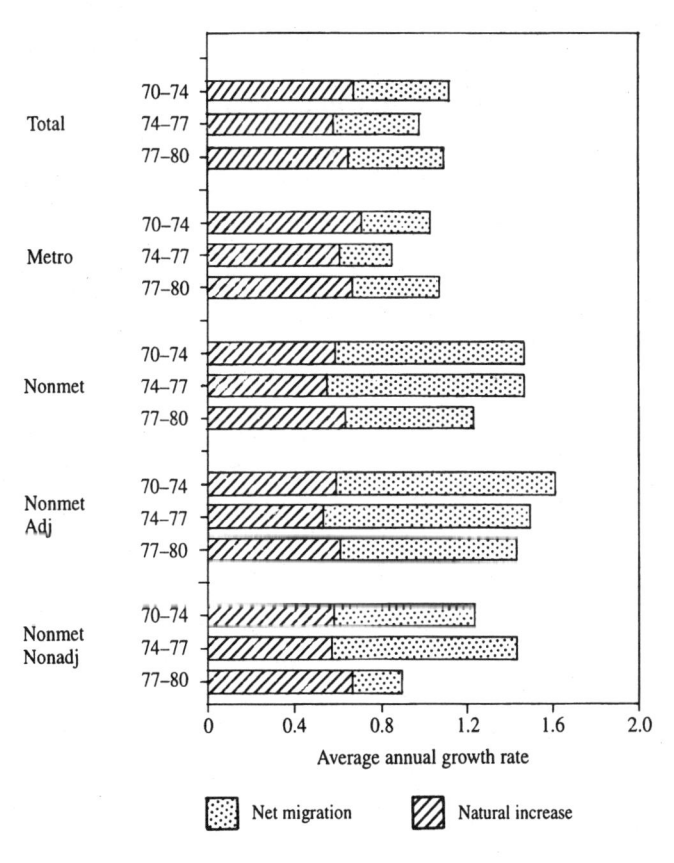

Fig 5.3 Components of growth by 1980 metropolitan status

These findings indicate that the age structure of nonmetropolitan areas became younger over the course of the decade; relative to metropolitan areas, death rates were lowered and birth rates rose. Census data confirm that while the mean age for those in metropolitan counties increased 1.8 years from 1970 to 1980 (rising from age 31.6 to 33.4); in nonmetro counties the increase was only 1.3 years (from 32.8 to 34.1) (US Bureau of the Census, 1981). These changes have helped to offset the drop in net migration by the late 1970s, to the point that natural increase is the largest component of growth for nonmetro areas in 1977–80. An examination of the age composition of metro to nonmetro and nonmetro to metro streams in 1970–75 by Lichter *et al.* (1979) showed that while the 'turnaround' stream has an older age structure, the greater retention of young people in nonmetro areas has helped to offset the impact of the new migrants. Indeed, the two streams are so similar that the net impact on the age structure on either destination is minimal. The natural increase findings presented here, however, would tend to confirm Long and DeAre's (1980) conclusion that the momentum of growth to nonmetro areas should continue, even if migration slows, since part of the effect of the turnaround is the retention of young families. In this way it appears that the growth differential between metro and nonmetro areas will continue to narrow.

Regional trends in nonmetropolitan migration

The turnaround literature has cited other ways in which the differences between metro and nonmetro areas have lessened, such as in lifestyle, socioeconomic status and income. These changes come about as urban expansion made nonmetro areas more accessible, while the decentralisation of manufacturing and service jobs has led to the convergence of the economic roles of metro and nonmetro areas. Functional explanations for the turnaround have identified regions of the country which have benefited from this economic decentralisation as well as from increased energy extraction and recreational activity.

Nonmetropolitan migration is shown on a regional basis in Table 5.4 in order to examine the endurance of these trends over the course of the 1970s. Counties are grouped by the 26 economic subregions developed by Beale (Beale and Fuguitt, 1978b), as shown in Figure 5.4. Net outmigration is seen in all but four of the areas in the 1960s. By the 1970–74 period net outmigration occurred in only three areas, all of which were agricultural (the Central Corn Belt, Mississippi Delta and Northern Great Plains). Other agricultural areas tended to have below average net inmigration, such as the Dairy Belt, Southern Corn Belt, Coastal Plain Tobacco and Peanut Belt, the Old Coastal Plain and the Southern Great Plains. Areas which received above average migration included regions typified by urban expansion (the Northern Metropolitan Belt); retirement/amenity migration (Upper Great Lakes, Ozark–Ouachita Uplands, Florida Peninsula, the Southwest and Hawaii); and energy extraction activities (southern Appalachian Coal, the Rockies, Blue Ridge/Smokies, and East Texas/Coastal Plain).

Table 5.4 Annual net migration rates for nonmetropolitan counties by region[a]

Region	1960–70	1970–74	1974–77	1977–80
US nonmetropolitan total	−0.64	0.88	0.92	0.60
Northern New England–St. Lawrence	−0.62	0.73	0.77	0.04
Northeastern Metropolitan Belt	1.07	2.23	1.78	1.28
Mohawk Valley and New York–Pennsylvania Border	−0.35	0.42	0.24	−0.49
Northern Appalachian Coal Fields	−0.78	0.71	0.62	0.22
Lower Great Lakes Industial	−0.08	0.36	0.03	0.12
Upper Great Lakes	−0.15	1.70	1.43	0.38
Dairy Belt	−0.35	0.74	0.89	0.91
Central Corn Belt	−0.75	−0.17	0.02	−0.58
Southern Corn Belt	−0.47	0.58	0.54	0.37
Southern Interior Uplands	−0.18	1.02	1.31	0.69
Southern Appalachian Coal Fields	−2.19	1.18	2.27	0.14
Blue Ridge, Great Smokies, and Great Valley	−0.45	1.54	1.52	1.09
Southern Piedmont	−0.57	1.00	0.44	0.65
Coastal Plain Tobacco and Peanut Belt	−1.38	0.63	0.78	0.30
Old Coastal Plain Cotton Belt	−1.33	0.49	0.42	−0.08
Mississippi Delta	−2.31	−0.65	−0.14	−0.89
Gulf of Mexico and South Atlantic Coast	−0.66	0.23	1.23	1.08
Florida Peninsula	2.48	6.97	3.69	5.46
East Texas and Adjoining Coastal Plain	−0.09	1.39	1.19	1.97
Ozark–Ouachita Uplands	0.46	2.24	1.82	1.14
Rio Grande	−1.69	0.48	1.37	1.09
Southern Great Plains	−1.45	0.08	0.41	0.05
Northern Great Plains	−1.60	−0.20	0.13	0.33
Rocky Mountains, Mormon Valleys, and Columbia Basin	−0.56	1.53	1.58	1.50
North Pacific Coast (including Alaska)	0.07	1.41	1.98	1.70
Southwest (including Hawaii)	0.32	2.79	2.12	3.08

[a] Nonmetropolitan status as of 1980

By the mid-1970s some of these trends have continued while others have ebbed. In the 1974–77 period regions with energy-extractive activities showed increased net inmigration (Southern Appalachian Coal, Gulf of Mexico/South Atlantic, the Rio Grande). But the net flow to these areas slowed by 1977–80, as it did in some areas associated with retirement/amenity migration (Upper Great Lakes, Ozark–Ouachita), though the Florida Peninsula continued to have high net inmigration. Agricultural areas with the exception of the Dairy Belt continued to have below average (but mainly positive) net migration rates in the 1977–80 period. Areas which showed above average migration throughout the 1970s are the Northern Metro Belt, the Southern Interior Uplands, the Blue Ridge/Great Smokies/Great Valley, the Florida Peninsula, East Texas/Coastal Plain, Ozark–Ouachita, the Rockies, the North Pacific and Alaska, and the Southwest and Hawaii. This group includes the five regions with positive net migration rates in the 1960s; the other four (the Southern Interior Uplands, Blue Ridge, East Texas, and Rockies) did not experience a

Fig 5.4 Beale economic subregions of the United States. 1, Northern New England–St Lawrence; 2, Northeastern Metropolitan Belt; 3, Mohawk Valley and New York–Pennsylvania Border; 4, Northern Appalachian Coal Fields; 5, Lower Great Lakes Industrial; 6, Upper Great Lakes; 7, Dairy Belt; 8, Central Corn Belt; 9, Southern Corn Belt; 10, Southern Interior Uplands; 11, Southern Appalachian Coal Fields; 12, Blue Ridge, Great Smokies, and Great Valley; 13, Southern Piedmont; 14, Coastal Plain Tobacco and Peanut Belt; 15, Old Coastal Plain Cotton Belt; 16, Mississippi Delta; 17, Gulf of Mexico and South Atlantic Coast; 18, Florida Peninsula; 19, East Texas and Adjoining Coastal Plain; 20, Ozark–Ouachita Uplands; 21, Rio Grande; 22, Southern Great Plains; 23, Northern Great Plains; 24, Rocky Mountains, Mormon Valleys, and Columbia Basin; 25, North Pacific Coast (including Alaska); 26, the Southwest (including Hawaii)

turnaround until the 1970–74 period. It would seem that while these regions have entered a period of sustained nonmetropolitan growth, others such as the Mohawk Valley, Northern New England, and several of the agricultural regions experienced a short-lived boom in the mid-1970s which levelled off as the widespread nature of the trend moderated. Economic characteristics that provide further information to analyse these regional differences in migration rates will be examined below.

Analysis of county characteristics

In order to understand more fully the factors that may continue to drive non-metropolitan growth or that may have produced short-lived migration trends, we have examined county-level economic characteristics. While more recent data would be useful to examine how employment trends affected migration over the course of the decade, 1980 figures are as yet unavailable. However, counties can be categorised by their characteristics as of 1970. In this way we can see how counties, defined by their economic characteristics at the outset of the decade, fared throughout the 1970s.

Table 5.5 shows annual net migration rates throughout the decade by selected economic characteristics. Counties characterised by a high employment in agriculture experienced lower rates of migration throughout the decade. This effect was more moderate in the 1974–77 period, when even counties that had 30 to 40 per cent employment in agriculture experienced positive net migration. High outmigration among heavily agricultural counties returned in the 1977–80 period however. Counties which were not characterised by a high degree of manufacturing employment in 1970 had higher rates of growth throughout the 1970s, with those with a moderate degree of such employment having the greatest amount of net inmigration. As indicated

Table 5.5 Annual net migration rates by selected county characteristics for nonmetropolitan counties[a]

Selected county characteristics	1970–74	1974–77	1977–80	Number of counties
US nonmetropolitan total	0.88	0.92	0.60	2390
% employed in agriculture				
0–4.9	1.05	1.05	0.62	443
5–9.9	0.99	0.96	0.72	535
10–19.9	0.97	0.96	0.76	712
20–29.9	0.30	0.68	0.23	370
30–39.9	–0.39	0.06	–0.56	221
40+	–0.88	–0.79	–1.57	109
% employed in manufacturing				
0–4.9	0.37	1.03	0.54	391
5–9.9	1.36	1.32	1.12	329
10–19.9	0.91	1.18	0.73	530
20–29.9	0.96	0.94	0.58	535
30–39.9	0.72	0.66	0.32	361
40+	0.74	0.46	0.41	244
% employed in mining				
0–4.9	0.90	0.85	0.56	2078
5–9.9	0.90	1.30	1.18	155
10+	0.66	1.51	0.66	157
% employed in entertainment and personal services				
0–4.9	0.80	0.78	0.32	1048
5–9.9	0.84	0.94	0.69	1199
10+	1.93	1.82	1.87	143
% employed in the military				
0–4.9	0.93	0.94	0.64	2344
5–9.9	0.51	0.10	–0.24	21
10+	–1.38	0.94	–0.31	25
State college				
No	0.86	0.96	0.65	2227
Yes	1.03	0.67	0.30	163

[a] Nonmetropolitan status as of 1980

in the regional tables, counties with some degree of mining employment had high rates of inmigration during the 1974–77 period, but this effect declined for those with a high (10 per cent and up) degree of mining employment by 1977. Counties with at least 10 per cent of their employment in entertainment and personal services had very high rates of inmigration throughout the period. Two variables explored by Beale and Fuguitt (1978a), military employment and the presence of a senior state college, showed fluctuating effects throughout the decade. While counties with a high degree of military employment showed population losses in 1970–74, this process reversed in the 1974–77 period. By 1977–80, however, the small number of counties with over 5 per cent military employment were experiencing negative net migration. Presence of a state college was found to be a powerful explanation of the nonmetro turnaround in the 1970–74 period. However, the relationship was reversed after 1974, with counties having a state college experiencing only half the rate of net migration of the nation as a whole.

These factors driving migration throughout the 1970s are analysed further using multiple regression analysis. By estimating a multivariate model we can examine the effect of county characteristics in a combined fashion. This analysis utilises many of the variables developed by Heaton *et al*. (1981) and Beale (1977). The dependent variable is the annual net migration rate in the three time periods for the 2390 nonmetropolitan counties and their equivalents, using the 1980 metropolitan definition. Independent variables include county characteristics in 1970, such as percentage employed in agriculture, percentage black, and dummy variables measuring military employment over 5 per cent and presence of a state college. Measuring deconcentration are a dummy variable for counties adjacent to an SMSA and a dummy for counties having a centre of at least ten thousand people in 1970. The three amenity variables are similar to those developed by Heaton *et al*. (1981). These are mild temperature, as measured by the ratio of the average January temperature to the average June temperature; presence of water, as measured by the sum of two standardised variables measuring water presence and the log of the area of inland water; and recreational development, measured by summing three log-transformed standardised variables: percentage employed in entertainment, recreation and personal services; the number of hotels and motels per capita; and the proportion of seasonal housing units in 1970. Interactions among the three amenity variables were also examined in order to see if the presence of these characteristics in combination made a further contribution to migration. Weighted regression analysis was used to reduce the measurement error resulting from unequal error variance (with errors being greater in smaller counties). The results were found to be similar to that using unweighted regression.

The regression results are presented in Table 5.6. Columns 1, 2 and 3 can be compared for changes in the effects of the independent variables over the course of the decade. Counties characterised by a high degree of agricultural employment had lower net inmigration throughout the decade. Nonmetropolitan counties with a high percentage black were found by Beale (1977) to lose population in the early turnaround period of 1970–75. The regression indicates that this pattern continued throughout the decade. Areas with a high

Table 5.6 Regression of net migration rates on selected county characteristics for nonmetropolitan counties[a]

Selected county characteristics[b]	(1) 1970–74	(2) 1974–77	(3) 1977–80	(4) 1974–77	(5) 1977–80
% employed in agriculture	−0.262** (−0.032) (0.003)	−0.200** (−0.023) (0.003)	−0.265** (−0.033) (0.003)	−0.105** (−0.012) (0.003)	−0.172** (−0.021) (0.003)
% black	−0.306** (−3.28) (0.222)	−0.275* (−2.80) (0.210)	−0.285** (−3.09) (0.238)	−0.163** (−1.66) (0.204)	−0.173** (−1.88) (0.234)
Military employment	−0.139** (−1.64) (0.193)	−0.051** (−0.567) (0.182)	−0.106** (−1.26) (0.206)	−0.001 (−0.015) (0.172)	−0.056** (−0.668) (0.196)
Presence of state college	0.052** (0.325) (0.110)	−0.023 (−0.138) (0.104)	−0.019 (−0.119) (0.118)	−0.042* (−0.246) (0.097)	−0.037 (−0.231) (0.111)
Largest place in county 10 000+	−0.073** (−0.256) (0.075)	−0.115** (−0.383) (0.071)	−0.100 (−0.356) (0.080)	−0.089** (−0.295) (0.066)	−0.074** (−0.264) (0.075)
Adjacency to SMSA	0.062** (0.183) (0.064)	0.019 (0.054) (0.060)	0.175** (0.526) (0.068)	−0.042* (0.118) (0.056)	0.152** (0.458) (0.064)
Presence of water	−0.111 (−0.145) (0.076)	−0.047 (−0.058) (0.072)	−0.136* (−0.179) (0.081)	−0.001 (−0.001) (0.067)	−0.088 (0.116) (0.076)
Recreational development	0.273** (0.301) (0.050)	0.316** (0.327) (0.047)	0.069 (0.076) (0.053)	0.212** (0.219) (0.044)	−0.037 (−0.040) (0.050)
Mild temperature	0.838** (3.48) (0.132)	0.838** (3.28) (0.124)	0.830** (3.46) (0.140)	0.528** (2.07) (0.132)	0.520** (2.17) (0.150)
Mild temperature x development	−0.043 (−0.096) (0.100)	−0.111* (−0.233) (0.094)	0.185** (0.411) (0.105)	−0.095* (0.199) (0.087)	0.202** (0.447) (0.099)
Mild temperature x water presence	0.159** (0.418) (0.152)	0.139* (0.344) (0.143)	0.245** (0.641) (0.162)	0.075 (0.186) (0.134)	0.179** (0.468) (0.152)
Water presence x development	0.158** (0.099) (0.011)	0.092** (0.054) (0.010)	0.045* (0.028) (0.012)	0.033* (0.002) (0.010)	−0.014 (−0.009) (0.011)
Net migration rate 1970–74	—	—	—	0.370 (0.342) (0.018)	0.371 (0.360) (0.020)
R^2	0.100	0.111	0.336	0.490	0.414

* significant at 0.05% level
** significant at 0.01% level
Key: standardized coefficient
   (unstandardized coefficient)
   (standard error)
[a] Based on weighted regression analysis where each county is weighted by population size
[b] See text for explanation of variables

degree of military employment also had lower net inmigration rates but this tendency had lessened by 1974–77. The positive effect of the presence of a state college in the county was not significant by 1974. This finding tends to support that of Beale (1977), who concluded that the relative role of the state college in nonmetropolitan growth dropped off after 1970 when other factors became more salient.

The regression indicates that patterns of population deconcentration appear to have shifted over the course of the decade. The tendency for counties with larger population centres to have lower migration rates was strongest in the middle period. The equation for the 1970–74 period also shows a positive effect for counties adjacent to an SMSA. This variable is not significant in the 1974–77 period but had a strongly positive effect in 1977–80. This finding confirms the hypothesis that while deconcentration continued in the late 1970s, the attraction of the more remote areas appears to have waned.

The results for the amenity variables reveal that the types of nonmetro areas that were receiving migrants changed over the course of the decade. In the 1970–74 period the single effect of water presence was not significant, but the interactions of water presence with mild temperature and with recreational development were both strongly positive, as were the single effects of these two variables. This would indicate that in the early part of the 1970s the presence of water alone did not draw migrants, but that amenity development and more temperate areas of the country were a large part of the explanation for nonmetropolitan growth. In the 1974–77 period presence of water alone was not significant, and the coefficients of the interaction variables were somewhat lower. The single-order effects of mild temperature and recreational development continue to be high in the middle period. The equation for the last period shows that only mild temperature of the single-order effects was significant at the 0.01 per cent level, along with the interaction of mild temperature with the other two amenity variables. In particular, the interaction of development with mild temperature had a high positive coefficient in comparison with the earlier two periods. This finding confirms the results in Table 5.4 showing that migration to some of the northern nonmetropolitan regions identified as amenity areas, such as the Upper Great Lakes and the Ozarks, had fallen off by 1977–80, while areas such as the Florida Peninsula and the Southwest continued to have high positive net migration.

In columns 4 and 5 of Table 5.6 the net migration rate for the 1970–74 period is added as an independent variable. By controlling for the migration of the previous period it is possible to see what factors are associated with changes in migration rates over the period, i.e. which had an effect on migration in the later period net of the trends found in the early 1970s. It is seen by comparing columns 2 and 4 and columns 3 and 5 that the independent variables have a continued effect on migration in the same direction as the earlier equations.

To summarise, the regression results indicate that many of the factors cited as explanations for the turnaround in the early 1970s have shifted in importance by the end of the decade. Areas characterised by a high degree of agricultural employment, a high proportion of black population or a high degree of military employment tended to lose population throughout the 1970s, and the presence of a state college did not appear to draw people to nonmetropolitan areas after 1974. While areas most remote from urban centres appear to be recipients of much turnaround migration up until 1977, after this point adjacency to a SMSA became a more salient factor. Amenity variables continued to explain nonmetro migration throughout the 1970s, but the types of areas receiving such migration appear to have shifted by the end of the decade. In particular, areas with both mild temperatures and recreational development appear to have been most successful in continuing to draw migrants by the late 1970s.

Conclusions

The most important conclusion to be drawn from this analysis is that the turnaround from negative to positive net migration in nonmetropolitan areas was sustained throughout the 1970s. This shift was pervasive; it occurred in nearly all regions of the country and at all levels of population concentration. It is certain that the impact of the turnaround has been greatest in nonmetropolitan areas. In many cases, communities which had a small, declining population base experienced an influx of new residents, an expanding economic sector and novel development in the 1970s. Our analysis has confirmed that amenity and recreational characteristics continued to attract migrants by the end of the decade, particularly in warmer climates, indicating that people may have been continuing to act upon a preference for rural areas. Recent research by Long and DeAre (1982) confirms that jobs and household income both grew in nonmetropolitan areas in the 1970s. As an unprecedented shift in metropolitan migration patterns, the importance of the turnaround should not be minimised.

The evidence presented here nevertheless indicates a slowdown in the growth of nonmetropolitan areas in the late 1970s. An even sharper decline was found for nonmetropolitan net migration rates, as natural increase returned to its traditional position as the most important component of nonmetropolitan growth (Johnson and Purdy, 1980). We have seen that many of the explanations for the turnaround given in past literature appear to have had short-term effects, such as energy-extractive activities, presence of a state college, and military employment. In addition, the widespread nature of the turnaround in the 1970-74 period, when growth was found even in heavily agricultural and declining regions of the country, had dropped off by the late 1970s. There is much evidence that areas identified as 'turnaround regions', especially the more northern amenity areas such as the Upper Great Lakes, may have experienced a short-lived migration boom which has now passed. The appeal of the most remote rural areas including those with small population centres appears to have ebbed by 1977, at the same time that nonmetro counties adjacent to a metropolitan area showed the highest net migration rates.

Does this evidence indicate an affirmative answer to the question put forth in this paper – is this the end of the turnaround? The 'discovery' of the upsurge in nonmetropolitan growth by Beale (1975) was at first dismissed by many as some kind of statistical aberration, as it did not fit into the current theories of metropolitan settlement. While the validity of the turnaround eventually gained acceptance, the evidence presented here could feasibly imply that the phenomenon was short-lived and anomalous, like the baby boom of the 1950s. The preference for rural areas may have arisen out of dissatisfaction with urban life in the late 1960s, but was a faddish and temporary trend. There also may be a limit to the ability of rural areas to accommodate newcomers. If development occurs as a result of heavy inmigration, these areas may lose their original appeal. And energy costs may have become a prohibitive factor for many who considered such a move after the oil embargo of 1973–74, as witnessed by the slowdown in growth for more remote and colder regions.

The evidence is stronger, however, that the validity of the turnaround and the evidence for a slowdown in nonmetro growth by the end of the decade may be incorporated into new theories of urban–rural migration. These theories provide a rationale for the movement into and out of metropolitan areas which help to describe how settlement patterns are changing in developed, 'post-industrial' societies. Innovations in communication and transportation technology have led to a decline in the importance of distance from an urban centre for both individuals and industries (Wardwell, 1980; Long, 1982). These developments mean that individuals are able to act on a long-held preference for rural living and that firms may escape the high costs of metropolitan location. Wardwell (1977) has outlined how this functional explanation of the shift in the importance of metro areas has led to an equilibrium in metro/nonmetro settlement patterns.

> Such an equilibrium may take the form of regularised streams of migration in both directions, approximately equal in total volume and roughly similar in composition. Equilibrium might thus be indicated by the comparability of these streams rather than by any cessation of lessening of total movement.... Were such an equilibrium hypothesis found to be viable with additional data analysis, we would be in a position to explain the recent turnaround in migration patterns in part as a temporary and stabilising return to equilibrium, following slight movement beyond the limit, or as temporal fluctuations about that limit in a condition of long-term equilibrium already achieved.

Evidence on the age structure of the two migration streams over the decade of the 1970s indicates that they have been becoming more similar, as have the rates of natural increase. In this way it is seen that there is a tendency towards balance in the interchange between metro and nonmetro areas, as evidenced by the continuing pattern of deconcentration throughout the 1970s, including a resurgence of growth in smaller metropolitan areas.

The full picture of how growth trends in the 1970s conform to population distribution theories will not emerge until some time has passed for reflection and further analysis. But the findings presented here would tend to support the thesis that nonmetropolitan growth, as a part of the continuing deconcentration of the country as a whole, will continue past the 1970s. Evidence of a slowdown in fact may only corroborate the notion that migration between nonmetro and metro areas is tending towards equilibrium.

Acknowledgements

An earlier version of this paper was presented at the annual meetings of the Population Association of America, Pittsburgh, Pennsylvania, 15 April 1983. This research has been supported by the College of Agricultural and Life Sciences, University of Wisconsin – Madison, and the Economic Development Division, US Department of Agriculture, through a co-operative agreement, and by the Center for Demography and Ecology, University of Wisconsin–Madison, through Training Grant HD07014 from the Center for Population Research of the National Institute of Child Health and Human Development. Computing was facilitated by NICHD Center Grant HD05876. I am indebted to Glenn Fuguitt, Calvin Beale, Karl Taeuber and Halliman Winsborough for their helpful comments, and to Karen Weed, Chris Fox, Ward Patton and Kris Zehner for help in the preparation of the manuscript.

References

Beale, C. L. 1975: *The revival of population growth in nonmetropolitan America.* Washington, DC: Economic Research Service, US Department of Agriculture.

Beale, C. L. 1976: A further look at nonmetropolitan population growth since 1970. *American Journal of Agricultural Economics* 58, 953–8.

Beale, C. L. 1977: The recent shift of United States population to nonmetropolitan areas, 1970–75. *International Regional Science Review* 2, 113–22 (*see* this volume, Chapter 2).

Beale, C. L. and Fuguitt, G. V. 1978a: The new pattern of nonmetropolitan population change. In Taeuber, K. E., Bumpas, L. L. and Sweet, J. A. (eds), *Social Demography.* New York: Academic Press, 157–77.

Beale, C. L. and Fuguitt, G. V. 1978b: Population trends in nonmetropolitan cities and villages in subregions of the United States. *Demography* 15, 605–20.

Dillman, D. 1979: Residential preferences, quality of life, and the population turnaround. *American Journal of Agricultural Economics* 61, 960–6.

Heaton, T. B., Clifford, W. B. and Fuguitt, G. V. 1981: Temporal shifts in the determinants of young and elderly migration in nonmetropolitan areas. *Social Forces* 60, 41–60.

Johnson, K. M. and Purdy, R. L. 1980: Recent nonmetropolitan population change in fifty-year perspective. *Demography* 17, 57–70.

Lichter, D. T., Heaton, T. B. and Fuguitt, G. V. 1979: Trends in the selectivity of migration between metropolitan and nonmetropolitan areas: 1955 1975. *Rural Sociology* 44, 645–66.

Long, J. F. 1982: A theory of population redistribution. Paper presented at the Annual Meeting of the Population Association of America, San Diego.

Long, L. H. and DeAre, D. 1980: *Migration to nonmetropolitan areas: Appraising the trend and reasons for moving.* Washington, DC: US Bureau of the Census, Special Demographic Analysis 80–82.

Long, L. H. and DeAre, D. 1982: The economic base of recent population growth in nonmetropolitan settings. Paper presented at the Annual Meeting of the Association of American Geographers, San Antonio.

McCarthy, K. F. and Morrison, P. A. 1979: *The changing demographic and economic structure of nonmetropolitan areas in the United States.* Rand Report R-2399-EDA. Santa Monica: The Rand Corporation.

Phillips, P. D. and Brunn, S. D. 1978: Slow growth: a new epoch of American metropolitan evolution. *Geographical Review* 68, 274–92.

Tucker, C. J. 1982: Metropolitan decentralisation: United States in the 1970s. In Brunet, Y. (ed.), *Urban exodus, its causes, significance and future*. Proceedings of the Conference, 13–15 April, Department of Geography, University of Montreal, pp. 47–62.

United States Bureau of the Census 1981: *Geographical mobility: March 1975 to March 1980*. Washington, DC: Current Population Survey Series P-20, no. 368.

Voss, P. R. and Fuguitt, G.V. 1979: Turnaround migration in the Upper Great Lakes Region. *Madison, Wisc.: Applied Population Laboratory*, Population Series 70–12.

Wardwell, J. M. 1977: Equilibrium and change in nonmetropolitan growth. *Rural Sociology* 42, 156–79.

Wardwell, J. M. 1980: Toward a theory of urban–rural migration in the developed world. In Brown, D.L. and Wardwell, J. M. (eds), *New directions in urban–rural migration: The population turnaround in rural America*. New York: Academic Press.

Appendix: Procedure for adjustment for error of closure

$$P_{ijt} = Q_{ijt}\,[(10-t)\,Q_{ijt}10 + t\,P_{ij}10]/10\,Q_{ij}10$$

for $\quad t = 0, 1.25, 2.25, ..., 9.25, 10;$
$\quad\quad i = 1, 2, ..., n$ and
$\quad\quad j = 1, 2, ..., 51$

where $\quad n$ = the number of counties in state j;
$\quad\quad P_{ijt}$ = the intercensal estimate for county i in state j at time t;
$\quad\quad P_{ijt}$ = the postcensal estimate for county i in state j at time t;
$\quad\quad P_{ij}10$ = the 1 April 1980 census count for county i in state j;
$\quad\quad Q_{ij}10$ = the provisional 1 April 1980 postcensal estimate for county i in state j; and
$\quad\quad Q_{ij}0$ = $P_{ij}0$ = the 1 April 1970 census count for county i in state j, including corrections made subsequent to the release of the official population counts.

SECTION TWO
INTERNATIONAL TRENDS

6 D. R. Vining Jr and T. Kontuly,
'Population Dispersal from Major Metropolitan Regions: An International Comparison'

From: *International Regional Science Review* 3, 49–73 (1978)

Introduction

Around the beginning of the 1970s, a number of the major centres of population concentration in the industrial nations began to experience a decline in the in-movement of population from the more remote and peripheral regions of those nations. This decline has continued, with but transient reversals in a few countries, to the writing of this paper, and in many places has gone so far as to create a net flow of population out of these major conurbations back into the peripheral and predominantly rural regions. The suddenness of this development, and the even more intriguing fact that it has occurred simultaneously across nations separated by vast distances, must stand as a major challenge to geographers and regional scientists. We seem to be confronted with no less than a breakdown in Clark's 'law of concentration' – of all the laws of population geography, surely the most enduring:

> A general description of what is happening in the mod0ern industrial world can be given in one sentence, vast though its consequences may be. The macro-location of industry and population tends towards an ever-increasing concentration in a limited number of areas; their micro-location, on the other hand, towards an increasing diffusion, or 'sprawl.'
>
> (Clark, 1967, p. 280).

This paper has the single objective of documenting the breakdown of this law in the industrial nations. There are, of course, variations from country to country in the rates of decline in the net migration of persons into the major core regions, but the overall impression is one of regularity, such that they seem to be moving in lockstep. Our object is to convey this regularity without suppressing totally local deviations that may alert one or more of our readers to the cause of the overall regularity itself.

Most of our analysis will be conducted on regions which are aggregations of the basic political subdivisions of each country. Where possible, we have adopted a regionalisation scheme that has been either officially adopted by the government or planning agency of the country in question or proposed by a prominent student of its geography. The overriding principle governing the adoption of these regionalisation schemes was that the metropolitan regions be large enough to contain all conceivable spillover of population from their central cities. That is, we have purposefully overbounded these regions in anticipation of the objection that the decline of migration into the major metropolitan regions is simply an extension of their 'functional fields' beyond their official boundaries. Thus, in most countries, the core regions contain between 20 and 30 per cent of the territories of those countries, a much larger area than is commonly assigned them.

Needless to say, there are numerous plausible and intuitively appealing ways of aggregating administrative subdivisions into larger regions. The reader may suspect that were one of these alternative schemes employed, the phenomenon here touted as a 'discovery' would not appear in the data, that in short, the phenomenon of deconcentration is an artefact of the scheme adopted. Rather than attempt to protect our thesis by demonstrating that it holds up under all alternative aggregation schemes, which would only end up overwhelming the reader with statistical detail, we have been especially careful to document our data sources, most of which, particularly for the later years, can be found in a reasonably large research library and are, therefore, available to be experimented upon by the reader. We have conducted numerous such experiments ourselves and have found surprisingly little variance in our results. Migration into the major conurbations of Europe and Japan has fallen (or remained steady, as the case may be), no matter how one draws the boundaries around them. Yet we are most anxious to know if any one of our readers can show this 'invariance property' in the data not to hold.

We shall not attempt here an explanation of the phenomenon of deconcentration. Several will doubtlessly leap to the mind of the reader as he studies our graphs and tables, and he may therefore grow impatient with our suppression of all causal explanation in this paper. We prefer to let the facts stand by themselves, unadorned by theoretical discussion. Only the most facile and obvious of theories will be disposed of here (i.e. the recession theory of some urban economists) and then only as an incidental adjunct to our primary descriptive task.

There is already a large literature documenting the decline of the major metropolitan regions of North America, in particular the North American core region of the northeastern United States (excluding northern New England) and southern Ontario and Quebec (Beale, 1977; Beaujot, 1978; Berry and Dahmann, 1977; Bourne, 1977-78; McCarthy and Morrison, 1977; Sternlieb and Hughes, 1977; Tucker, 1976; Vining and Strauss, 1977). Because of the availability of this literature, we shall confine ourselves here to a comparative study of the nations of Europe and East Asia, which have yet to be analysed in any detail.[1]

For ease of exposition, we have chosen to treat the countries of our study *seriatim*. To provide a context for the most recent developments, we first describe the historical trends in regional concentration since the nineteenth century for each country and thereby identify its major points of concentration.

We then turn to a more detailed study of the trends in interregional migration flows in that country since World War II, with particular emphasis on the flow of population into these major points of concentration since 1970. In several countries where the data are sparse, however, we simply point out how these data are consistent with our thesis.

The evidence

Japan: the pure case of deagglomeration

Japan is the purest and simplest instance of what we are claiming is a universal mode of evolution in the population geography of the industrial nations. With the advent of the Meiji era in 1867, there began a steady flow of population from the peripheral islands of the Japanese archipelago (Kyushu, Shikoku, and, more recently, Hokkaido) and from the peripheral parts of the main island of Honshu (Chugoku, Hokuriku, and Tohoku) towards the great alluvial plains of Pacific Honshu stretching back from the port cities of Tokyo, Osaka, and Nagoya and constituting the Kanto, Tokai, and Kinki regions, or what is now called the Tokkaido megalopolis.

There were, of course, variations in the rate of net flow into these regions from decade to decade (World War II even saw large-scale return migration to the peripheral regions), and there were also local cities of some size in the peripheral regions that grew at rapid rates over this period, e.g. the industrial cities of northern Kyushu. There was also throughout the modern era movement into the frontier area of northern Japan, particularly Hokkaido. But the overall tendency for the Japanese population to 'pile up' in the Pacific plains is the single most important feature of that country's modern population geography. The growth of the Pacific regions was particularly rapid in the post-war period and, within that period, in the 1960s. In certain years of the early 1960s, over a half million persons net, or one-half of 1 per cent of the entire Japanese population annually, moved into these regions.

This century-long movement of population towards the Pacific coast of central Honshu came to an end in the 1970s. In the first year of this decade, around 430 000 persons moved into the Pacific regions *net*; in 1976, the latest year for which we have data, just 9000 persons did *net*. This sudden and precipitous drop in net migration into Japan's three major metropolitan regions is shown in Figure 6.1.[2] Furthermore, it is not only the peripheral regions on either side of the Pacific regions of central Honshu that are experiencing a corresponding rise in net migration, but the peripheral islands as well. Thus, the decline in net migration into the regions of central Honshu cannot be dismissed simply as a further extension of these regions to either end of Honshu.

These are important differences, both within the metropolitan regions and within the six peripheral regions, but the overall pattern in interregional migration flows over the post-war period is a simple one: a steady rise in net flows into the metropolitan regions until the early 1960s, followed by a decline in the middle 1960s, though the numbers are still very large, followed by a second rise in the late 1960s, and then the precipitous and dramatic decline of the

1970s. We shall observe this general pattern for a number of other countries of our study.

Two of the three Pacific regions have actually become net exporters of population. Kinki, site of three of the six largest metropolises of Japan (Osaka, Kobe, and Kyoto), was importing net from 100 000 to 200 000 persons annually between 1954 and 1970. In 1974, it *exported* net 18 000 and in 1976, 34 000. The third and largest of these regions, Kanto, site of the Tokyo and Yokohama metropolises, by contrast has continued to import population on a net basis, though at a much reduced rate. The reason for the relatively slower decline of the eastern portion of the Tokkaido megalopolis, as compared to its western portion, is probably due to the fact that the Kanto plains are from three to four times larger than the Nobi and Nara basins and Osaka plains that are the principal foci of settlement in the western end of the megalopolis, and have, therefore, considerably more empty land left of the type suited to urban settlements than exists in the western end.

Sweden: the replacement of natives by immigrants

Two major themes dominated the population geography of Sweden up to 1950, both of which have their parallel in Japan. The first was the movement of population out of dispersed rural settlements into cities. At the regional scale, this has meant the expansion of the metropolitan regions surrounding the three largest cities of Sweden (Stockholm, Göteborg, and Malmö) and the decline of the predominantly rural regions, such as the Southeast and Mid-Sweden Forest regions. The second was the colonisation of the frontier regions to the north, a theme equally prominent in Japan, where the northernmost island of Hokkaido, inhabited largely by aboriginals until the Meiji era, rapidly expanded its share of population up to around 1950. Thus, in both Sweden and Japan, we observe population expansion in the largely uninhabited frontier areas and in the densely populated regions containing these countries' major metropolises, as well as population stagnation or decline in the older and long-settled rural regions, such as the Southeast in Sweden or the southernmost island of Kyushu in Japan. This is a pattern that we shall observe again in the other Nordic countries.

After 1950, however, the northern counties of Sweden joined the more established rural counties to the south in experiencing a declining share of the national population. In a sense, the pull of the urban regions of the south became so great that even the frontier began to lose population (note again the similarity to Japan, where the growth in Hokkaido's share of the national population was arrested in the 1950s). The 1960s, again as in Japan, saw particularly rapid expansion in the populations of the metropolitan regions; maximum net inmigration from the rural regions was recorded at the beginning and end of this decade (Figure 6.2). The three metropolitan regions, covering 22 per cent of Sweden's land area, contained almost 70 per cent of its population in 1970.

With the advent of the 1970s, however, net migration into the metropolitan regions from elsewhere in Sweden began an unprecedented decline (Figure 6.2).

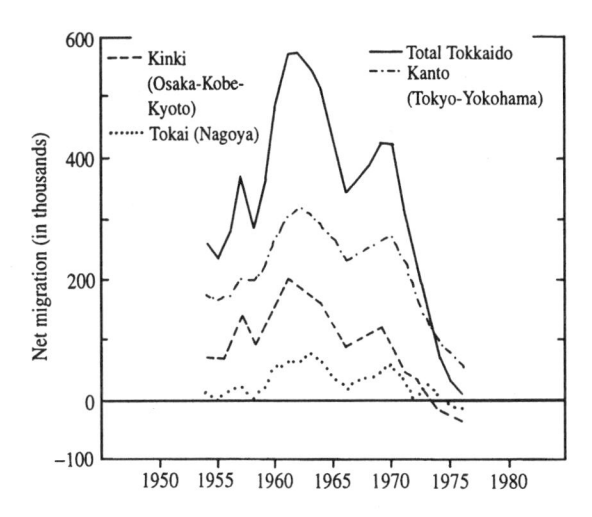

Fig. 6.1 Annual net internal migration into core regions of Japan, 1954–76

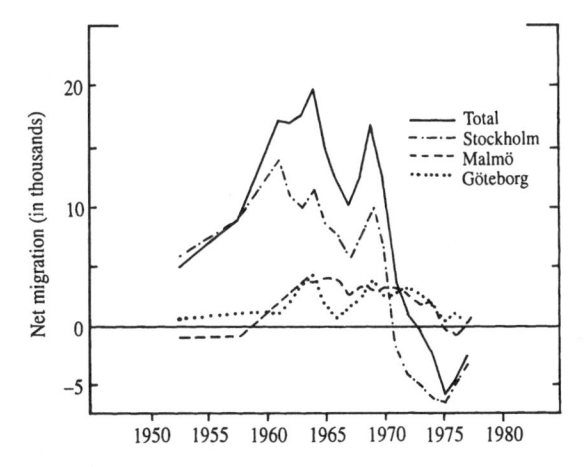

Fig. 6.2 Annual net internal migration into core regions of Sweden, 1951–77

The decline was most severe for the Stockholm region, but by 1975, all three metropolitan regions had begun to experience negative or close to negative net inmigration from the other regions of Sweden. The overall similarity of this pattern to that of Japan is striking; indeed, Figure 6.1 is almost a carbon copy of Figure 6.2, except for scale. The coefficient of correlation between the 'total' series in these two figures is 0.96.

The sizes of the metropolitan regions as here defined preclude the possibility that what we are observing in Figure 6.2 is simply a growth of the major urban agglomerations of Sweden beyond their conventionally defined boundaries. Twenty-two per cent of Sweden's land area is contained in these regions, and between them they cover a large part of the southern third of Sweden, excluding only the Southeast. The northern regions are in the true sense of the word Sweden's 'peripheral' regions and can by no stretch of the imagination be classified as the outermost hinterlands or exurbs of the metropolises of southern Sweden. Yet after two decades of net outflows averaging around

72 Differential Urbanization

10 000 persons a year, there is now a net inflow into these regions from the south.

There are two additional sources of regional growth that could theoretically counteract the recent trends in interregional net migration: natural increase and immigration from abroad. The former, at least, seems more likely to reinforce these trends. Since migrants are predominantly young, an area experiencing positive net inmigration will see its age distribution shift towards the younger and more fertile age groups. If age-specific fertility and mortality rates are more or less the same across regions, as they tend to be in the industrial nations of northern Europe and Japan, regions of positive net inmigration will have above average rates of natural increase and regions of negative inmigration below average rates of natural increase.[3] Alonso (1973) has called this phenomenon the 'demographic multiplier' (p. 10). The growth of the peripheral regions induced by positive net inmigration of the middle 1970s, therefore, eventually should be reinforced by above average rates of natural increase, though shifts in rates of natural increase appear to lag those in net migration rates for unknown reasons (Johnson and Vining, 1976).

Foreign immigration, on the other hand, appears to be counteracting, rather than reinforcing, the trends in interregional domestic migration in many countries. There has been speculation in the United States (Glazer 1975, p 242) that the accelerating loss of native population from the old industrial centres of the Northeast and Midwest is being compensated for by the inmigration of foreigners from abroad. In Sweden, we have direct evidence of this tendency. Not only do the metropolitan regions capture a disproportionate share of the foreign immigration into Sweden, but this share has risen as domestic net migration into these regions has fallen and gone negative (Table 6.1). Foreign migration has more than compensated for the loss of natives from these regions. Of course, the converse of this is that in the event of widespread migration of foreigners out of Sweden, say through the action of the government, the metropolitan regions would lose disproportionately. Nonetheless, it is quite possible that a major consequence of the partial abandonment of the major metropolitan regions by the native populations of the industrial nations will be to create a vacuum that will be filled by migrants from abroad.

Table 6.1 Domestic and foreign immigration into the metropolitan regions[a] of Sweden, 1973–77

Year	Net domestic migration	Net foreign immigration	Metropolitan regions' share of total net foreign immigration	Metropolitan regions' share of Swedish population
1973	−208	−9 732	—	0.693
1974	−2 607	5 944	0.714	0.693
1975	−5 980	12 504	0.768	0.692
1976	−4 600	15 914	0.797	0.692
1977	−2 477	19 369	0.845	0.692

Sources: *Belfolkningsförändringar 1975 Del 1 Församlingar, kommuner och A-regioner* (Statistiska Centralbyrån, Stockholm), Table 7, p. 128; *Statistiska meddelanden - Be 1977: 2* (Statistiska Centralbyrån, Stockholm), Table 1, pp. 8–9; *Statistiska meddelanden - Be 1978: 3* (Statistiska Centralbyrån, Stockholm), Table 1, pp. 8–9
[a] The Greater Stockholm, Malmö, and Göteborg regions

Norway: dispersal in a recession-proof economy

As in Sweden and Japan, over the last century there have been two major directions of population flow at the broad regional scale within Norway: one towards the frontier areas of the north and the other towards the region surrounding the country's largest city, Oslo. Nord-Norge, like Hokkaido in Japan and Norrland in Sweden, grew faster than the rest of the country up until 1950, as did Østlandet, site of Norway's major urban agglomeration, Oslo. During the 1950s, however, the direction of net flow became largely fixed in a single channel directed at the capital and its hinterlands.

Where Norway differs is in its economy. The two recessions of the 1970s that struck the economies of every other major industrial nation were avoided in Norway (OECD, 1975, p. 5; 1976a, p. 5). This fact makes Norway an ideal test case for a theory put forward at least once in print (Kain, 1975) and on a number of occasions in conversation by urban economists, namely, that the decline in the migration of persons into the core regions of the industrial nations during the 1970s is due to the depressed economic conditions of this period throughout the industrial world. If the recession were the main cause of the decline in net migration into the core regions of the industrial nations, then we would not expect to find this decline in a country that did not experience the recession. However, we find the same sharp drop in the rate of net migration into the region of Østlandet as was observed for the capital regions of Sweden and Japan (Figure 6.3).[4] Moreover, in all three countries, the decline begins about the same year, 1970. In 1973, there was actually a small net outflow from the Østlandet region. Unlike in Japan and Sweden, net flows into this region did not continue their decline after 1973 but turned upwards again; they nonetheless remain in 1976 below their level of the 1950s and 1960s (Figure 6.3). Note, however, that the Østlandet region, like the Stockholm region in Sweden, has continued to capture a disproportionate share of the net foreign immigration into Norway (Table 6.2).

Table 6.2 Net internal migration and net foreign immigration into the Østlandet[a] region, 1971–76

Year	Net internal migration	Net foreign immigration	Østlandet's share of total net foreign immigration	Østlandet's share of population
1971	2 221	3 883	0.587	0.491
1972	416	2 114	0.478	0.491
1973	–397	1 516	0.440	0.490
1074	266	2 830	0.677	0.189
1975	1 153	2 877	0.603	0.488
1976	2 899	2 614	0.535	0.488

Source: *Flyttestatistikk 1976* (Statistiska Centralbyrå Oslo), Table 3, pp. 19–20

Italy: the recession theory revisited

While the Scandinavian populations have drifted south towards the heartland of Europe, the Italian population has drifted northwest, also towards Europe's centre.

74   Differential Urbanization

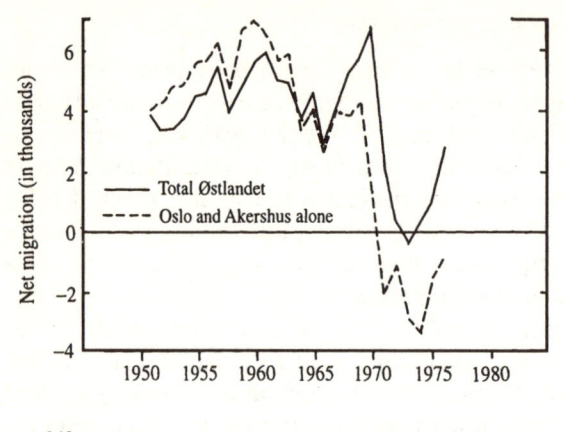

Fig. 6.3 Annual net internal migration into core regions of Norway, 1951–76

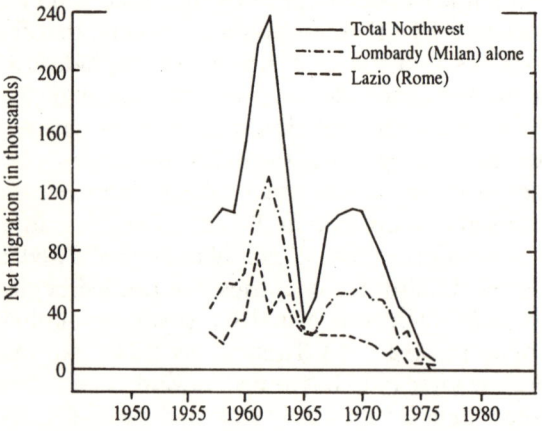

Fig. 6.4 Annual net internal migration into core regions of Italy, 1957–76

Since the unification of Italy, internal movements have followed the same direction, south to north, and east to west: from the regions of the south and from the islands (especially Sicily) to the centre regions (e.g. Lazio-Rome and Tuscany); to the northwest (e.g. Lombardy, Liguria, and Piedmont); and from the northeast (e.g. Venetia) to the northwest. During 1962–69, net immigration from the south and islands to the north amounted to 798 000 and from the same area to the centre, 247 000. Altogether, the south and islands have had a net loss through internal migration of 1 045 000 or about 130 000 each year

(Nangeroni, 1974, p. 1097).

Figure 6.4 shows net migration into the northwest and Lazio-Rome regions from the other regions of Italy over the years 1957–76. It has a form remarkably similar to that of Figures 6.1 and 6.2: peaks in the early and late 1960s, a trough in the middle 1960s, and a sharp falling off through the first half of the 1970s.[5] A unique feature of the Italian series is the depth of the trough in the middle 1960s, which nearly equals that of the current period. The comparable depth of these two troughs gives strength to the claim that the recession is causing the decline in net migration into the metropolitan regions, since there was a world recession in the middle 1960s as there was in the middle 1970s.

Figures 6.1, 6.2, and 6.4 undoubtedly owe part of their common shape to a close synchrony in the business cycles of Japan, Sweden, and Italy. Left unexplained by this theory (in addition to the anomaly of Norway) is the failure of net migration into the metropolitan regions to recover during the economic expansion of 1972 and 1973. In our view, therefore, aggregate economic conditions can explain only part of the reduction of migration into metropolitan regions in the 1970s.

Owing to the widespread return of Italian workers from northern Europe in the 1970s, the majority of whom have their origins outside of the Northwest and Centre regions, immigration has tended to reinforce, rather than counteract, the internal migration trends during this period. The Northwest and Centre regions' share of this return migration is much below their share of the population as a whole (*see* Table 6.3), in contrast to Sweden, where the metropolitan regions have captured a disproportionate share of the foreign movement into that country. Rates of natural increase in the Northwest and Centre regions also remain lower than elsewhere in Italy (Golini, 1974, p. 3).

Table 6.3 Net internal migration and net foreign immigration into the Northwest and Central regions of Italy, 1971–76

Year	Net internal migration	Net foreign immigration	Northeast and Central regions' share of total net foreign immigration	Northwest and Central regions' share of population
1971	112 901	22 323	a	0.466
1972	92 645	9 401	0.103	0.466
1973	73 176	18 914	0.322	0.465
1974	60 349	14 537	0.128	0.465
1975	33 366	16 375	0.200	0.465
1976	25 807	11 033	0.231	0.465

[a] Total net foreign immigration into Italy was negative for this year
Sources: *Annuario Statistico Italiano* 1973 (Istituto Centrale di Statistica, Rome), Table 15, p. 15; ibid., 1974, Table 23, p. 49; ibid., 1975, Table 23, p. 51; ibid., 1976, Table 20, p. 40; ibid., 1977, Table 14, p. 25

France: the revival of Le Désert Français

France and Great Britain are virtually alone among the countries of Western Europe in not keeping a continuously updated record of their interregional population flows. Data on internal migration in these countries are collected only during the census years. Therefore, the evidence that we will present here on deconcentration in these two countries will be less definitive than was the case for the countries studied so far. However, the importance of these two countries in the industrial world is such that we are forced in a paper pretending to be a comparative study of the industrial nations to make do with what data are available.

76 Differential Urbanization

France is well known for the growth of population in and around its capital, Paris. The Paris region has more than tripled its share of the national population in the last hundred years; whereas 1 out of 17 Frenchmen lived in the Paris region in 1851, 1 out of 5 does today. Yet what migration data we have for France suggests that the *internal* flow that fed this growth has largely dissipated. Figure 6.5 shows a decline in net migration from elsewhere in France into the Paris region and into the Paris region and the surrounding Paris Basin between the periods 1954–62 and 1962–68, and a net outflow in the period 1968–75. More generally, the backward, relatively underdeveloped western half of France (see Beaujeu-Garnier, 1974, Figure 6.1) began for the first time in the period 1968–75, to draw population from the highly industrialised and dynamic eastern half, particularly its northern part. The southeast, consisting of the Mediterranean and Centre-East regions, has exerted a steadily growing attracting force on the remainder of the country over the entire period 1954–75.

Though there was a net outflow of *natives* from the Paris region during the first half of this decade, probably for the first time in the peacetime history of France, foreign immigration into the region remains strong. Over 200 000 persons *net* entered the Paris region from abroad over the period 1968–75, or a little over 30 000 persons a year, which more than compensates for the net loss of natives (Table 6.4). The French situation then is remarkably similar to that of Sweden, where the net outflow of natives from its three largest metropolitan regions has been more than equalled by a net inflow of foreign immigrants (Table 6.1).[6]

Table 6.4 Net internal migration and net foreign immigration into the Paris region, 1954–75

Period	Net internal migration	Net foreign immigration	Paris region's share of total net foreign immigration	Paris region's share of total population at mid-period
1954–62	335 000	375 000	0.279	0.177
1962–68	68 000	297 000	0.244	0.184
1968–75	–140 000	200 000	0.258	0.187

Sources: *La population de la France* (Paris: Institut National d'Études Démographiques, 1974), Table 9, p. 229; *Recensement Général de la Population de 1975, premières estimations, sans doubles comptes par département et région* (Paris: INSEE, 1975), Table 3, p. 8, and p. 6.

Great Britain: an apparent exception

The population geography of Great Britain, the oldest of the industrial nations, has a different character from that of the more recently industrialised nations studied at the beginning of this paper. In the other countries studied here, the declining regions have traditionally been those of below average density, an exception being the north-eastern area of France. In Great Britain, we are as apt to find a region of above average density losing population as one with below average density. In fact, many of Great Britain's densely populated areas have been net exporters of population for close to half a century, in

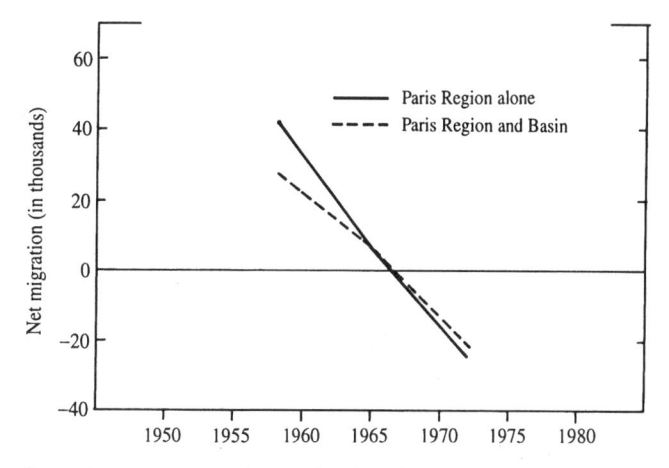

Fig. 6.5 Annual average net internal migration into the Paris region and Basin, 1954–62, 1962–68, and 1968–75 (shown at the midpoint of these periods)

particular the industrialised and urbanised counties in the north of England and in south Wales 'that were focal areas of inward migration in the nineteenth century but areas of economic depression, or only sluggish expansion, and thus of outward movement, in the period since the First World War' (Osborne, 1964, p. 139).

Great Britain is similar to the other countries studied here in the continuous and unbroken expansion of its capital region around London. In the twentieth century, the growth of the Southeast became part of a more general 'drift south' of the British population. Since 1921, the Southeast, together with the adjacent regions of East Anglia and the Southwest, both net losers throughout the nineteenth century, have seen their share of the British population steadily increase. The south of England, then, is Great Britain's major area of population concentration, and it is here, therefore, that we should look for a replication of the pattern found elsewhere: a sudden decline, beginning around 1970, in the net migration of natives into the capital regions of the industrial nations.

British data on internal migration are even more meagre than the French. Interregional migration tables exist for just two periods, 1965–66 and 1970–71, neither of which reveals an abatement in the 'drift south' of the British population (Central Statistical Office, 1973, p. 90, 1976, p. 40). Though there was a net outflow from the Southeast in both of these periods, the net inflow into its adjoining regions, East Anglia, the Southwest, and East Midlands, more than compensated for this loss. In fact, for the south as a whole, the net inflow was greater in the latter of these two periods. Of course, there is nothing inconsistent in this with the pattern observed elsewhere in Europe and Japan. Net inflows into the major core regions of Japan, Norway, Italy, and Sweden all increased between 1965 and 1970. In no country, with the possible exception of France, have we observed a decline in net inflows into these regions over the last half of the 1960s.

78 *Differential Urbanization*

The 1976 census, which would have supplied us with a third interregional migration table for the period 1975–76 and thereby given us some indication of the degree to which Great Britain conforms to the general pattern already observed for such countries as Japan and Sweden, was unfortunately cancelled (Clarke, 1975). Preliminary and provisional population statistics, however, suggest that Great Britain may be an exception to that general pattern. Though the Southeast had an absolute decline in population between 1971 and 1975, when the Southeast is combined with its neighbouring regions, East Anglia and the Southwest, the capital region as a whole is still growing. In fact, these three regions, with 42 per cent of the population, captured 61 per cent of the growth of the total British population between 1971 and 1975, though this growth was drastically lower than it was in the 1960s, in part because of a higher rate of emigration of Britishers abroad. In the period 1966–71, they captured 49.9 per cent of the national growth and in 1961–66, 55.6 per cent (Central Statistical Office, 1977, p. 35). Though the latest figures are provisional and subject to error, it is difficult to reconcile them with a net flow of population out of the south in the period 1971–76, or even with a decline in net inflows into the south.

The two Germanies: dispersal from the Ruhr and Saxony Regions

West Germany stands apart from the rest of Europe in having, according to one author, 'a highly favourable distribution of population. Other European countries are often dominated by a single metropolis or have only a few metropolitan regions that constantly grow at the expense of the less populous or less prosperous districts. This rather even population spread is one beneficial dividend of Germany's traditional regionalism, a factor that has militated against the dominance of any one city or district' (Kirby, 1974, p. 55).

This account ignores the large concentration of population and industry in the Ruhr district and along the Rhine Valley in general, 'one of the densest conurbations in the world ... the most impressive concentration of heavy industries one sees on the continent' (Gottmann, 1969, p. 596). The percentage of population living in Nordrhein-Westfalen, the *Land* containing the heart of the Ruhr district and the richest and most densely populated of the German states, increased by a third between 1871 and 1975, while in most other states (excluding the city-states of Bremen and Hamburg), population shares either remained stable or declined over this period. Furthermore, after World War II, the principal axes of population flow in West Germany were from north and southeast to the states of the Rhine Valley (Nordrhein-Westfalen, Rheinland-Pfalz, Hessen, Saarland, and Baden-Württemberg), predominantly states of above average density. Thus, though there may be a more even spatial distribution of urban centres in West Germany and a city size distribution less dominated by the very largest cities than we observe elsewhere in Europe, historically we find the pattern found in the rest of Europe: population flows from regions of below average density to regions of above average density.

What is exceptional about West Germany, and it is here that the rather uniform distribution of equal-sized cities may be having its effect, is that the concentration trend there grew progressively weaker over the post-war period (Figure 6.6), whereas in Japan and elsewhere in Europe, it showed signs of slackening only in the 1970s. Thus, in West Germany, there was a reversal in the direction of net flows between the densely and sparsely populated states much earlier than was the case in Sweden, Japan, France, Norway, and Italy.[7]

In contrast to Sweden and France, where foreign immigrants are disproportionately drawn to the core regions, foreign immigrants into West Germany distribute themselves more or less in proportion to the states' populations. In fact, the share of total net foreign immigration captured by the state of Nordrhein-Westfalen is slightly below its share of the total population of West Germany, though in the 1950s it drew a disproportionate share of this immigration (Table 6.5).

Table 6.5 Net internal migration and net foreign immigration into Nordrhein-Westfalen, West Germany, 1950-75

Period	Net internal migration[a]	Net foreign immigration	Nordrhein-Westfalen's share of total net foreign immigration	Nordrhein-Westfalen's share of total population, at mid-period
1950–61	915 000	1 085 400	0.443	0.270
1961–70	−265 000	706 900	0.288	0.281
1971	−10 500	111 000	0.258	0.279
1972	−24 800	83 000	0.251	0.278
1973	−22 400	99 000	0.258	0.278
1974	−14 800	15 500	[b]	0.278
1975	−9 352	−44 174	[b]	0.277

Sources: *Statistisches Jahrbuch 1963 für die Bundesrepublik Deutschland* (Statistisches Bundesamt, Wiesbaden), Table 6, p. 42; ibid., 1964, Table 6, p. 43; ibid., 1965, Table 6, p. 42; ibid., 1966, Table 6, p. 36; ibid., 1967, Table 6, p. 36; ibid., 1968, Table 5, p. 33; ibid., 1969, Table 5, p. 33; ibid., 1970, Table 5, p. 34; ibid., 1971, Table 5, p. 34; ibid., 1972, Tables 2,3, pp. 52–3; ibid., 1973, Tables 2,3, pp. 64–5; ibid., 1974, Tables 2,3, pp. 63-4; ibid., 1975, Tables 4.15, 4.16, pp. 78–9; ibid., 1976, Tables 4.16, 4.17, pp. 76–7; ibid., 1977, Table 3.35, p. 77
[a] Net migration from West Berlin included
[b] Total net foreign immigration was negative for this year

East Germany, like West Germany, since the early 1960s has seen a net flow of population out of its industrial heartland, the Southeast or Saxony region, which contains eight of East Germany's 10 cities over 100 000 and over 40 per cent of its population (Figure 6.7). (Data are not available to us for the 1950s, but Benderman 1964, p. 223 states that as late as 1960 this region absorbed 65 per cent of internal migrants annually.) However, here the flow has been strongly directed to the capital city of Berlin and the surrounding Centre region rather than to the more sparsely populated North and Southwest regions (Figure 6.7). In the continued strong growth of its capital, East Germany bears a strong resemblance to the other Eastern European countries studied here, Hungary and Poland, which are discussed below.

80 Differential Urbanization

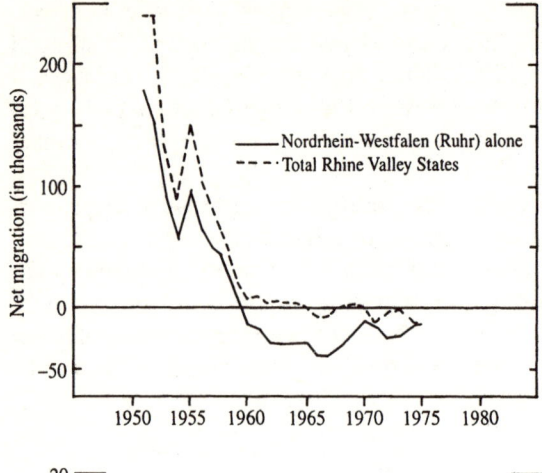

Fig. 6.6 Annual net internal migration into core regions of West Germany, 1951–75 (excluding net migration from West Berlin)

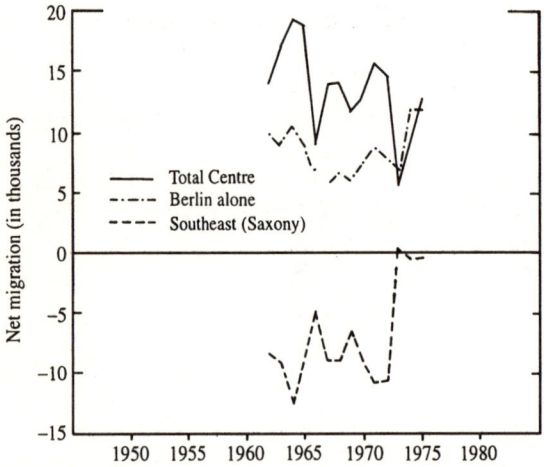

Fig. 6.7 Annual net internal migration into core regions of East Germany, 1961–75

Belgium and the Netherlands: dispersal from the Randstad and Brussels regions

A net movement out of the heavily populated western region of the Netherlands, which contains its major conurbation, Randstad Holland, into the more sparsely populated East, North, and South was first observed in the early 1960s (bearing a certain resemblance to the cases of East and West Germany) and has continued up to the present (Figure 6.8). Net movement into the Greater Brussels region from the remainder of Belgium was interrupted and reversed only in the early 1970s (Figure 6.9). Both countries are rather small (the Netherlands is approximately the same size as Nordrhein-Westfalen and it together with Belgium is appreciably smaller than the Paris Basin). Deconcentration, therefore, has less significance in these two countries than in the other countries of our sample, where long-term movements over large distances are being reversed. Nevertheless, the similarity of pattern is worth noting.

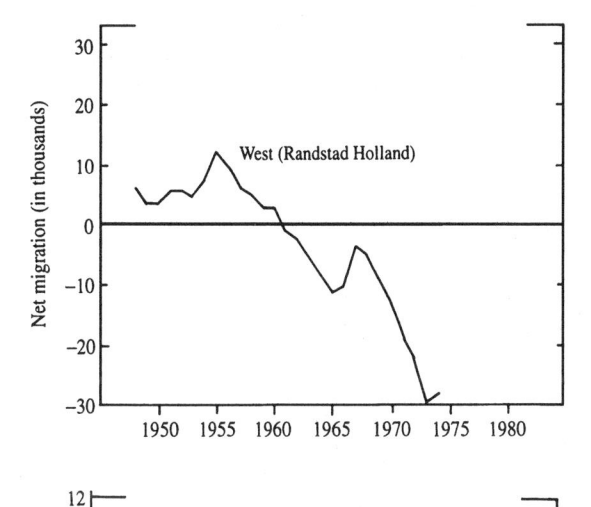

Fig. 6.8 Annual net internal migration into core regions of the Netherlands, 1948–74 (omitting net migration from ZIJ Polders and moves of unknown destination and origin)

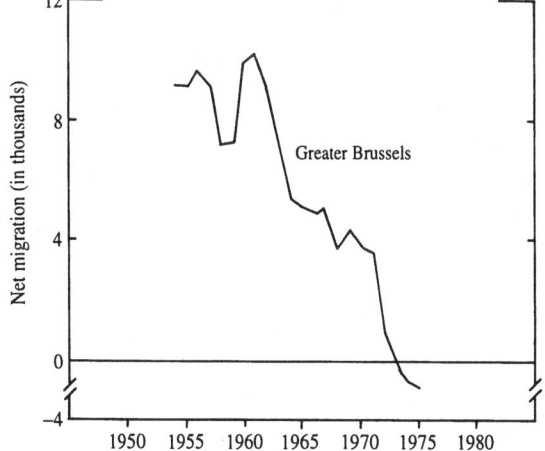

Fig. 6.9 Annual net internal migration into core regions of Belgium, 1954–75 (1963 figure not available)

Denmark and New Zealand: dispersal from the Copenhagen and Auckland regions

Interregional migration statistics are not readily available for these two countries, but what data we have been able to gather are consistent with the pattern described in this paper thus far. Both New Zealand and Denmark show a recent net outflow from their regions of traditional net inflow and concentration (the eastern islands of Denmark, which contain its capital, Copenhagen, and the North Island of New Zealand, which contains the majority of its cities, including the capital, Wellington, and largest city, Auckland). In Denmark, the eastern islands experienced a net outflow of -3602 and -1234 in 1974 and 1975 (Danmarks Statistik 1976, pp. 44–5; 1977, pp. 26–27). Although we have no internal migration data for the 1960s, other investigators report that the direction of net flow was from west to east during this period (Biraben and Duhourcau, 1973); thus, the 1974–75 figures appear to represent a reversal of long-standing trends. The 1976 census for New Zealand revealed that the

82   Differential Urbanization

South Island had a net inflow of 10 000 in the period 1971–76, to be contrasted with a net outflow of –16 000 in the period 1966–71. The North Island continues to have greater proportionate rate of growth, owing to differences in rates of natural increase between the two islands (Department of Statistics, 1977, p. 59).

The exceptions: the rapidly industrialising nations of Poland, Spain, Finland, Hungary, South Korea, and Taiwan

Until the 1970s, there was little to distinguish these countries from the countries studied so far in this paper. In all of them there was a steady growth of a few high-density core regions at the expense of the low-density peripheral regions. In Hungary, Finland, South Korea, and Taiwan, the regions surrounding their capital cities have completely dominated their urban hierarchies, while in Spain and Poland the major points of concentration have been more dispersed, with the capital regions in competition with other large metropolitan regions for the flow out of the predominantly rural regions.

However, whereas the migration data of the countries of Western Europe and Japan reveal an attenuation in the attraction of their major metropolitan regions since 1970, the migration statistics of these six countries do not exhibit this trend. There was a fairly rapid decline in net migration into the Budapest area (the Central Industrial region consisting of Budapest and Pest counties) between 1960 and 1970, but since 1970 the level of net migration has remained more or less constant. In 1975, it actually increased (Figure 6.10). There has been a similar constancy in the level of net migration into the Helsinki region since 1970; a fairly large drop occurred in 1974, but more data are needed before we can call this anything more than a random variation (Figure 6.11). Figure 6.11, however, does exhibit an intriguing isomorphy with Figure 6.3 for Norway.[8] Net flows into the major core regions of Poland (Warsaw, the industrial district of Upper Silesia, and the port city of Gdánsk) actually grew over the 1970s (Figure 6.12). Net flows into the major metropolitan regions of Spain[9] (the regions surrounding its three largest cities, Madrid, Barcelona, and Valencia or, more generally, the north-eastern quadrant of Spain) likewise showed no signs of systematic decline over the 1970s, at least until the sharp contraction in economic activity in 1975–76 (OECD, 1976b) (Figure 6.13). South Korea and Taiwan, which have the lowest per capita incomes of the countries of our sample, as well as the highest growth rates in these per capita incomes, show very large net flows into the regions surrounding their capital cities, Seoul and Taipei, continuing unabated up through 1976 (Figures 6.14 and 6.15).

Thus, in contrast to the more developed countries of Western Europe and Japan, there has yet to be a downturn of any duration in the net migration of persons into the capital regions of Hungary, Finland, South Korea, and Taiwan, and into the capital and other large metropolitan regions of Poland and Spain . Possibly, the later industrialisation of these nations and their larger agricultural and peasant populations will explain this difference. Of the countries that maintain population registers, these six are the least industrialised

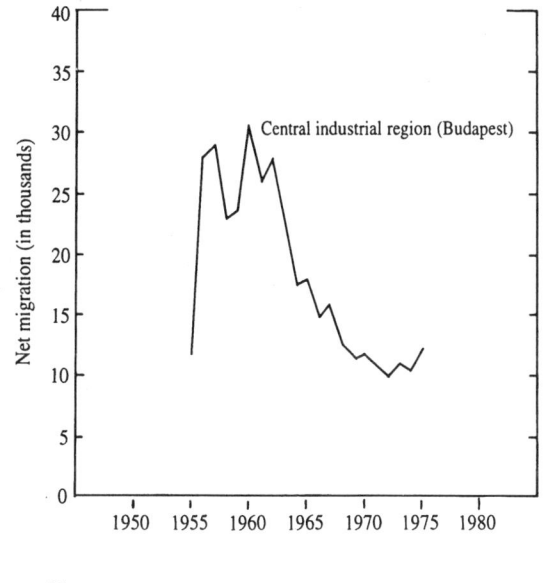

Fig. 6.10 Annual net internal migration into core regions of Hungary, 1951–75

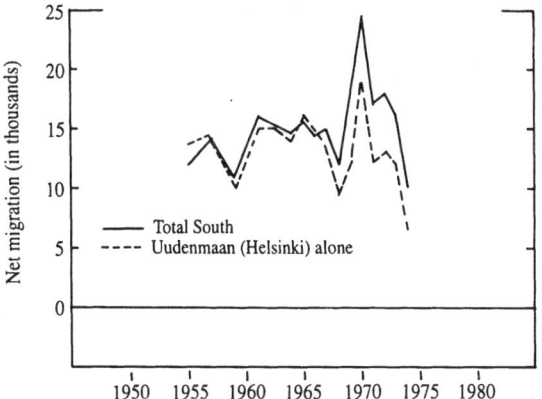

Fig. 6.11 Annual net internal migration into core regions of Finland, 1955–74

and the closest in the economic respect to the nations of the so-called Third World, where it is at least our impression from the popular press that the concentration of population has continued apace through the 1970s.

Summary and some concluding remarks on the causes of deconcentration

Of the 18 countries studied here, 11 (Japan, Sweden, Italy, Norway, Denmark, New Zealand, Belgium, France, West Germany, East Germany, and the Netherlands) show either a reversal in the direction of net population flow from periphery to core or a drastic reduction in the level of this net flow. In the first seven of these 11 countries, this reduction or reversal first became evident in the 1970s; in the last four, its onset was recorded in the 1960s. Six countries (Hungary, Finland, Spain, Poland, Taiwan, and South Korea) have yet to show

84 Differential Urbanization

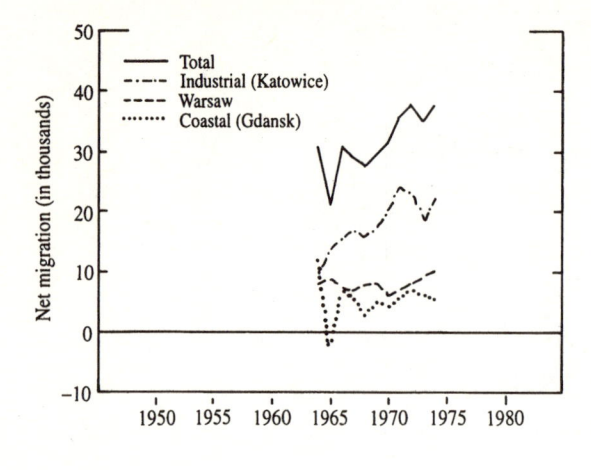

Fig. 6.12 Annual net internal migration into core regions of Poland, 1963–74

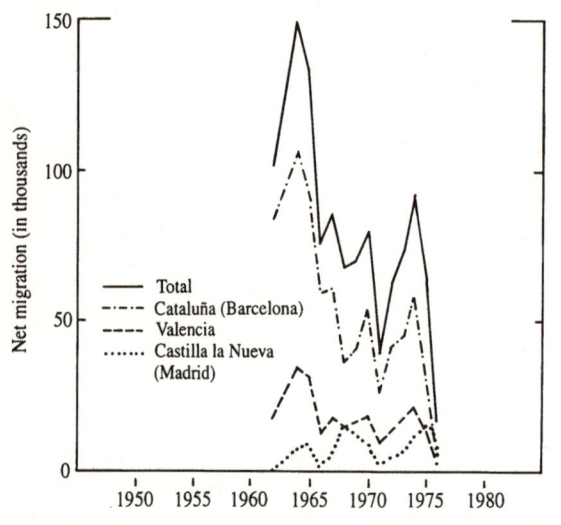

Fig. 6.13 Annual net internal migration into core regions of Spain, 1962–76 (excluding net migration from the Canary Islands and overseas dependencies, 1974–76)

an attenuation in the movement of persons into their core regions. Some possibly unreliable British data likewise fail to reveal a slackening in the growth of the regions surrounding London.

Three additional minor discoveries were described in this paper. First, migration continues heavy into the capital regions of three of the Eastern European countries that publish annual migration data (Poland, Hungary, and East Germany). However, the low rates of natural increase in these regions have blunted their expansion. Second, though domestic migration into the capital regions of France and Sweden has declined dramatically, foreign immigration into these regions remains at a high level. Third, net domestic migration into the core regions of Sweden, Japan, and Italy, countries separated by vast distances, fluctuates from year to year in a remarkably similar manner.

The reader should not conclude from the decline of net domestic migration into the largest conurbations of western Europe and Japan that migration

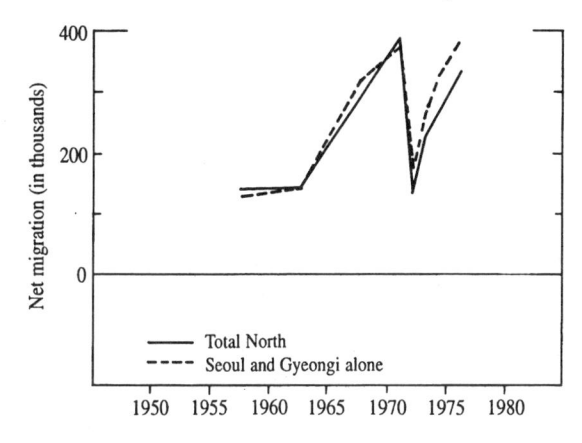

Fig. 6.14 Annual net internal migration into core regions of South Korea, 1955–76 (data for 1975 unavailable; annual averages for 1955–60, 1961–65, and 1966–70 shown at midpoints)

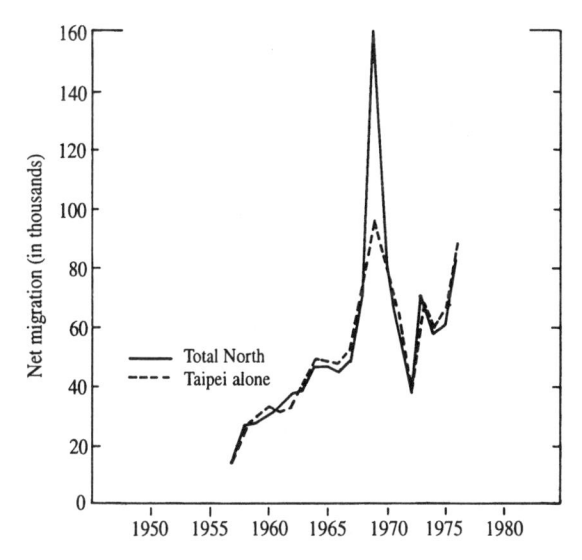

Fig. 6.15 Annual net internal migration into core regions of Taiwan, 1957–76 (including net foreign immigration)

towards cities in general has declined. It is our impression that people in moving back to the peripheral regions have tended to concentrate in a limited number of small and medium-sized cities there. Tachi (1971, p. 19) has called this tendency 'dispersed concentration'. It would be interesting to know how general a tendency this is in the developed countries and whether there are signs of its slackening as well.

Although we have purposely avoided speculation as to the possible causes of the phenomenon of deconcentration, we were able to raise doubts concerning one candidate, the recession theory. Cyclical variations in aggregate economic activity can explain neither the decline of net migration into the Oslo region in Norway (a country which did not experience the two recessions of the 1970s) nor the failure of this decline to be halted in Italy and Japan during the economic expansions of 1972 and 1973. The short-run business cycle undoubtedly explains part of the observed fluctuations in net migration into the metropolitan regions. Else, how does one explain the remarkable synchrony in these fluctuations in Italy, Sweden, and Japan in the 20-year period, 1955–75, or the very sharp decline in net migration into Spain's metropolitan regions in 1976? However, there is also a longer-run downward trend in the time series of these fluctuations that is apparent to the eye as well as to more rigorous statistical tests (Kontuly, 1978, Chapter 6).

Sundquist has recently attributed population dispersal in the advanced nations to decentralisation policies adopted by their governments (Sundquist, 1975). Dispersal has occurred in countries both with (Sweden, France) and without (the USA, Japan) strong decentralisation policies (Sundquist, 1975, pp. 228 , 248; Swain and Logan, 1975, p. 746). Dispersal appears not to have occurred in Great Britain, whose government has been trying to stem the 'drift south' of the British population for four decades (Osborne, 1964, p. 139; Sundquist, 1975, Chapter 2; De Jong, 1977). We have seen little evidence of deconcentration in Eastern Europe (with the possible exception of East Germany whose industrial heartland, the Saxony region, like the Ruhr district in West Germany, has exported population net since the early 1960s), though state policy to deter the concentration of population in large agglomerations is universal there and the means to enforce it more effective than in most other regions of the world (Fuchs and Demko, 1977). De Jong (1977) has found that while government policies in the Netherlands have undoubtedly reinforced and enhanced dispersal away from the western provinces, they did not initiate this trend nor have they directed this movement into targeted growth areas in the periphery.

The data of this paper suggest a developmental theory of the phenomenon of deconcentration. Diseconomies of metropolitan scale perhaps set in only at an advanced stage in a country's economic development, a stage not yet reached by six of the countries in our sample. Of course, migration out of the metropolitan regions may not occur even in the presence of these diseconomies, if undeveloped areas are limited in extent or only exist near to the current areas of concentration. Thus, in long-industrialised nations such as Great Britain, where the industrial and capital regions do not coincide and cover between them the preponderance of the national territory and where the largest expanses of developable rural land are in adjoining regions to the capital, the capital combined with its surrounding regions continues to attract population from the remainder of the country. Where development possibilities do exist in regions remote to the capital and other major metropolitan regions, the diseconomies of metropolitan scale eventually express themselves in the outmigration of persons from these core regions. *When* these diseconomies appear seems to be a function of what stage of economic development the country as a whole is in. At least this is how we would sketch out a theory of metropolitan decline with the data of this paper before us.

The development of this theory is clearly the next step. It will have the somewhat unusual feature (in regional science and geography) of having to be consistent not only with the past but also with the present, for the data base employed here is continually being updated so that one can

Developed World: International Trends 87

never rest easy with a theory that fits the data published to date. The availability of a continually updated data base provides a discipline to theoretical work in this area that is largely missing in theoretical regional science as a whole.

Notes

1 *See*, however, Kuroda (1969) and Alexandersson and Falk (1974).
2 During the rapid growth of these regions during the 1960s, the Japanese demographer, Toshio Kuroda (1969), stood virtually alone in foreseeing these developments.
3 Documentation of this fact for the countries of northern Europe and Japan may be found in Johnson and Vining (1976, p. 324); Drewett *et al.* (1976, p. 7); INSEE (1976, p. 7–8); and Statistiska Centralbyrån (1976, p. 128). In the countries of Eastern and Mediterranean Europe, however, we find the opposite pattern: rates of natural increase in the metropolitan (inmigration) regions are lower than in the rural (outmigration) regions. *See* Latuch (1973); Central Statistical Office (Hungary) (1972, p. 26); Staatliche Zentralverwaltung für Statistik (1976, p. 400); and Golini (1974, p. 3).
4 Part of this decline is a statistical artefact. Data for 1970 include migration occurring in previous years but not registered until the population census of 1970, and net migration figures for that year are therefore artificially high.
5 The abnormally high peak in the early 1960s is probably due to a change in the registration system in 1961. According to Salvatore (1977, p. 398): 'The 1961–62 migration figures are inflated as a result of the abolishment in February 1961 of the 1939 law regulating internal migration.... [Overreporting] for these years ... may be on the order of 20 percent.'
6 Since 1974, however, net foreign immigration into France (and presumably into the Paris region as well) has declined precipitously. In 1976, there was a net outflow. *See* Chi (1977).
7 Part of this reversal may be explained by the policy adopted after the war of temporarily settling German refugees from elsewhere in Europe in the rural areas of Germany, areas located predominantly in the states of Bayern, Niedersachsen, and Schleswig-Holstein, rather than in the devastated urban areas, which were located predominantly in the Rhine Valley states. As soon as the industrial centres of the Ruhr and Rhine Valley began to be reconstructed, however, these settlements broke up and their inhabitants moved into the traditional industrial heartland of West Germany. In no other country were refugees such an important component of interregional population flows in the early 1950s. Thus, the bulge in Figure 6.6 at the beginning of the 1950s may be due to the large flow of refugees out of their temporary camps in rural Germany and might well not have occurred in their absence. For more details of the effect of refugees on the internal distribution of population, *see* George (1972, pp. 526–7).
8 That isomorphy is possibly due more to similar defects in the registration systems of the two countries than to anything real. *See* note 3.
9 The capital regions of Eastern Europe generally have very low rates of natural increase relative to the rural and peripheral regions, however, which have tended to slow the expansion of these regions. For references, *see* note 3 above.

References

Alexandersson, G. and Falk, T. 1974: Changes in the urban pattern of Sweden 1960–1970: the beginning of a return to small urban places? *Geoforum* 18, 87–92.

Alonso, W. 1973: *National interregional demographic accounts: A prototype.* Monograph no. 17, Institute of Urban and Regional Development, University of California, Berkeley, Calif.

Beale, C. 1977: The recent shift of United States population to nonmetropolitan areas, 1970–75. *International Regional Science Review* 2, 113–22 (*see* this volume, Chapter 2).

Beaujeu-Garnier, J. 1974: Toward a new equilibrium in France? *Annals, Association of American Geographers* 64, 113–25.

Beaujot, R. 1978: Canada's population: growth and dualism. *Population Bulletin* 33 (April), 1–47.

Benderman, G. 1964: Regionale besonderheiten der Bevölkerungsbewegung in der DDR: Dargestellt am Beispiel des Jahres 1960. *Petermanns Geographische Mitteilungen* 108, 221–7.

Berry, B. and Dahmann, D. 1977: Population redistribution in the United States in the 1970s. *Population and Development Review* 3, 443–71.

Biraben J. and Duhourcau, F. 1973: La redistribution géographique de la population de l'Europe Occidentale de 1961 à 1971. *Population* 28, 1158–69.

Bourne, L. 1977–78: Some myths of Canadian urbanisation: reflections on the 1976 census and beyond. *Urbanism Past and Present* 5 (Winter), 1–11.

Central Statistical Office (Great Britain), 1973: *Abstract of regional statistics 1972.* London.

Central Statistical Office (Great Britain), 1976: *Regional Statistics* no. 11, 1975. London.

Central Statistical Office (Great Britain), 1977: *Regional Statistics* no. 12, 1976. London.

Central Statistical Office (Hungary) 1972. *Statistical yearbook 1971.* Budapest.

Chi, D. 1977: Bilan Démographique 1976: estimations provisoires. *Économie et Statistique* 86 (Feb.), 69–73.

Clark, C. 1967: *Population growth and land use.* New York: St. Martin's Press.

Clarke, J. 1975: The abortive 1976 census. *Area* 7, 81–2.

Danmarks Statistik. 1976 and 1977: *Statistisk Årbog 1976 and 1977.* Copenhagen.

De Jong, G. F. 1977: *The demographic impact of regional population redistribution policies in the Netherlands and Great Britain.* Working Paper No. 1977-08 (Pennsylvania State University: Population Issues Research Office).

Department of Statistics (New Zealand). 1977: *New Zealand official yearbook 1977.* Wellington.

Drewett, R., Goddard, J. and Spence, N. 1976: Urban Britain: beyond containment. In Berry, B. J. L. (ed.), Urbanisation and Counter-Urbanisation. *Urban Affairs Annual Review*, Vol. 2, 43–80, Beverly Hills, Calif.: Sage Publications.

Fuchs, R. and Demko, G.. 1977: Spatial population policies in the socialist countries of Eastern Europe. *Social Science Quarterly* 58, 60–73.

George, P. 1972: Questions de géographie de la population en République Fédérale Allemande. *Annales de Géographie* 447, 525–37.

Glazer, N. 1975: Social and political ramifications of metropolitan decline. In Sternlieb, G. and Hughes, J. (eds), *Post-industrial America: Metropolitan decline and inter-regional job shifts.* Center for Urban Policy Research, Rutgers University, New Brunswick, NJ.

Golini, A. 1974: *Distribuzione della popolazione, migrazioni interne e urbanizzazione in Italia*. Istituto de Demografia. Facoltà di Scienze, Statistiche Demografiche ed Attuariali, Università di Roma, Rome.

Gottmann, J. 1969: *A geography of Europe*, 4th Edition. New York: Holt, Rinehart & Winston.

INSEE. 1976: *Recensement général de la population de 1975: Premières estimations*. Paris.

Johnson, P. and Vining, D. 1976: A note on the equilibrium Hoover index associated with regional migration and natural growth patterns in Japan, 1955–1974. *Journal of Regional Science* 16, 337–44.

Kain, J. 1975: Implications of declining metropolitan population on housing markets. In Sternlieb, G. and Hughes, J. (eds), *Post-industrial America: Metropolitan decline and inter-regional job shifts*. Center for Urban Policy Research, Rutgers University, New Brunswick, NJ.

Kirby, G. 1974: Germany, Federal Republic of (in part). *Encyclopaedia Britannica, Macropaedia*, 15th edition, vol. 8, pp. 45–56.

Kontuly, T. 1978: The actual or imminent reversal of regional population concentration in the highly industrialised countries, Ph.D. Dissertation, University of Pennsylvania.

Kuroda, T. 1969: *A new dimension of internal migration in Japan*. English Pamphlet Series no. 69, Ministry of Health and Welfare, Institute of Population Problems, Tokyo.

Latuch, M. 1973: The role of internal migrations in contemporary population growth in big cities in Poland. *Studia Demograficzne* 34, 35–47.

McCarthy, K. and Morrison, P. 1977: The changing demographic and economic structure of nonmetropolitan areas in the United States. *International Regional Science Review* 3, 123–42.

Nangeroni, G. 1974: Italy. *Encyclopaedia Britannica, Macropaedia*, 15th edition, vol. 9, 1085–114.

OECD. 1975. *Economic Surveys – Norway*. Paris.

OECD. 1976a. *Economic Surveys – Norway*. Paris.

OECD. 1976b. *Economic Surveys – Spain*. Paris.

Osborne, R. 1964: Migration trends in England and Wales. In Beaver, S. and Kosinski, L. (eds), *Problems of Applied Geography* 11. Warsaw: PWN – Polish Scientific Publishers.

Salvatore, D. 1977: An econometric analysis of internal migration in Italy. *Journal of Regional Science* 17, 395–408.

Staatliche Zentralverwaltung für Statistik (East Germany). 1976: *Statistisches Jahrbuch 1976*. East Berlin.

Statistiska Centralbyrån (Sweden). 1976: *Belfolkningsförändringar 1975 Del Församlingar Kommuner och A-regioner*. Stockholm.

Sternlieb, G. and Hughes, J. 1977: New regional and metropolitan realities of America. *Journal of American Institute of Planners* 43, 227–40.

Sundquist, J. 1975: *Dispersing population: What America can learn from Europe*. Washington, DC: Brookings Institution.

Swain, H. and Logan, M. 1975: Urban systems: a policy perspective. *Environment and Planning A* 7, 743–55.

Tachi, M. 1971: The inter-regional movement of population as revealed by the 1970 census. *Area Development in Japan* 4, 13–24.

Tucker, C. 1976: Changing patterns of migration between metropolitan and nonmetropolitan areas in the United States: recent evidence. *Demography* 13, 435–43.

Vining, D. Jr. and Strauss, A. 1977: A demonstration that the current deconcentration of population in the United States is a clean break with the past. *Environment and Planning A* 9, 751–8 (*see* this volume, Chapter 3).

7 S. G. Cochrane and D. R. Vining Jr,
'Recent Trends in Migration between Core and Peripheral Regions in Developed and Advanced Developing Countries'

From: *International Regional Science Review* 11, 215–43 (1988)

Introduction

With the front-page headline 'Population's Rise in the Northeast Reverses a Trend', the *New York Times* (1985) announced that Rogerson and Plane (1985) had found that the annual net migration loss from the north-east United States had slowed since 1980. Not only had net outmigration in this core region decreased in recent years, but the total population of the region had begun increasing again in 1981 after almost a decade of decline. Net migration in the north-east was positive in 1983 for the first time since 1970 (by at least one measure). The strong shift during the 1970s of interregional population movements in the United States towards the south and west appears to have subsided. These statistics were the first indication that reconcentration of population is taking place since Vining and Pallone (1982) put forward the thesis that the century-long migration towards the high-density core regions was over in the developed world.

Plane (1984) suggested that the pattern of core–periphery dispersal of population should begin to reverse itself through an increase in outmigration from the peripheral areas as 'newly expanding urban areas become more integrated in the national economic system and rise in the functional national urban hierarchy'. Net migration into the south and west of the United States indeed began to decrease in the first half of the 1980s.

Along with these new interregional migration trends, the location of growth within regions in the United States appears to be shifting as well. The 1984 United States population estimates confirm results from earlier studies (Engels and Forstall, 1984) that growth of the nonmetropolitan population since 1980 has been slower than that of the metropolitan population, again reversing the situation of the 1970s (Engels and Forstall, 1985).

Similar trends have been reported recently for several countries of Western Europe. A slowdown of deconcentration occurred in Denmark during the 1980s (Illeris, 1984), although not returning to the concentration of former times. Internal migration trends favouring the peripheral regions of Norway in the early 1970s were reversed later in that decade (Hansen, 1978, 1988). In Japan, Yamaguchi (1983) observed a process of population reconcentration as well.

These recent findings suggest that it is appropriate to update the net migration figures from Vining and Pallone (1982) for as many countries as possible in order to identify any trends that may appear across several countries. Vining and Pallone examined internal migration statistics for 22 countries, grouped into five categories and covering from the 1950s to 1978–79. Data have since become available up to 1984–86 for 17 of these 22 countries. The 17 countries represent each of the five categories originally defined by Vining and Pallone: (1) *North-western Europe*: Belgium, Denmark, the Federal Republic of Germany, France, and the Netherlands. This article updates the data for all these countries. (2) *North America*: Canada and the United States. Recent data are available for both countries. (3) *Japan, New Zealand, and the periphery of Western Europe*: Finland, Iceland, Italy, Norway, Spain, Sweden, the United Kingdom, Japan, and New Zealand. Data after 1979 are available for all except Iceland, the United Kingdom,[1] and New Zealand. (4) *Eastern Europe*: Czechoslovakia, the German Democratic Republic, Hungary, and Poland. Recent data are available for Czechoslovakia and the German Democratic Republic. (5) *Two advanced developing countries of East Asia*: the Republic of Korea and Taiwan. Data for both of these countries are updated.

The definitions used in this article for core and peripheral areas are those of Vining and Pallone. The core regions intentionally are defined to be overbounded to make sure that they are fully contained within the specified boundaries. The net migration rate is regional net internal migration during time period t divided by total regional population at the beginning of t, multiplied by 1000; note that net migration includes internal but not international migration, except in a few cases where data sources do not allow such a separation. The multiplication by 1000 is done to generate a measure which is always a whole number or is almost always greater than one in order to make it easier to understand the figures. The net internal migration rate, thus, is equivalent to percentage times 10.

Before we present the new findings, the migration trends to 1978 described by Vining and Pallone (1982) are summarised briefly. The countries of northwestern Europe experienced a long-term decline in net inmigration to their core regions during the 1950s and the 1960s and net outmigration to their peripheral regions during the 1970s. The decline in net inmigration to the core regions was more abrupt in Canada and the United States and amounted to very substantial losses to their core regions in the 1970s. The countries on the periphery of Western Europe plus Japan and New Zealand experienced stable or rising net inmigration to their core regions between the 1950s and the 1960s, followed by a sharp drop in net inmigration beginning around 1970 (but not to the point where sustained net outmigration was observed). The Eastern European countries had sustained, though moderate, net inmigration to their core regions without exhibiting any real decline. Net inmigration to the core regions of both Taiwan and the Republic of Korea was very large and showed no signs of abatement in 1978.

Recent migration trends

A brief examination of the trends from the late 1970s to 1984–86 follows. The data are summarised in graphs depicting the trends since 1970. Core regions are marked with an asterisk. Space does not allow the printing of all of the data, which include regional growth rate differentials as well as migration for each country; however, the complete data tables are available upon request.[2]

North-western Europe

The net migration flows of Denmark, the Netherlands, and Belgium have already reversed their previous trends of net outmigration from their core regions. The core regions of each of these countries have had net migration rates near zero since 1981. West Germany continues to show net outmigration from the core region, but at quite moderate rates. Census figures for France show increasing net outmigration from the core region for the censal periods 1968–75 and 1975–82. Recent indications, however, are that net outmigration from the Paris region and basin has decreased quite significantly.

In the Federal Republic of Germany (Figure 7.1), the moderate net outmigration from the core area of the west (Rhine–Ruhr) region to the periphery that began in the 1960s continues through 1985. The southern region of Bavaria had the largest net migration rate of all the regions through 1983. All the other peripheral regions (north, central, and Baden-Württemburg) had modest positive net internal migration rates between 1978 and 1982. Baden-Württemburg surpassed Bavaria with net migration rates of 2.1 in 1984 and 2.4 in 1985. During this same period, the net migration rates of the central and north regions dropped to −1.2 and −0.9, respectively, in 1985. The total population of the country as a whole has been stable between 1970 and 1985. Bavaria is the only region that has shown consistent population growth since 1970.

In Belgium (Figure 7.2), a similar migration trend was evident between 1970 and the early 1980s, with the southern, less prosperous region of Wallonia experiencing moderate net inmigration from the core Brussels metropolitan area and the northern region of Flanders. The rate of net inmigration in Wallonia during the late 1970s was approximately double that of the early 1970s. Several years of data are not available, but net migration to the core region apparently became positive again in the early 1980s and net migration in Wallonia has been near zero. It should be noted that the net migration rates for all three regions have been moderate, between −1 and +1, for the last decade. Flanders, even with a slightly negative net migration rate, remains the only region of the three that has grown consistently at an annual rate greater than the country as a whole. Flanders has the highest birth rate and the lowest death rate of Belgium's three regions.

Migration trends in Denmark (Figure 7.3) were similar to those in the Federal Republic of Germany and Belgium through the 1970s, with the core eastern islands region (Copenhagen) experiencing net outmigration to the peripheral regions of Fyn and Jutland. The core region had small but positive net migration rates, however, from 1981 to 1983. The entire country has exhibited general stability since 1980, with all net migration rates between −1 and +2. Even though the eastern islands region experienced net inmigration in the early 1980s, its net migration rate was −0.2 in 1984 and −0.4 in 1985, and the total population of the region has declined slightly since 1979. Fyn and Jutland both have above average population growth rates, due to higher birth rates.

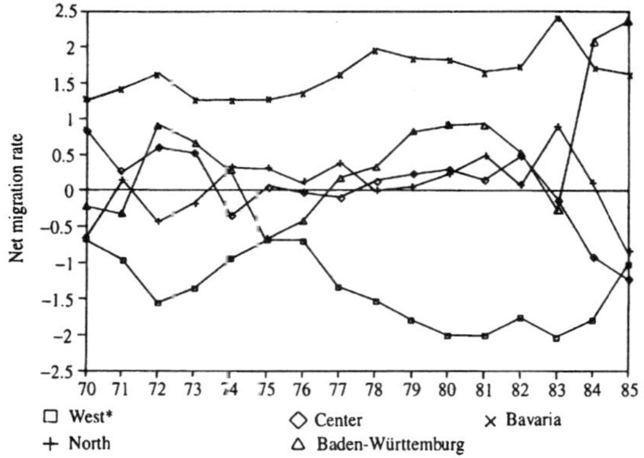

Fig. 7.1 Changing migration trends in the Federal Republic of Germany, 1970–85 Note: in this and subsequent figures core regions are designated by an asterisk

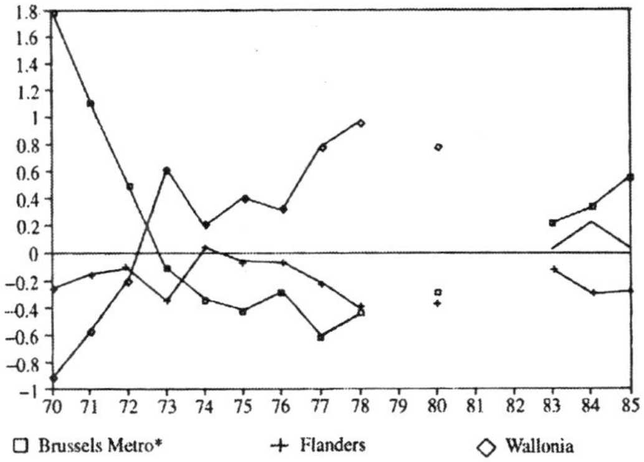

□ West*   ◇ Center   × Bavaria
+ North   △ Baden-Württemburg

Fig. 7.2 Changing migration trends in Belgium, 1970–85

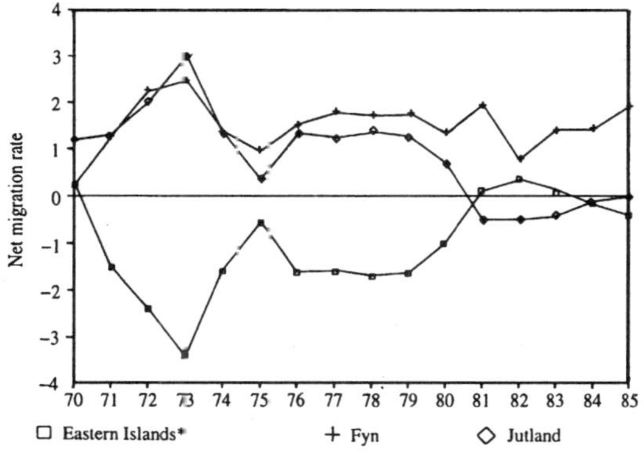

□ Brussels Metro*   + Flanders   ◇ Wallonia

Fig. 7.3 Changing migration trends in Denmark, 1970–85

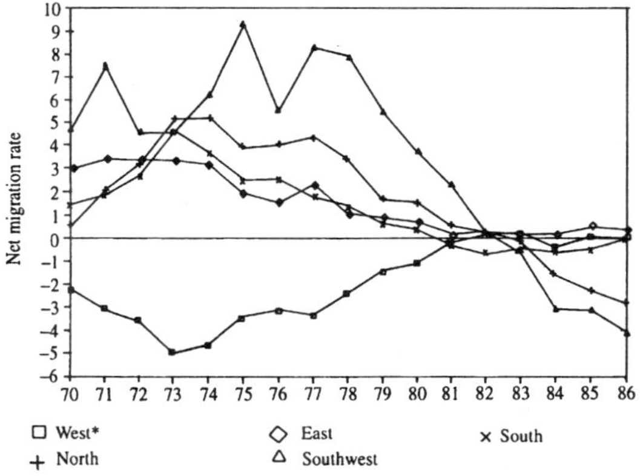

□ Eastern Islands*   + Fyn   ◇ Jutland

Fig. 7.4 Changing migration trends in the Netherlands, 1970–86

□ West*   ◇ East   × South
+ North   △ Southwest

94 Differential Urbanization

A similar convergence of net migration rates towards zero is seen in the Netherlands (Figure 7.4), but it is much more pronounced because the net migration rate in the south-west region was as high as 9.4 in 1975. From the mid-1960s to the mid-1970s, all of the peripheral regions had positive net migration rates at the expense of the core western region. Beginning in 1978, there was a rapid convergence towards zero, with the core region recording slightly positive rates in 1982, 1983, 1985, and 1986.[3] Net outmigration began in the southern region in 1981 and in the north and south-west in 1983.

Migration data for France (Figures 7.5 and 7.6) are available only for intervals of approximately seven years, with the last period ending in 1982.

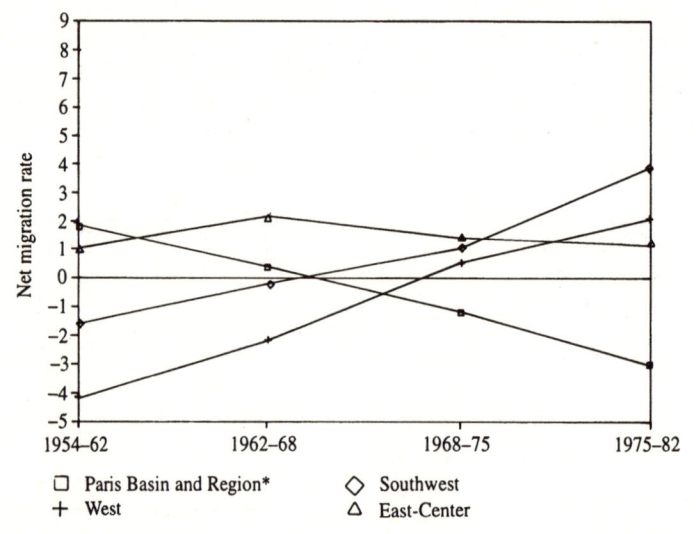

☐ Paris Basin and Region*   ◇ Southwest
+ West   △ East-Center

Fig. 7.5 Changing migration trends in France, regions 1–4, 1954–62, 1962–68, 1968–75, and 1975–82

+ Mediterranean   ◇ North   ☐ East

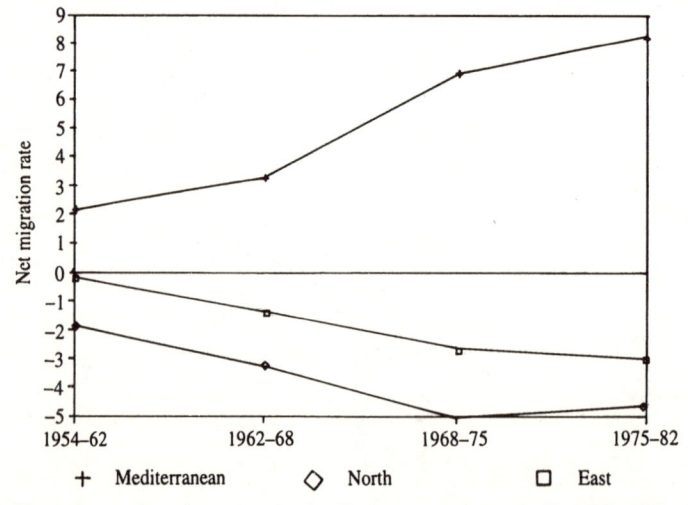

Fig. 7.6 Changing migration trends in France, regions 5–7, 1954–62, 1962–68, 1968–75, and 1975–82

France is distinguished from the other countries of north-western Europe in that the net migration rates of its regions are diverging away from zero, with the Paris, north, and east regions showing increasingly negative rates and the Mediterranean, west, and south-west regions showing increasingly positive rates. Although the total population of the Paris region continues to increase (at a rate below the national rate), its negative annual net migration rate during the period 1975–82 (−3.0) was more than twice the annual rate during the previous period 1968–75 (−1.2). The west, south–west, and Mediterranean regions continue to experience net inmigration at the expense of the northern and eastern regions of the country. A recent study, however, indicates that the slowdown in net outmigration from the core seen in other countries seems to be appearing in France as well (Courgeau, 1986). This result, obtained from a survey of employment, needs to be corroborated. It could indicate that net outmigration from the core region of France may have already reached its peak.

North America

Figures available for the United States for five-year periods during the 1970s and for annual periods for 1980–84 indicate a decrease in net outmigration from both the north-east and midwest regions since 1980, and decreasing net inmigration to the south (Figure 7.7).[4] Bureau of the Census reports for net internal migration actually show a net migration rate of −1.2 for the west region in 1983, although it rose again to 5.6 in 1984. These data reflect the findings from Internal Revenue Service data reported by Rogerson and Plane (1985), with the exception of the midwest.

Perhaps more useful figures for the USA are a longer time series for annual total migration figures (including international). They are calculated as a residual, the difference between population growth and natural increase. These figures (Figure 7.8) show net outmigration from the north-east decreasing steadily since 1978, turning positive in 1983 with a net total migration rate of 1.1, and dropping to −0.4 in 1984. The north-east's population began growing again in 1981 following five years of decline, although the region's percentage growth rate was still 0.64 less than the national percentage growth rate. Net outmigration from the midwest continually increased during the late 1970s, with its net total migration rate changing from −2.5 in 1976 to a low of −7.9 in 1982. It has since increased to −3.8 in 1984. Rogerson and Plane's (1985) observation that the south and west may have passed their peak periods of net inmigration and growth has been confirmed by the most recent data. The net total migration rate for the west increased from 6.9 to a high of 12.7 between 1970 and 1979, then decreased to 9.4 in 1984. Net migration to the south peaked two years later in 1981 with a net total migration rate of 12.4, declining to 6.7 in 1984.

A general, if less detailed, description of United States net total migration trends is based on annual averages of 10-year migration figures gathered by the decennial census and by intercensal estimates for 1980–87 based on the residual method. The results, although more generalised, repeat the pattern of the annual figures from the residual method (Figure 7.9).

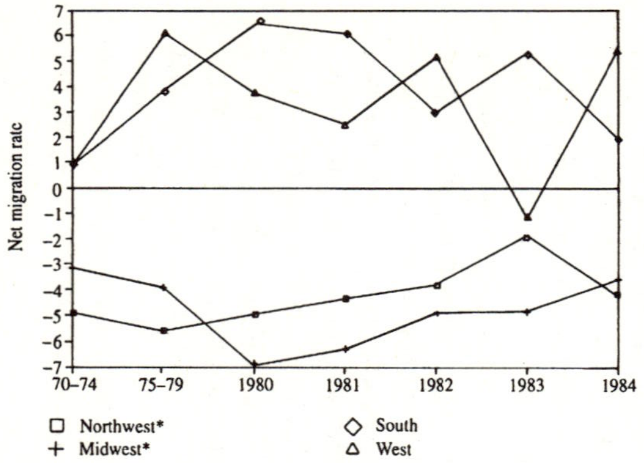

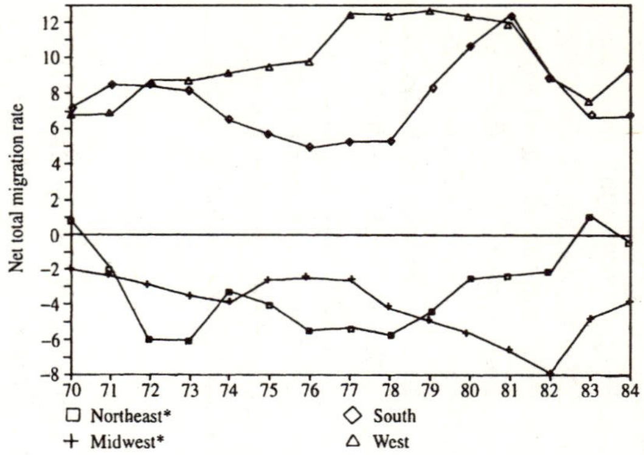

Fig. 7.8 Annual net total migration estimates for the USA, 1970–84

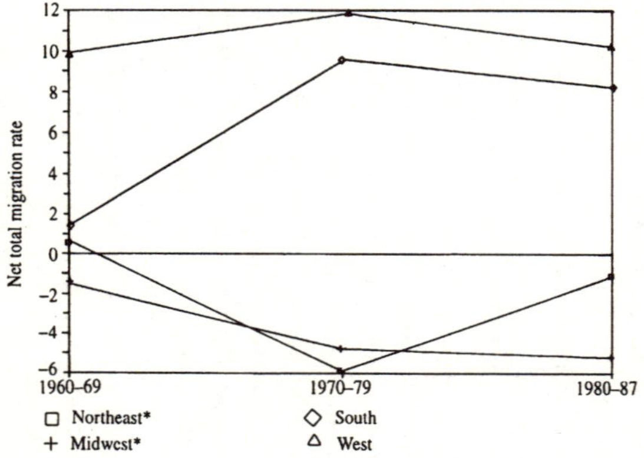

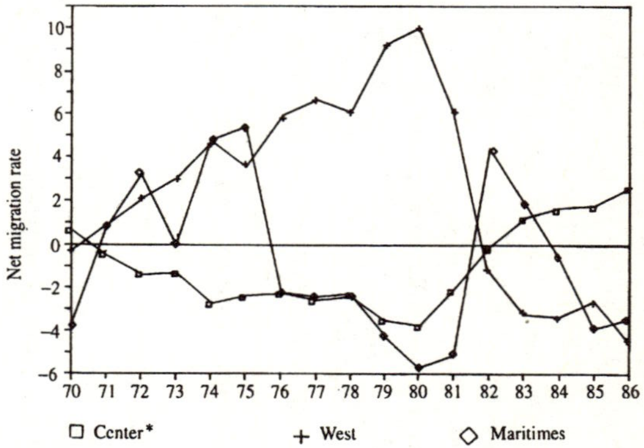

Fig. 7.10 Changing migration trends in Canada, 1970–86

Fig. 7.7 Net internal migration in the USA, 1970–84

Fig. 7.9 Annual net total migration for the USA, 1960–69, 1970–79, and 1980–87

Recent trends in Canada (Figure 7.10) are similar to those of the USA, but the reversal of net outmigration from the core is complete. The core region of Canada (Ontario and Quebec) experienced steadily increasing net outmigration after 1971 with a maximum negative net migration rate of −3.8 in 1980. The trend then reversed, with positive net migration rates of 1.2, 1.7, 1.8, and 2.6 in the years 1983 to 1986. The opposite trend is seen in the west region where the net migration rate gradually increased to a peak of 10.0 in 1980, followed by a sudden drop between 1982 and 1986 to −4.5. The speed with which net migration first rose and then declined in the west can almost certainly be explained by the performance of the oil industry. In 1976, the population of Alberta accounted for only 29 per cent of the west region, but during the period 1976–81, inmigration to this oil-producing province accounted for 48 per cent of total inmigration to all of the provinces of the west (Statistics Canada, 1985, p. 78). Net migration in the maritimes has been anomalous in comparison to the other regions, increasing in the early 1970s to a peak of 5.4 in 1975, then decreasing to −5.7 in 1980, increasing again to 4.4 in 1982, and declining again to −3.5 in 1986. In 1984 the migration pattern of Canada returned to that previously seen in 1970, with a net gain of migrants in the core region and a net loss in the two peripheral regions. What is most dramatic is that in 1985 every province and territory in the two peripheral regions experienced net outmigration and in 1986 only British Columbia and Yukon had positive net migration rates. In the core region, however, Ontario gained 45 278 net migrants in 1986 and Quebec had a net loss of only 3686 migrants, which is a very small loss compared with the net loss of 46 427 in 1977.

Peripheral West Europe and Japan

This group had a significant drop in net inmigration to the core regions beginning only after 1970, with maximum net flows into the core regions and away from the peripheral regions occurring during the 1960s. Since 1980 the rates of net migration have shown a slight divergence away from zero, with the core areas again attracting migrants from the periphery. In no way, however, are the net flows to the core regions approaching the previous levels of the 1960s.

Finland (Figure 7.11) is the only country in this category which has never reversed the negative net migration rates in any of its peripheral regions during the study period, although the net flows have been greatly reduced from their maximum in 1970 and nearly approached zero in 1977. Between 1979 and 1984, the north region again experienced increasing net outflows to the core region.

Norway (Figure 7.12) shows a similar trend, with net flows out from the north region approaching zero in 1979, but decreasing to a net migration rate of −9.7 in 1985, the greatest level of net outmigration in the north region since 1970. The net migration rate in the core (east) region has increased steadily between 1979 and 1985 from 0.1 to 3.1. The south region, which consistently had the highest net migration rate since 1971, had a net migration rate of only 0.9 in 1985, its lowest level since 1966. This recent decline in the south is probably due to a slowdown of the oil industry which followed a decline in oil prices.

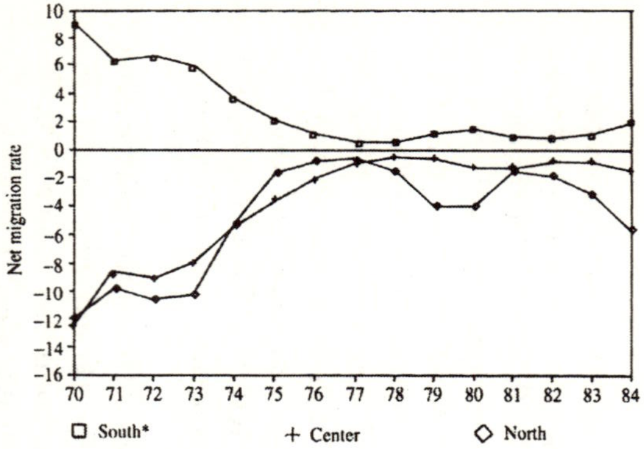

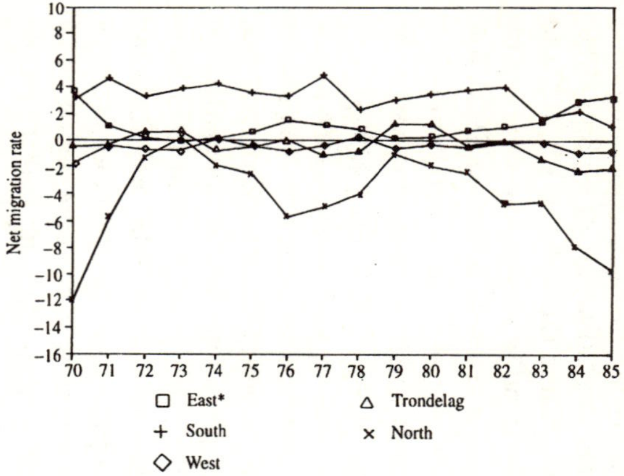

Fig. 7.12 Changing migration trends in Norway, 1970–85

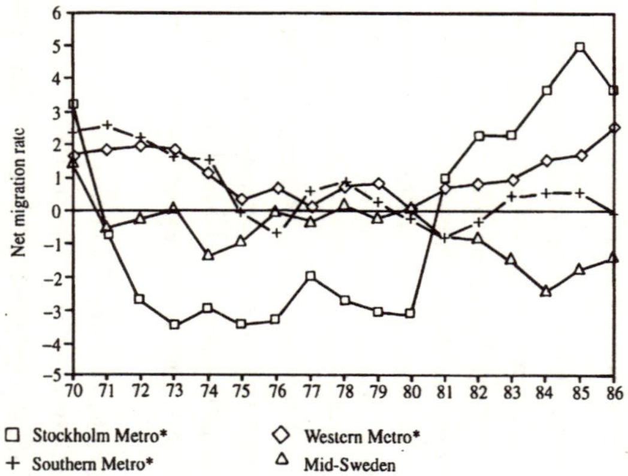

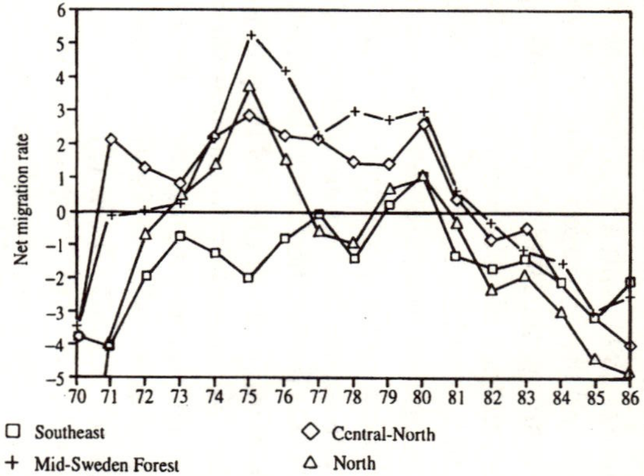

Fig. 7.14 Changing migration trends in Sweden, regions 5–8, 1970–86

Fig. 7.11 Changing migration trends in Finland, 1970–84

Fig. 7.13 Changing migration trends in Sweden, regions 1–4, 1970–86

A much more abrupt change has occurred recently in Sweden (Figures 7.13 and 7.14), reversing the trend of net flows out from the core seen during the 1970s. From 1971 to 1980 the Stockholm metropolitan region had a net outflow of persons to other regions. In 1981 this trend reversed and continued increasingly positive through 1985, declining slightly in 1986. The opposite is seen in Sweden's peripheral regions, which exhibited generally positive net migration rates in the 1970s, peaking in 1975. Net migration dropped abruptly and continuously for these regions beginning in 1981, with all peripheral regions having negative net migration rates between 1982 and 1986. The Stockholm and southern metropolitan regions have had positive net migration rates since 1981 and 1983, respectively. The net migration rate of the third core region, the western metropolitan area, was zero in 1980 and has since been positive and increasing.

No such sharp change is seen in Japan (Figures 7.15 and 7.16) after 1980. Its net migration rates remain relatively small and steady when compared to their pre-1970 rates. The migration patterns of Japan most closely resemble those of Finland. Net migration out of the peripheral regions dropped almost to zero in the mid-1970s, but the three core regions have since been gaining gradually in their attraction of migrants. The net migration rate of the Kanto region (containing Tokyo) has increased from a low of 1.7 in 1976 to 4.2 in 1986.[5] The net migration rate of Tokai, another core region (containing Nagoya), was 0.7 in 1985 and 0.9 in 1986, its first positive net migration rates since 1974. Overall, since 1980 the three core regions of Japan have begun to attract more migrants once again as net outmigration from all of the peripheral regions has increased.

The net migration rates for Italy (Figure 7.17) also continue to hover near the zero mark, but differ from Japan and other countries in this category in that Italy is still in the long process of convergence that began in the 1960s. The net migration rate of the core (north-west) region became negative for the first time in 1982. In 1985, the core region's net migration rate stood at −0.8, identical to that of the south region, although in 1986 the net migration rate of the north-west increased slightly to −0.3 and the south declined to −1.4. The net migration rate for the south has increased from a low of −13.0 in 1961. Moreover, the population of the north-west has been declining since 1979, whereas the growth of the south has been the highest in the country since 1971, in part because of higher birth rates and returning international migrants (Fielding, 1982).

Vining and Pallone (1982) described a rapid decline in net inmigration to the four core regions of Spain (Figures 7.18, 7.19 and 7.20) in 1974 and a rapid decline in net outmigration from the peripheral regions of the south and west. At the time it was difficult to call this a long-term trend, given the recession which struck the country around this time. More recent data, however, confirm this trend to 1985. The net migration rates of all of the regions continue to converge towards zero, with some reversal of migration trends. Two of the four core regions, Cataluña and Vascongados y Navarra, have had net outflows of migrants since 1980 and 1978, respectively. As of 1982 all of Spain's peripheral regions have had positive net migration rates with the exception of Castilla la Vieja, Asturias, and Baleares. Population estimates following the 1981

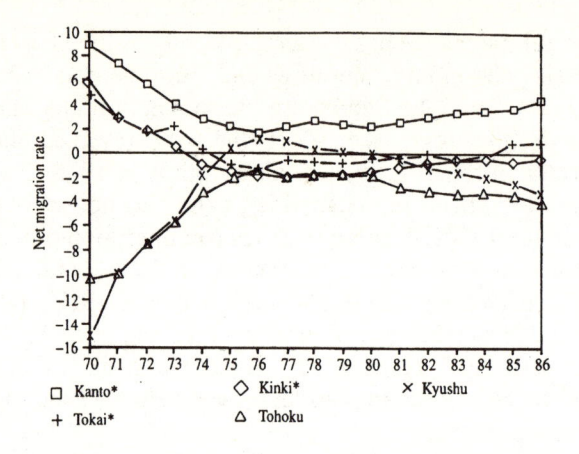

Fig. 7.15 Changing migration trends in Japan, regions 1–5, 1970–86

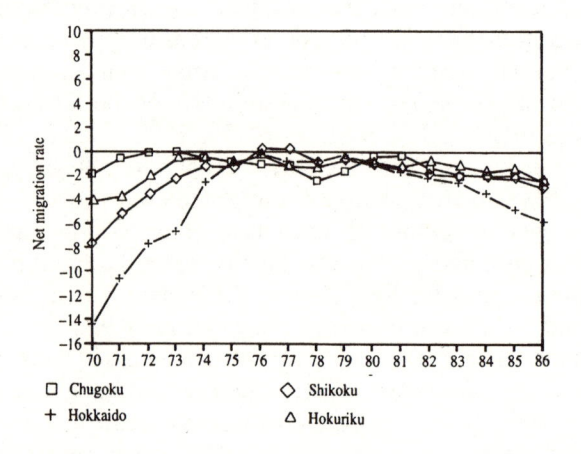

Fig. 7.16 Changing migration trends in Japan, regions 6–9, 1970–86

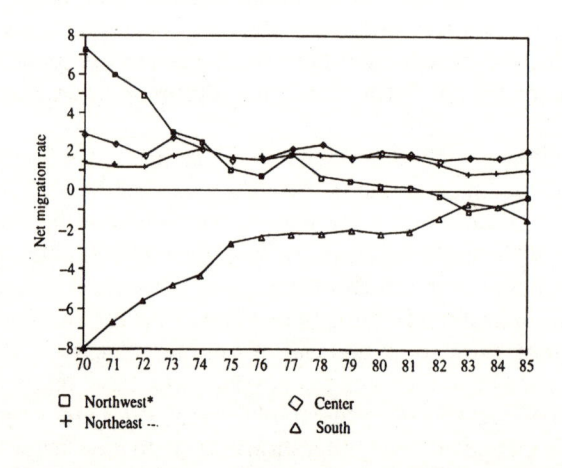

Fig. 7.17 Changing migration trends in Italy, 1970–85

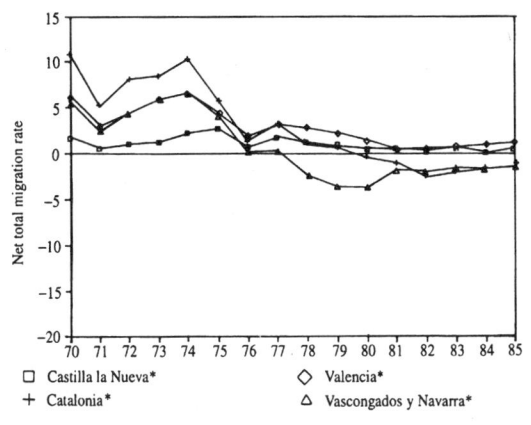

Fig. 7.18 Changing migration trends in Spain, regions 1–4, 1970–85

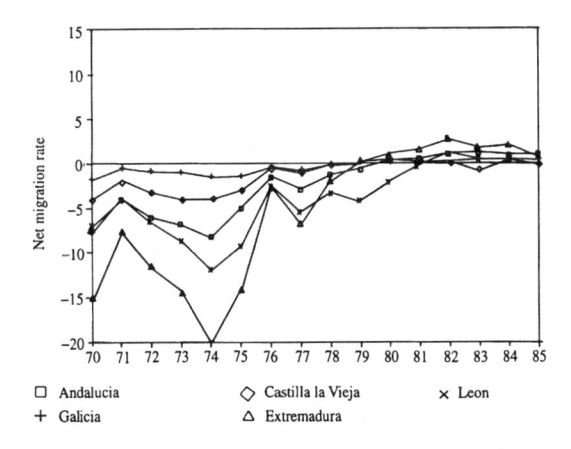

Fig. 7.19 Changing migration trends in Spain, regions 5–9, 1970–85

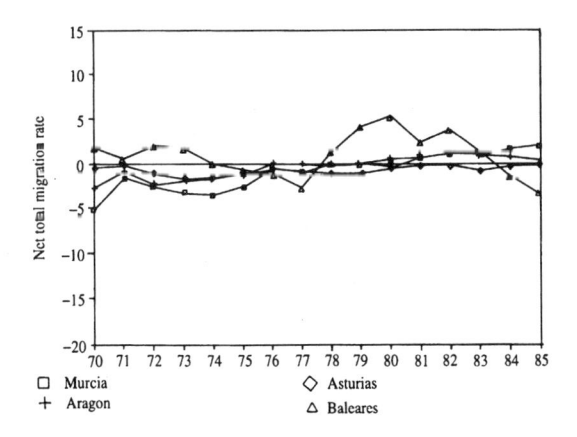

Fig. 7.20 Changing migration trends in Spain, regions 10–13, 1970–85

census reflect these recent net migration trends, with current estimates for the core regions reduced up to 1.9 per cent over earlier estimates and those for the peripheral regions increased up to 4.8 per cent over previous estimates.

In the United Kingdom, changes of registration with the National Health Service Central Register were used by Ogilvy (1980) to estimate migration flows during the 1970s. His data show reduced net outmigration in the mid-1970s for several of the northern peripheral areas (Scotland, Yorkshire and Humberside, and the north district). The core region (East Anglia, south-east, and south-west areas) had a small net outmigration in 1974–75, which became increasingly positive in the latter half of the decade, while the peripheral regions, except for Wales, all showed net outmigration from 1975 to 1979. Later data from the Central Statistics Office (1982, p. 60; 1983, p. 51; 1984, p. 51) shows the south-east (London Metropolitan) area with positive net migration in 1980–82. Champion (1987a, 1987b) adds that recent trends in net migration have been primarily responsible for the switch from population decline to growth in Greater London since 1983.

East Europe

There is little change in the constant rate of net migration towards the core areas in the German Democratic Republic (Figure 7.21) or Czechoslovakia (Figure 7.22). The high rate of net migration into the East Berlin region achieved in the mid-1970s has continued with a net migration rate of 17.6 in 1986, the highest since these figures became available in 1962. Net migration towards Prague continues, but at a more moderate rate than toward East Berlin. The net total migration rate of Czechoslovakia's core region has fluctuated between 1.5 and 6.8 during the period 1970 to 1986, and none of the three peripheral regions of Czechoslovakia has experienced positive net migration rates since 1976.[6] Neither country shows any sign of a systematic or sustained decline in net migration towards their core regions.

Taiwan and the Republic of Korea

The core (north) region of Taiwan (Figures 7.23 and 7.24) continues to attract net migrants from all of the other regions of the island, but its net migration rate has declined from 19.8 in 1978 to 10.9 in 1986. During the same period, net outmigration from the centre region (containing Taichung City) subsided, with its net migration rate changing from −5.7 in 1978 to −0.6 in 1986. The other peripheral regions still have net migration rates much less than zero, and, unlike the centre region, they do not appear to be approaching zero. The net migration rate of the core (north) region declined by 51 per cent between 1978 and 1983, then increased slightly but steadily between 1983 and 1986. Although Taiwan has slowed its migration to the core, it cannot be said that its migration pattern has reversed.

During the period 1978–81 it appeared that net migration trends in the Republic of Korea (Figure 7.25) might be beginning to converge towards zero, but net migration rates to the core area during the 1982–84 period returned to

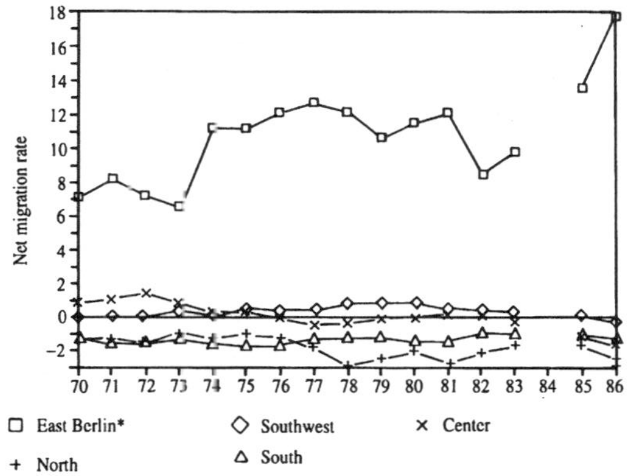

Fig. 7.21 Changing migration trends in the German Democratic Republic, 1970–86

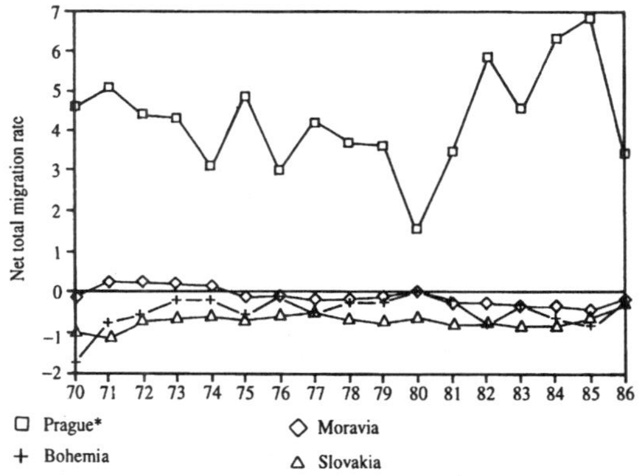

Fig. 7.22 Changing migration trends in Czechoslovakia, 1970–86

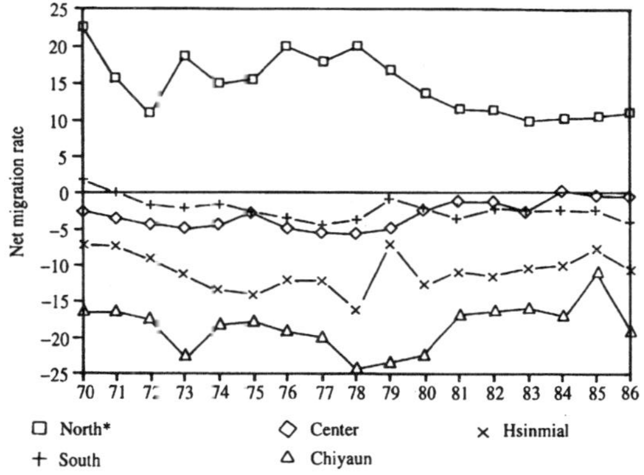

Fig. 7.23 Changing migration trends in Taiwan, regions 1–5, 1970–86

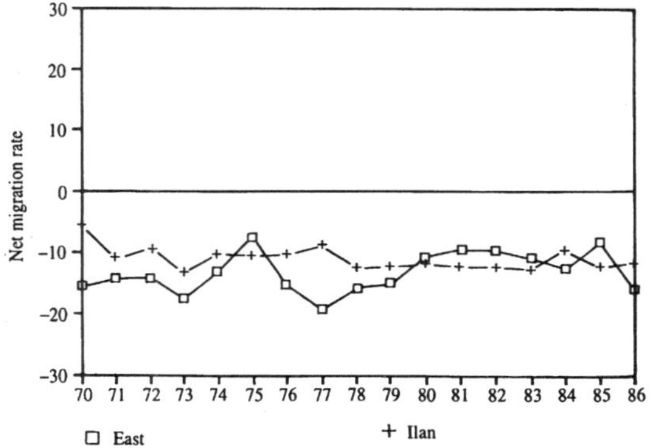

Fig. 7.24 Changing migration trends in Taiwan, regions 6 and 7, 1970–86

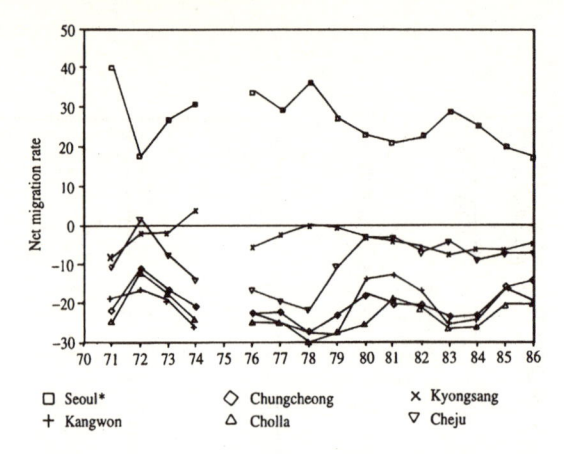

Fig. 7.25 Changing migration trends in South Korea, 1970–86

levels seen in the mid-1960s. The Seoul region attracted 427 000 net migrants in 1983, nearly as many as in 1978, which was the highest year since 1969. Then during the period 1984–86, the net migration rate for the Seoul region declined again to 17.2, its lowest level since 1972 and about half the rate of 36.3 seen in 1978. The general trend of net migration into the Seoul region has been downward since 1978, but the fluctuations in the rates make it difficult to determine whether the country is making a break with its past migration patterns.

Discussion

North-western Europe

The countries of north-western Europe show the most moderate rates of net migration examined in this article. Only France continues to have increasingly divergent rates, with the Paris Basin, the east, and the north regions losing increasingly more people to the other regions. Recent data, however, indicate that net outmigration from the Paris region may be subsiding. Net migration rates in the Federal Republic of Germany are fairly steady, with the core region still having a net loss of migrants, primarily to the southern regions of Bavaria and Baden-Württemburg. Migration trends in the smaller countries of north-western Europe – Belgium, Denmark, and the Netherlands – have reversed the earlier pattern of net flows from the core region to the periphery. The core regions of each of these smaller countries have had net migration rates near zero since 1980.

When this turnaround occurred between 1978 and 1983, the total number of migrants decreased in Denmark by 5.6 per cent, while the total number of migrants in the Netherlands increased by 4 per cent. A decreasing supply of new dwellings may have contributed to the decline of migration levels in Denmark, although growth in the primary and tertiary economic sectors in the eastern islands region appears to have led to net migration into the core region in the early 1980s (Champion, 1988; Illeris, 1984). The Netherlands has been particularly hit by unemployment in recent years. Although total movement has increased, perhaps as job seekers search for opportunities in other areas, no single region dominates the internal migration pattern. Internal migration surveys in Australia have shown consistently that, during the 1980s, unemployed persons have had rates of mobility that are twice as high as those who are employed (Smailes and Hugo, 1985). A similar finding was made for the United States for the late 1960s (Herzog and Schlottmann, 1984). This may also have been the case in the Netherlands, as long as unemployment benefits are portable throughout the country.

A detailed study of migration between functional urban regions in the Federal Republic of Germany (Kontuly and Vogelsang, 1988) indicates that net migration towards the southern regions of the country in the 1980s spans all age groups. This finding led the authors to 'speculate that counterurbanisation will not reverse in the Federal Republic in the near future'. In the United Kingdom migration patterns have also been consistent across age groups; however, there is evidence that the core region of the UK is losing less population through internal migration (Champion, 1988).

Peripheral West Europe and Japan

The peripheral European countries and Japan show larger rates of net migration during the past decade than the countries of north-western Europe. Although the rates were converging towards zero in the mid-1970s, a slight divergence generally was seen during the late 1970s to mid-1980s. Norway, Finland, and Japan are examples whose core regions had net migration rates approaching zero in the mid-1970s, but which are now increasing slightly. The core regions of Sweden have also attracted more net migrants since 1981, following a decade of net outmigration. The trends in Spain and Italy lag behind those of the other peripheral European countries by about a decade, perhaps because of the large size of their own peripheral regions (the Mezzogiorno of southern Italy, Andalucia and Galicia in Spain) and their relatively wide historical disparities in regional development, but they are following the general trend of the other countries in this category.

Hansen (1988) offers two reasons why Norway was able to reduce outmigration from the peripheral regions, but not to the point of completely reversing net inmigration to the core (east) region. He argues that the urbanisation process in Norway is still active and the hierarchy of urban places is not yet complete. The added prosperity of oil production, however, provided for an expansion of the welfare state and regional development programs until the decline in oil prices of recent years.

North Sea oil production and expanded regional and rural development policies also contributed to the growth of peripheral regions in the United Kingdom in the 1970s (Champion, 1988). The reduction in the flow of migrants from the core to periphery in the early 1980s coincided with a cutback in regional policy measures and the oil price decline, as well as an increase in inner-city development programs (Champion, 1988).

Regarding Italy, Champion (1988) attributes some importance to the policy of infrastructure improvement in contributing to the decline in net outmigration from the south. He also presents evidence that the north-east and centre regions of Italy may continue to have dynamic economies in the future and attract a positive flow of net migrants, where:

> New dispersed forms of industrial development are being grafted on to the urban network with its high density of small- and medium-size towns, aided by a tradition of middle-class enterprise and self-reliance. By effectively bypassing the life-cycle model of metropolitan development based on cores and rings, there seems to be emerging here more quickly than elsewhere a new type of dispersed settlement which has all the necessary benefits of agglomeration economics without the need for a high level of concentration.
>
> (Champion, 1988, p. 12)

United States and Canada

In the United States, as in north-western Europe, a change is seen in the net migration trends of the past two decades. Net outmigration has decreased in the north-east and midwest regions since 1979–80. Internal migration statistics do not show a return to zero net migration for the core regions as in the smaller western European countries, but total net migration rates are near zero for the north-east in 1983 and 1984. Net migration in the United States may be approaching a new stability as efficiencies of agglomeration begin to decrease in the south and west and service industries take hold in the north-east.

Changing net migration patterns in Canada are similar to, but more dramatic than, those of the United States. The drop in net migration to the west region is much more abrupt than seen in the United States, owing to the dependence on oil for much of the growth of the Canadian west during the 1970s. The general pattern shows migration trends converging towards zero in 1982, followed by net movement towards the core through 1986.

The general migration trends indicated by the Bureau of the Census estimates have been studied in greater detail by Morrill (1988) and by Plane and Isserman (1983). Morrill studied all possible pairs of state-to-state migration flows and grouped the states into migration 'regions', which happen to be quite similar to the broad census regions used to define core and periphery in this article. Patterns of flows were relatively volatile, but a general core-to-periphery structure was found to be valid for the United States, although data for 1980–85 suggest a part-way return to a less dramatic core-to-periphery redistribution of population. Both studies discovered that although California and Florida are the major attracting states in the west and south regions, they have an important role in redistributing flows from the north-east. The census figures presented here reflect this pattern with large flows between the west and the south regions in recent years.

Taiwan and Korea

Taiwan had shown signs of convergence toward zero during the early 1980s, but, although the net migration rate for the core (north) region declined by

50 per cent since 1970, the core still dominates all other regions. The government has provided incentives and encouragement to many industries to locate outside the Taipei municipality, which 'has resulted in a high concentration of manufacturing companies and their official buildings in a few major urban areas such as the Taipei metropolitan area and Kaoshiung City' (Tsai, 1978). Although Kaoshiung Municipality has had a positive net migration rate since 1984, the net migration rate for the entire south (Kaoshiung) region has remained negative, standing at −4.4 in 1986. All of the other peripheral regions except the central region continue to have rates below −8. The concentration of the south region's net inmigration within the municipality of Kaoshiung reflects the recent spatial pattern of manufacturing growth (Todd and Hsueh, 1988), in which the government's program of industrial decentralisation has led to 'excessive local polarisation' in rapidly growing peripheral areas. Todd and Hsueh also conclude that, although the peripheral regions are benefiting from a relative rise in manufacturing and tertiary activities, the agglomeration economies in the core region still dominate the country. Their shift-share analysis of the spatial growth of manufacturing led them to state that 'the advantage of industrial inertia renders a total shift outcome that does no other than to underscore the core–periphery division of Taiwan'.

In the Republic of Korea, development policies seem to have had a significant impact on the movement of industrial activities from Seoul to its suburbs and some effect on the movement of industry or the creation of new industry in the large cities of the peripheral regions, such as Pusan and Taegu (Kim, 1988). The Seoul region is still the predominant region, with its net migration rate increasing from 20.9 in 1981 to 28.9 in 1983, then declining to 17.2 in 1986. These wide fluctuations make it difficult to predict if South Korea will soon break from its migration pattern of the past (Figure 7.25). South Korea has had plans since 1977 to control Seoul's population growth to obtain a more appropriate national population distribution pattern (Kim and Donaldson, 1979). Government policies have tried to promote deconcentration away from Seoul (Kim, 1988; Rondinelli, 1984) but at the same time have promoted concentration in urban centres in areas adjacent to the capital city within the Seoul metropolitan area (Smith *et al.* 1983).

Net migration rates in the core region of South Korea have been of the order of twice the rates seen in the core region of Taiwan, even though the countries have followed similar paths of development through export promotion of finished products. There seems to have been a key difference in the methods they employed regarding the spatial distribution of development (Kim, 1988; Todd and Hsueh, 1988). Korean policymakers have focused on controls to inhibit core region development. In Taiwan, however, government efforts to promote industrial decentralisation have combined a policy of dispersal of new large-scale industries with the development of major infrastructure projects that have linked all regions of the country. Early efforts in promoting agricultural development and labour-intensive industry also may have helped to keep Taiwan's net migration toward the core region at a relatively lower rate (Oshima, 1986). On the other hand, an important similarity between these countries is that growth in the peripheral regions appears to be confined to a limited number of central places. Perhaps only when the rural and urban economies are more closely linked and development reaches the smaller-size cities will a turnaround in the core–periphery migration trends appear as in the countries of Western Europe and North America.

Conclusion

Rogerson and Plane (1985) observed that the large loss of net migrants from the United States industrial core is a phenomenon of the 1970s, and that 'the imminent demise of the industrial core states, predicted by many, is far from certain'. The data indicate that this is true not only for the United States but for north-western Europe, as well, where many of the core regions have recovered from a period of net migration loss. Rogerson and Plane suggest that 'the high growth rates of areas of the United States that are rich in high-tech jobs may already be slowing because the inmigration of firms to these areas is bidding-up wages and causing the growth areas to lose some of their advantage over the industrial core'.

Counterurbanisation has been a major theme of European migration literature in recent years. Fielding (1982, p. 13) concluded from a study of migration trends in France and other European countries that:

> 'Urbanisation', used in the sense of a positive relationship between net migration and settlement size, has ceased in almost all of the countries of Western Europe in the period 1950–80. ... In seven out of the nine countries in which it can be shown that urbanisation had ceased, the metropolitan and principal industrial cities showed, during the 1970s, signs of net migration loss, and rural regions containing small and medium-sized towns showed signs of net migration gain.

Do the trends of the early 1980s for Belgium, Denmark, the Netherlands, Norway, Sweden, and the United Kingdom, as well as Canada, Japan, and the United States, indicate an end to counterurbanisation in these regions of the world? As Fielding said, there is an immense complexity of regional and local situations in Western Europe, which makes conclusions from empirical analysis difficult; but contrary to the period of the 1970s, many of the metropolitan regions of Europe, the United States, and Japan are losing fewer migrants or are gaining population through net internal migration. Between the core and peripheral regions of some of these countries, a new balance has been achieved with little or no net migration to the core region.

It has been shown clearly that all countries go through major shifts in settlement patterns as their economies undergo shifts from agriculture to industrialisation and from industrialisation to post-industrial activities (Alonso, 1968; Fielding, 1982; Mera, 1973; Plane, 1984; Vining and Pallone, 1982; Williamson, 1965). Net outmigration to peripheral regions at the expense of the core regions seems only to be, however, a temporary phenomenon that occurs while an economy adjusts to the new spatial location requirements of post-industrial economic activity. If this is true, the Paris region of France and the Rhine–Ruhr region of the Federal Republic of Germany might be expected to decrease their flows of net outmigration as their economies adjust further to postindustrialism. Over a decade ago, when net outmigration trends from the United States core regions were becoming apparent, Mera (1973, p. 320) wrote:

> The relative advantage of urban concentration is greater in less developed countries than in developed countries. However, this statement should not be taken as implying that the advantage of urban concentration disappears once a country reaches a certain stage of development. The findings of the United States and Japan show that there are still observable advantages in concentration.

The countries of eastern Europe must still be excluded from the pattern seen in western Europe, Japan, and the USA, as they show no likelihood of shifting their trends from a pattern of steady net inmigration to their core regions.

It is more difficult to generalise about the net migration trends of the rapidly developing countries of the Republic of Korea and Taiwan. In recent years, the core regions of both countries have shown, in general, declining trends in their levels of net inmigration. The fluctuations in the net migration rate of the Seoul region, the recent levelling off of the net migration rate of the Taipei region, and the lack of any sustained net inmigration to any peripheral region of either of these countries make it impossible to conclude that either South Korea or Taiwan will see a reversal in migration patterns in the near future. Aggressive infrastructure development programs in both countries (*see* Kim, 1988; Rondinelli, 1984; Todd and Hsueh, 1988) are auspicious signs that migration turnarounds may yet be seen in the future.

Acknowledgements

The authors are grateful for the assistance of Judith Z. Kalhacher, Vera J. B., David L. Brown, and Supdna M. Wilson in the preparation of various aspects of this paper.

Notes

1 Vining and Pallone reported data for Scotland only. Migration trends for the UK have been studied by Ogilvy (1980) using figures from the National Health Service Central Register. His results are briefly described in this article.
2 Working Paper no. 108a is obtainable from the Department of Regional Science, University of Pennsylvania.
3 The negative net migration rate for the west region in 1984 appears to be due to a large movement of migrants to the newly developed Zuidelijke IJsselmeerpolders area, which became Flevoland Province in 1985.
4 US Census Bureau reports provide estimates of net interregional migration figures based on sample surveys of civilian noninstitutional households and members of the armed forces living off base or with their families on base. These data must be interpreted with caution, however, because of the nature of the survey. Migration into the south may be biased downward because of the large number of military bases located in this region. Soldiers from other regions assigned to bases in the south may not be counted as inmigrants. Yet, when they are discharged and return to their home regions as civilians, they will be counted as migrants from the south to their home region. See Long (1983) for a further discussion of the effect of military populations on models of migration.

5 Within the Kanto region, the net migration rate of Tokyo prefecture was positive in 1985, the first time Tokyo had experienced net inmigration since 1966.
6 The net total migration rate may be regarded as a net internal migration rate because net international migration is quite small for Czechoslovakia.

References

Alonso, W. 1968: Urban and regional imbalances. *Economic Development and Cultural Change* 17(1), 1–14.
Central Statistics Office. Government Statistics Service. 1982–84: *Regional trends.* London: Her Majesty's Stationery Office.
Champion, A. G. 1987a: Counterurbanisation: The British experience. Paper presented at the annual meeting of the Association of American Geographers, Portland.
Champion, A. G. 1987b: Recent changes in the pace of population deconcentration in Britain. *Geoforum* 18 (4), 379–401.
Champion, A. G. 1988: *Recent changes in the pace and nature of population deconcentration: The European experience.* Paper prepared for The Dynamics of Centre–Periphery Relations: A Workshop on the Nation State, the Region and the Spatial Organisation of Power. Bergen, Norway.
Courgeau, D. 1986: Vers un ralentissement de la deconcentration urbaine en France. *Population et Sociétés* 200, 4.
Engels, R. and Forstall, R. 1984: *Growth in nonmetropolitan areas slows.* Washington, DC: United States Bureau of the Census, unpublished release.
Engels, R. and Forstall, R. 1985: Tracking the nonmetropolitan population turnaround to 1984. Paper prepared for presentation at the annual meeting of the Population Association of America, Boston.
Fielding, A. 1982: Counterurbanisation in Western Europe. *Progress in Planning* 17, Part 1, 1–52.
Hansen, J. 1978: Settlement pattern and population distribution as fundamental issues in Norway's regional policy. Paper presented to IGU Commission on Population Geography: Symposium on Population Redistribution Policies, Oulu, Finland.
Hansen, J. 1988: Norway in the eighties – the turnaround which turned round. Paper presented at the counterurbanisation session of the IGB annual conference, Loughborough, UK.
Herzog, H. W. Jr. and Schlottmann, A. M. 1984: Labor force mobility in the United States: migration, unemployment, and remigration. *International Regional Science Review* 9 (1), 43–58.
Illeris, S. 1984: Danish regional development during economic crisis. *Geografisk Tidsskrift* 84, 53–62.
Kim, S- U. and Donaldson, P. 1979: Dealing with Seoul's population growth: government plans and their implementation. *Asian Survey* 19 (4), 660–73.
Kim, W. 1988: Population redistribution policy in Korea. *Population Research and Policy Review* 7 (1), 49–77.
Kontuly, T. and Vogelsang, R. 1988: Explanations for the intensification of counterurbanisation in the Federal Republic of Germany. *Professional Geographer* 40 (1), 42–54.
Long, J. 1983: The effects of college and military populations on models of interstate migration. *Socio-economic Planning Sciences* 17 (5–6), 281–90.
Mera, K. 1973: On the urban agglomeration and economic efficiency. *Economic Development and Cultural Change* 21 (2), 309–24.

Morrill, R. L. 1988: Migration regions and population redistribution. *Growth and Change* 19 (1), 43–60.

New York Times 1985: Population's rise in the north-east reverses a trend, 7 April, p. 1.

Ogilvy, A. 1980: Population migration between the regions of Great Britain, 1971–79. *Regional Studies* 16 (1), 65–73.

Oshima, H. 1986: The transition from an agricultural to an industrial economy in east Asia. *Economic Development and Cultural Change* 34 (4), 783–809.

Plane, D. 1984: A systemic demographic efficiency analysis of United States interstate population exchange, 1935–1980. *Economic Geography* 60 (4), 294–312.

Plane, D. and Isserman, A. M. 1983: United States interstate labour force migration: an analysis of trends, net exchanges, and migration subsystems. *Socio-economic Planning Sciences* 17 (5–6), 251–66.

Rogerson, P. and Plane, D. 1985: Monitoring migration trends. *American Demographics* 7 (2), 27–9, 47.

Rondinelli, D. 1984: Land-development policy in South Korea. *Geographical Review* 74 (4), 425–40.

Smailes, P. and Hugo, G. 1985: A process view of the population turnaround: an Australian rural case study. *Journal of Rural Studies* 1 (1), 31–43.

Smith, W. R., Huh, W. and Demko, G. 1983: Population concentration in an urban system: Korea 1949–1980. *Urban Geography* 4 (1), 63–79.

Statistics Canada. 1985: *Canada Yearbook 1985*. Ottawa.

Todd, D. and Hsueh, Y-C. 1988: Taiwan: some spatial implications of rapid economic growth. *Geoforum* 19 (2), 133–45.

Tsai, H. 1978: Development policy and internal migration in Taiwan. Reprint, Department of Agricultural Extension, National Taiwan University, pp. 27–51.

Vining, D. and Pallone, R. 1982: Migration between core and peripheral regions: a description and tentative explanation of the patterns in 22 countries. *Geoforum* 13 (4), 339–410.

Williamson, J. 1965: Regional inequality and the process of national development: a description of the patterns. *Economic Development and Cultural Change* 13 (4), (Part 11): 3–84.

Yamaguchi, T. 1983: Population redistribution of Japan within the context of the national settlement system. *Proceedings of the Department of Humanities, College of Arts and Sciences, University of Tokyo* 78 (Series of Human Geography no. 8), 1–18.

8 R. Koch,

'"Counterurbanisation", also in Western Europe?'

From: *Informationen zur Raumentwicklung* 2, 59–69 (1980).
Originally published in German

Introduction

In the past the consequences of declining birthrates in demographic and regional policy discussion were an important topic of research in the countries of Western Europe,[1] but today, migration tends to become a more important factor affecting the spatial redistribution of the population in this region. Two factors are responsible for this.

On the one hand the birth and mortality rates in most of the countries and regions of Western Europe are almost in balance. This means that the size and structure of the population are affected more severely by migration than by birth and mortality rates. However, population development indicators in Germany and Austria show this to be only a temporary phenomenon. As birthrates decrease, the effect of migration on population development will inevitably diminish too.

On the other hand, regional scientists, mainly from English-speaking countries, are studying the occurrence of 'counterurbanisation'. This term, coined by the American geographer B. J. L. Berry, has not yet been translated correctly into German. Terms such as *'Entdichtung'* or 'deconcentration' do not fully reflect the meaning of the term. Originally counterurbanisation referred to the turnaround of migration flows in the USA during 1970 to 1975 compared to 1960 to 1970. Statistics of the US Bureau of Census proved that the growth of the large agglomerations declined during the 1960s and that they lost population because of out-migration during the early 1970s. This migration proved to be population movement from the big cities in the north to the smaller and medium-sized towns in the south and mid-west, not merely suburbanisation. Some American regional researchers regard the changes in migration during the last ten years as 'no less than a breakdown in Clark's law of concentration' (Vining and Kontuly, 1978, p. 49). Clark regarded the development of settlement in modern industrial countries as 'the macro-location of industry and population [which] tends towards an ever-increasing concentration in a limited number of areas; their micro-location, on the other hand, towards an increasing diffusion, or "sprawl"' (Clark, 1967, p. 280). To establish whether a similar breakdown of Clark's law occurred outside the United States, American researchers analysed migration flows in Western Europe. This proved to be a very stimulating exercise as an international comparison of migration trends.

At a seminar of the Council of Europe in autumn 1979 the hypothesis of counterurbanisation in Western Europe and the role of migration in the population development process were discussed. Against this background an attempt is made in this paper to identify the trends of population development in Europe between 1960 and 1975, and to investigate their possible consequences for different types of regions.

Interregional migration

During the years 1960–70, a significant out-migration of people from the peripheral regions occurred in Western Europe (Figure 8.1). The most significant losses were recorded in southern Italy, in Northern Ireland and Scotland, i.e. the peripheral regions of Western Europe. Migration losses shown by peripheral regions of individual countries were also remarkable, e.g. France (Lorraine), the Federal Republic of Germany (eastern Bavaria), and the Netherlands (Friesland). Substantial migration gains were concentrated in a few large urban agglomerations:

Developed World: International Trends 113

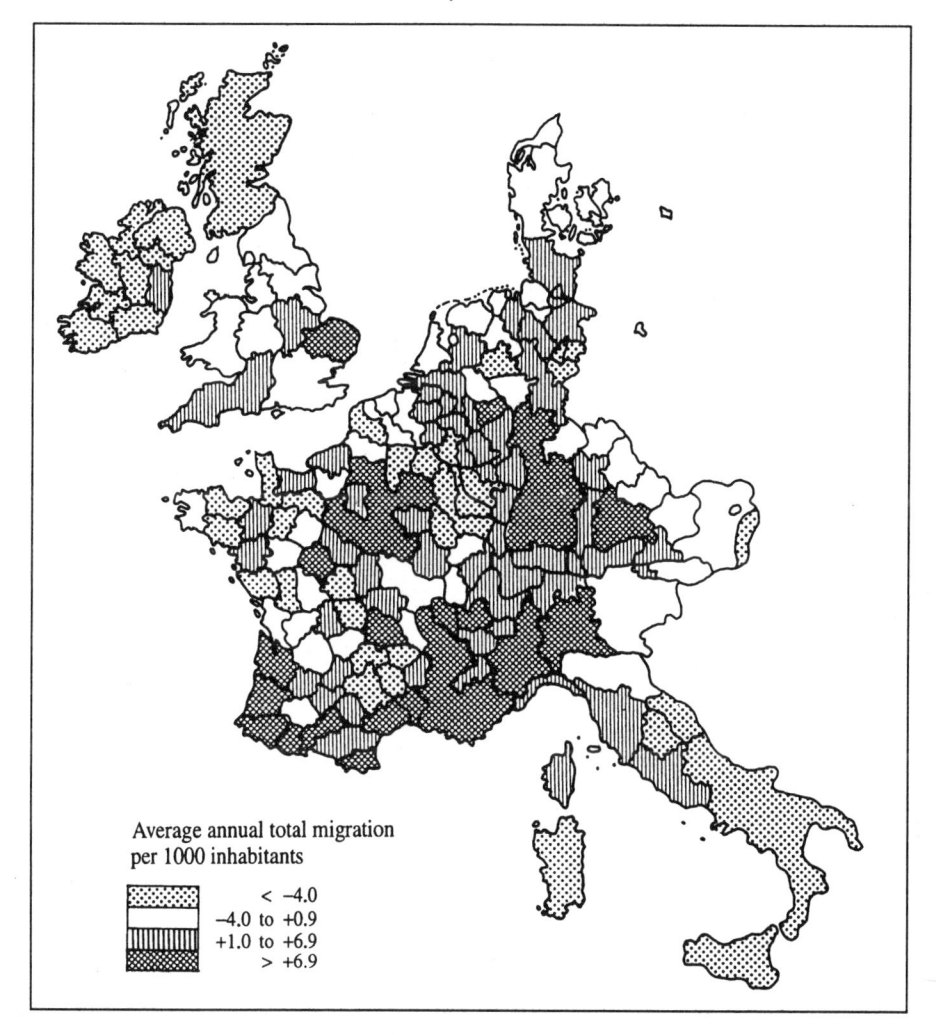

Fig. 8.1 Migration in Western Europe, 1960–70. (Source: European Research Institute for Regional and Urban Planning – ERIPLAN. Final report of interest group 2, 'migration', Brussels, 1978, p. 46)

Legend: Average annual total migration per 1000 inhabitants
- < −4.0
- −4.0 to +0.9
- +1.0 to +6.9
- > +6.9

- in France, the Paris region and the surrounding departments, the Rhone valley and the Marseille region;
- in the Federal Republic of Germany, the agglomerations along the Rhine and the Munich region;
- in Belgium, mainly the Brussels region.

In addition, the coal and steel areas (Ruhr and northern France) were already showing significant migration losses. In the south-east of England and the Randstad, the migration gains and losses were almost equal (Fielding, 1975, pp. 237–54).

114 *Differential Urbanization*

In recent years, a change has occurred in net migration. Large urban agglomerations, where in- and out-migration were in balance in the past, are now showing a net migration loss (Figure 8.2). The Paris region is still showing a slight migration gain, but there are only three regions showing significant migration gains, i.e. Munich, Lyon–Grenoble, and Marseille. Simultaneously, the migration losses of peripheral regions in these countries have decreased, most notably in southern Italy. In some areas which traditionally experienced out-migration such as Brittany and Southern Ireland, small migration gains are now observed for the first time. The highest losses are now occurring in the densely populated coal and steel areas of northern France and the Ruhr region (ERIPLAN, 1978, p.73).

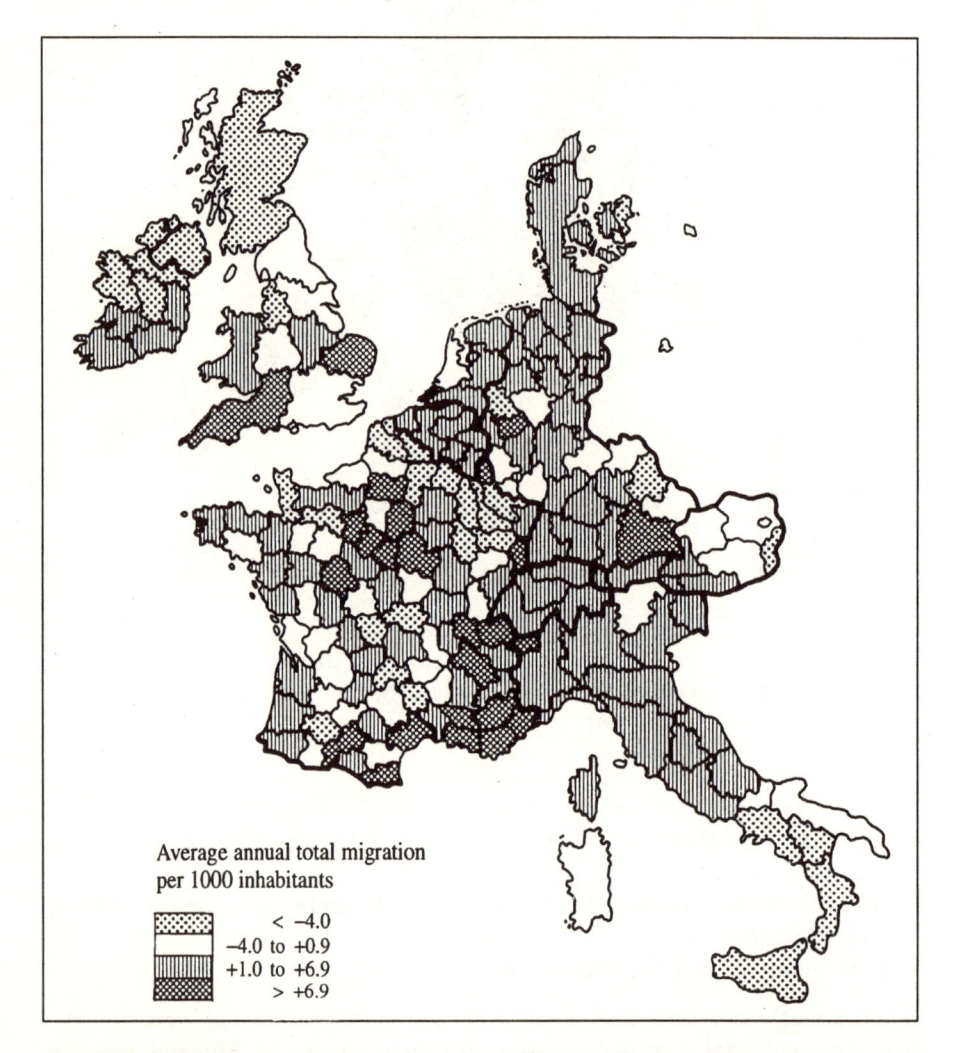

Fig. 8.2 Migration in Western Europe, 1970–75. (Source: European Research Institute for Regional and Urban Planning – ERIPLAN. Final report of interest group 2, 'migration', Brussels, 1978, p. 47)

However, these changes are not sufficient proof of the breakdown of Clark's thesis or of the beginning of 'counterurbanisation' in Western Europe. This would be a false inference to a very complex process on the basis of a simple indicator full of statistical weaknesses.

First, regional net migration was strongly influenced by international migration in both periods under study. Between 1960 and 1970 the migration of labour from the rural regions of Southern Europe to the densely populated regions of Western Europe reinforced the urbanisation process. The 1974–75 recession led to a substantial return of migrant 'guest-workers' to their home regions. This explains a large part of the migration losses of the hitherto expanding agglomerations as well as the reduced losses of the peripheral regions. The Stuttgart region, for example, one of the regions in the Federal Republic of Germany with a very high proportion of foreigners, gained about 32 000 foreign people every year between 1965 and 1973. Between 1974 and 1976 the average annual migration loss was about 23 000 persons.

Second, the reduced losses in some areas which traditionally experienced significant migration losses conceal the fact that there is still high out-migration of young people from these regions. Education facilities and job opportunities in these areas are still inferior to those in the large urban agglomerations. Eastern Bavaria showed a net loss of 3200 persons to the rest of the Federal Republic of Germany, but the migration loss of 18- to 24-year-olds were approximately 4200 persons. The Paris region, in spite of an overall migration loss of 23 417 persons per year during the period 1968–75, gained 11 691 employed persons per year. The real migration loss could be observed in the age group 65 and over. In this group there was a deficit of 16 242 inhabitants per year (Courgeau, 1978, pp. 533–4).

Third, it is unlikely that the decrease in migration loss in rural areas could be ascribed to an improvement of living and working opportunities. Rather, it could be argued that the employment opportunities for unskilled labour in urban areas deteriorated sufficiently during 1974 and 1975 to deter migration from these areas. Unemployment is possibly more acceptable in people's home regions, at least amongst older people, where their social contacts and some occasional employment may provide income in addition to the employment subsidies. A significant economic boom should also have an effect on this labour force's potential and could reinforce out-migration.

Fourth, inter-regional migration is losing importance compared to natural population growth. Migration has a great influence on population growth in regions where birth and deaths rates are evenly balanced. As in regions experiencing a decline in their birth rates, the number of regions showing an increase in their death rates are increasing. In many cases migration can no longer compensate for these losses. In some countries the volume of migration has decline significantly – by as much as 30 per cent in the Federal Republic of Germany between 1972 and 1976.

Finally, the selection of unsuitable regional units could easily lead to false conclusions. Vining and Kontuly (1977, p. 65) attempted to disprove the continued relevance of Clark's law in the case of the Federal Republic of Germany by analysing the migration balances of the federal state of North Rhine-Westphalia (Figure 8.2). Without doubt North Rhine-Westphalia is a densely populated state and the Ruhr area is one of the largest agglomerations in Europe. However, the Ruhr area, with its economic dependence on coal and steel, is not representative of the large urban agglomerations in the Federal Republic of Germany. The Ruhr area lost about 0.3 per cent of its population due to migration between 1970 and 1975, whilst the Munich region gained 5.8 per cent. The Paris region, according to Vining and Kontuly (1977, p. 41), extends from the Atlantic in the north to the surroundings of Lyon in the south enclosing a number of (urban/rural) departments which experienced migration gains and losses over the period. Statements about the urbanisation process in France based on this region need to be treated with caution.

Although out-migration from peripheral regions and in-migration into the core regions of Western Europe have slowed down somewhat, it would be an exaggeration to speak of a reversal of migration trends similar to those observed in the USA. The traditional basic patterns of migration are still present and become apparent when the migration of gainfully employed persons is analysed. They are concealed by the increasing migration of non-working people, especially of retirement migration.

Inter-regional migration of gainfully employed persons cannot be analysed without consideration of conditions in the regional labour market. Employment opportunities are still much better in large urban agglomerations for qualified people than in peripheral areas. In all the rural areas unemployment rates exceed the national average.

Generally, firms making use of highly qualified labour did not show any tendency to leave large urban agglomerations between 1970 and 1975. The headquarters of the larger national, and especially multinational, firms are still concentrated in the metropolitan areas. This differs significantly from the situation in the USA, where the out-migration of the population from the large urban agglomerations (Tucker, 1976, p. 435) is accompanied by the relocation of firms (Sternlieb and Hughes, 1977, p. 227). In Western Europe the positive external effects of the large urban agglomeration still predominate.

Trends in migration

Most regional forecasts in Western Europe are confined to the forecasting of natural population growth. This does not, however, reflect all the effects of current demographic trends. On the other hand the forecasting of migration has always been the most difficult part of regional forecasting. In view of changing migration trends forecasts based on existing models are mere rough extrapolations, a shot in the dark (Koch and Gatzweiler, 1978, p. 45).

Sometimes the imbalances in the labour markets are analysed only in terms of migration trends. The forecasting of total net migration from such a balance or imbalance does not take sufficient cognisance of the expected selective migration trends of the future. Thus the interpretation of labour market imbalances must include the comparison of labour force demand and supply which may offer important insights into future spatial structural problems.

The main regional problems of the next 10 to 20 years are not expected to result from the effects of population decline or lower population growth but from the consequences of the baby-boom of the 1960s. On the one hand there will be a decrease in the number of children, because of the decline in the birth rate, and a decrease in the number of persons retiring, while on the other hand the high birth rates of the 1960s will lead to a great increase in 15–64-year-olds. This development, which results from the population structure of the past, is expected to be the most important factor affecting the demography of Western Europe over the next 10 to 20 years. The economically active population is expected to grow substantially not only in southern and south-east Europe (up to 1990 this age group will grow by about 45 per cent), but also in Western Europe. It is expected that this group will increase by 8.1 per cent in the EEC, 26.8 per cent in Ireland, and 14.9 per cent in the Netherlands.

On the other hand a 1.6 per cent decrease in gainfully employed persons is forecast for Western Europe. The relative positions of regions with monostructural economies, especially peripheral regions, are expected to worsen. Job opportunities in the large urban agglomerations with strong tertiary sectors are expected to increase steadily, or would at least stand a better chance of dealing with structural problems.

Comparisons of labour supply and demand for 1990 on a regional scale show significant disparities. In 1990 there is expected to be a surplus of 11.4 million labourers in Western Europe. Compared with a real unemployment total of 2.8 million in 1974 this is a fourfold increase. The numbers of people searching for jobs will be highest in southern Italy and Northern Ireland (27 per cent). But in Central Europe high unemployment rates are also expected. Without out-migration or an increase in commuting, it is expected that an over-supply of labour will be an acute problem of labour markets in certain peripheral regions by 1990. Such regions include the north of Denmark, France, the Netherlands, and also the German regions of Ems-Osnabrück, Münster and East Bavaria.

The difference between the trends predicted for rural areas and those predicted for large urban agglomerations suggests the hypothesis that a large number of the unemployed people in the periphery will migrate to the large urban agglomerations. However, it is difficult to estimate the number of people likely to migrate to the large urban agglomerations in each country because the number of job opportunities in these large urban agglomerations will be less than the labour surplus in the rural areas. We may assume that young, well educated people in particular will leave the rural regions, whilst older people will tend to stay despite lower incomes and unemployment. This assumption is supported by the empirical evidence for the Federal Republic of Germany and Denmark.

The migration flow of young people towards the densely populated areas roughly equals, or even exceeds, that of people not depending on urban jobs, especially the retired. Thus the contradictions between the economic model and reality can be explained as follows:

- First, the selective migration patterns in Western Europe cannot sufficiently be explained simply by looking at net migrations.

- Second, in terms of the way in which the selective migration patterns in Western Europe are influencing the existing population and social structure of Western Europe, four types of regions can be identified. First, rural areas relatively isolated from large urban agglomerations, with unattractive job opportunities, insufficient infrastructure and without scenic beauty. These regions will probably continue to be areas of out-migration in the future. Taking into account the decline in the birth rate, the expected out-migration will greatly affect the population growth rate in these regions. This will result in an accelerated ageing of the peripheral population. Typical examples of such regions are Champagne-Ardennes and Limousin in France, northern Scotland in the United Kingdom or the Emsland in the Federal Republic of Germany. Second, rural areas closer to or better linked to large urban agglomerations, with scenic beauty and sufficient developed infrastructure. These regions could become popular recreation and housing areas. Many of the houses in these areas will be holiday housing, and a larger percentage of the houses will be built for retired people. Retirement settlements as in Florida or in Arizona are not yet known in Western Europe, but there is already a spatial concentration of retired people especially in the coastal regions such as south-west France or Provence, Côte d'Azur, and in the Bavarian Alps, Tessin (Cribier, 1970, p. 119).
- Third, large urban agglomerations which had expanded during the 1960s and which will continue to be centres of economic growth. However, their growth is expected to be much slower than in the early 1970s. These regions are expected to be popular migration destinations for the young, economically active age groups. Within these regions social and spatial segregation as well as urban sprawl are expected to continue. This will be the case in regions such as London, Paris, Lyon–Grenoble or Munich. However, the rate of population increase is expected to vary considerably in these urban regions owing to significant differences in the age structures of their populations and their fertility rates as well as the migration rates of the elderly.
- Fourth, large urban agglomerations such as the Ruhr area or the coal and steel regions of France, which have already experienced considerable structural problems. They are expected to attract no or relatively small numbers of young people in the economically active age groups because they offer relatively few job opportunities for qualified workers. However, growing numbers of the elderly are expected to leave these highly congested areas. Ultimately, this could result in a decline in the populations of these regions, depending on their natural growth rates.

Conclusions

The present and future regional population growth patterns in Western Europe do not indicate a continuation of the agglomeration process at the same rate as before 1970. Neither do they indicate a turnaround towards rural areas. Moreover, those who predicted an enormous growth of a few metropolitan areas such as the north-west European megalopolis of south Belgium, the Ruhr area, Hanover, Berlin and Warsaw (Hall, 1977, p. 107) and those who tried to demonstrate the breakdown of Clark's law of concentration are both expected to be proven wrong.

On the contrary, the demographic trends analysed above indicate an intensification of tendencies towards spatial division of labour between areas of production and consumption on the one hand and areas of recreation and consumption on the other hand (Koch, 1976, p. 9). A large proportion of the population, job opportunities, and infrastructure in Western Europe is already concentrated in a few regions. The German agglomerations of Hamburg, Rhine–Ruhr and Rhine–Main, for example, accounted for 31.5 per cent of the gross domestic product in 1970, although these areas represent only 5.6 per cent of the area of the country and 26.1 per cent of the population. In other countries of the EEC the degree of concentration is similar (see Table 8.1). This concentration is especially benefiting industrial and tertiary firms, because there it is associated with many positive external effects such as infrastructure development and the diversifying of the markets. Negative external effects did not significantly diminish the locational advantages of firms. However, the living environment of most inhabitants of large urban agglomerations has deteriorated.

Table 8.1 Concentration of population and economic power in Western Europe, 1972

State	Shares of regions with a population density of 500 inhabitants per square kilometre and over as a percentage of		
	area	population	GNP
Belgium	52.8	76.6	79.3
Federal Republic of Germany	24.5	50.1	50.4
France	4.5	26.2	36.7
Italy	14.2	26.5	32.1
Netherlands	36.3	67.7	69.3
United Kingdom	25.8	60.2	58.2
EG (without Denmark and Luxembourg)	15.8	44.8	50.0

As long as alternative job opportunities for qualified workers are lacking in rural areas, the population in the economically active age groups is to a large extent dependent on the availability of employment opportunities in the large urban agglomerations. Many of these people will own second homes which they will visit on weekends and during holidays. In the long run these homes can be regarded as a stepping-stone to out-migration from large urban agglomerations after retirement. Retired people can, however, leave the large urban agglomerations if they can afford it. Migration into regions with scenic beauty is preferred to return migration to the home region.

The volume of this selective migration and its change over time will depend on the increase in productivity and seasonal distribution of tourist migration.

Depending on these trends, the structural changes in regions with scenic beauty will increase. These factors, together with the efforts of the authorities to increase holiday settlements and expand infrastructure for tourism, could lead to serious ecological problems due to congestion.

EEC regulations and measures with direct or indirect and intentional or incidental spatial effects are increasing. They have a profound influence on the trends towards a growing spatial division of labour. Among the measures intended to influence the settlement structure are mutually supportive regional policy objectives of the EEC Regional Fund and the European Investment Bank. However, other measures often contradict these objectives and lead to growing disparities, e.g. the non-investment traffic policy and the policy on research and technology development. In one investigation, German researchers came to the conclusion (Kreuter, 1978, p. 949) that at least in the Federal Republic of Germany, EEC policies tend to reinforce the disparities in the rate of development in urban and rural areas. In the future the main function of the European regional policy has to be the creation of regional balance in terms of living conditions, income, and social services. Population change and the distribution of social groups at the supra-regional level are predetermined by the distribution of social and age groups in urban centres and their hinterlands at an intra-regional level. Both are important trends in the formulation of regional and urban planning policy, as well as in social and environmental policymaking. However, instead of concentrating on these issues, present political deliberations in Western European countries mainly revolve around the quantitative effects of demographic change. These discussions will ultimately lead to a scramble for more people, jobs, infrastructure and financial aid between regions and municipalities in the Federal Republic of Germany (Göb, 1977), while first priority needs to be given to the reduction in population decline in certain regions.

Notes

This article is a shortened version of a working paper presented by the author at the Seminar of the Council of Europe on the consequences of the present population trends on urban and regional development held in Strasbourg (18–20 September 1979).

1 The Council of Europe's programme of population studies, Appendix 5. Conclusions of the Seminar on the Implications of a Declining or Stationary Population, Strasbourg 1978, *Informationen zur Raumentwicklung*, Heft 2, pp. 59–69, 1980.

References

Clark, C. 1967: *Population growth and land use*. New York: St Martin's Press.
Cribier, F. 1970: Migrations de retraite en France: matériaux pour une géographie du troisième age. In *Bulletin de l'Association de Géographes Français*. 381, 119–122.
Courgeau, D. 1978: Les migrations internes en France de 1954 à 1975. *Population* 30 (3) 525–45.
ERIPLAN. 1978: *Final report of Interest Group 2, 'Migration'*. Brussels.

Fielding, A. 1975: Internal migration in Western Europe. In Kosinski, L. A. and Prothero, R. M (eds), *People on the move*. Methuen: London.
Göb, R. 1977: Die schrumpfende Stadt. *Archiv für Kommunalwissenschaften* 16, 149–77.
Hall, P. 1977: *Europe 2000*. European Cultural Foundation. London: Duckworth.
Koch, R. 1976: *Altenwanderung und räumliche Konzentration alter Menschen*. Bonn. Forschung zur Raumentwicklung, Bd 4.
Koch, R. 1977: Wanderung und Rezession: Wanderungen in der Bundesrepublik Deutschland 1974 und 1975. *Informationen zur Raumentwicklung* 12.
Koch, R. and Gatzweiler, H. P. 1978: *Migration and settlement in the Federal Republic of Germany*. IIASA Working Paper WP 78-39, Laxenburg, 1978.
Kreuter, H. 1978: Auswirkungen der EG-Verträge auf die räumliche Entwicklung in der Bundesrepublik Deutschland. *Informationen zur Raumentwicklung* 11/12.
Sternlieb, G. and Hughes, G. W. 1977: New regional and metropolitan realities of America. *Journal of the Institute of American Planners* 43, 227–41.
Tucker, C. J. 1976: Patterns of migration between metropolitan and non-metropolitan areas in the United States: recent evidence. *Demography* 10, 435–43.
Vining, D.R. and Kontuly, T. 1977: Population dispersal from major metropolitan regions: an international comparison. *RSRI Discussion Paper Series* 100, Philadelphia.
Vining, D. R. and Kontuly, T. 1978: Population dispersal from major metropolitan regions: an international comparison. *International Regional Science Review* 3 (1), 49–73 (*see this volume, Chapter 6*).

9 A. J. Fielding,
'Migration and Urbanisation in Western Europe since 1950'

From: *The Geographical Journal* 155, 60–9 (1989)

Introduction

This paper is concerned with population redistribution in the 14 major countries of Western Europe which lie to the west of a line drawn from the Baltic to the Adriatic. These countries are richly varied in their human geographies, and therefore in the spatial distributions of their populations. But they also differ in the manner in which their population distributions have changed over time. Furthermore, while many of the processes which affect population redistribution are present to some degree in each country, their importance differs from one country to another, as does the period over which they operate. This daunting complexity is summarised in Figure 9.1.

A population redistribution process, or 'prp', can be thought of as a bundle of relationships which produce as one of their outcomes an altered distribution of the population. Prp1 (Figure 9.1), for example, might stand for the restructuring of agricultural production, implying the decline in on-farm

122 Differential Urbanization

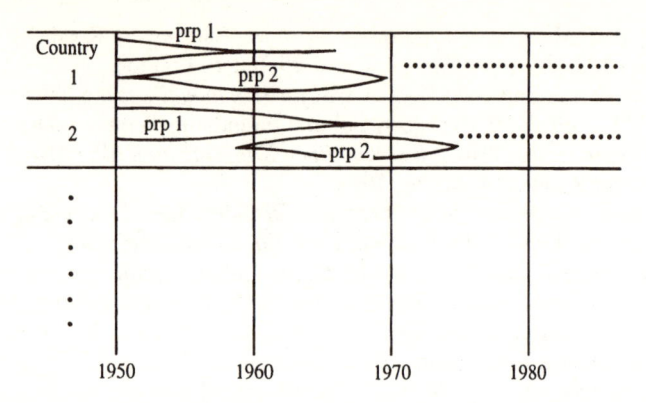

Fig. 9.1 The complexity of population redistribution processes in time and space

employment and the migration of young people from rural areas to the towns and cities. The timing and intensity of this process differs between countries; in Britain the restructuring process was already in its late stages by 1950, whereas in France it still had a long way to go and acted as a major determinant of population redistribution during the 1950s and 1960s. Very many population redistribution processes can be identified in post-war Western Europe; at any one time, however, it is likely that the effects of some will be cancelled out by those of others. Finally, the complexity of the whole can be increased by the coexistence of processes which result in spatial concentration at one spatial scale with those which result in spatial deconcentration at another (for example, overall urbanisation with local suburbanisation).

How then can one break into this complexity in order to separate the general processes from the particular and the more important from the less important? A first step is to ensure clarity and consistency in the analyses of the data on population redistribution. This calls for two decisions. The first is to concentrate attention on the migration component in population change. The justification for this is that the population redistribution effects of urban and regional differences in fertility and mortality over the period since 1950 have become far less important than the direct and indirect effects of migration (the latter arising from the selectivity of migration with respect to factors such as age, ethnicity and occupation). The second step is to operationalise the notions of spatial concentration (agglomeration, urbanisation) and deconcentration (deglomeration, counterurbanisation). This can be done by defining urbanisation as coincident with a significantly positive relationship between net migration rate and settlement size, and counterurbanisation as coincident with a significantly negative relationship between net migration rate and settlement size (where settlements are defined in labour market area terms rather than administratively). Where no significant relationship exists it is not necessarily the case that there is an absence of population redistribution: it simply means that such redistribution is, in aggregate terms, neutral with respect to settlement size (more on this below).

Urbanisation and counterurbanisation in Western Europe since 1950

By focusing on the net migration rate/settlement size relationship[1] certain general patterns emerge from the complexity alluded to above; in fact, the picture of population redistribution which results is remarkably clear and consistent. It is presented schematically in Figure 9.2 and is based upon calculations of the net migration rate/settlement size relationships for each of the 14 countries of Western Europe, for each of the three decades (1950s, 1960s and 1970s) and, wherever possible, for the early to mid-1980s; and at two (sometimes three) spatial scales. The detailed results of this research are reported elsewhere (Fielding, 1982, 1986a), but the main features are:

- that urbanisation was the dominant redistribution trend in the 1950s in all countries;
- that the relationship between net migration and settlement size began to break down in the 1960s. It broke down first in the countries of north-western Europe in the mid-1960s, but was sustained in the countries and regions of the southern and western European periphery through the 1960s and, in the case of Spain, into the 1970s;
- that by the 1970s most of the countries of Western Europe were recording a counterurbanisation relationship between net migration and settlement size. This was particularly so for the countries of the 'core' region of Western Europe, but was also the case for the regions of other countries which were located close to that core (for example, northern Italy);
- that counterurbanisation became less dominant in the early 1980s; it was not, however, replaced by urbanisation but by a situation in which no clear relationship between net migration and settlement size was in evidence. Only in West Germany and Italy did the counterurbanisation form of the relationship persist (Figure 9.2).

The transition from urbanisation to counterurbanisation: net migration and settlement size in France 1954–82

How did the change from urbanisation to counterurbanisation occur? The French census is uniquely helpful in this matter. It provides data for intercensal annual net migration rates by size category of urban agglomerations (plus rural communes), where the boundaries of the agglomerations are redefined at each census date (thus greatly reducing the problem of statistical underbounding). The results of these analyses, plotted in such a way as to take account of the importance of the different size categories, are presented in Figure 9.3.[2]

It can be seen that the curve was strongly positive for the late 1950s. The first signs of change came in the mid-1960s when the attractiveness of the Paris agglomeration suddenly diminished; the rest of the curve was, however, largely unchanged with high rates of rural depopulation continuing, especially in the smallest communes. In the early 1970s the curve changes shape quite dramatically: the dominant trend is now a gently sloping inverse relationship between net migration and settlement size, but with a much reduced residual rural depopulation. The late 1970s/early 1980s position is a further

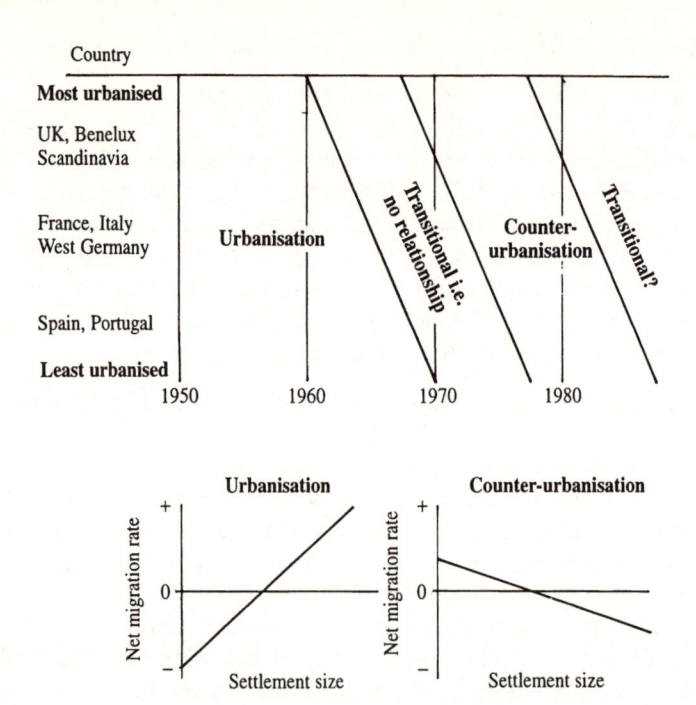

Fig. 9.2 Schematic representation of urbanisation and counterurbanisation in Western Europe, 1950 to present

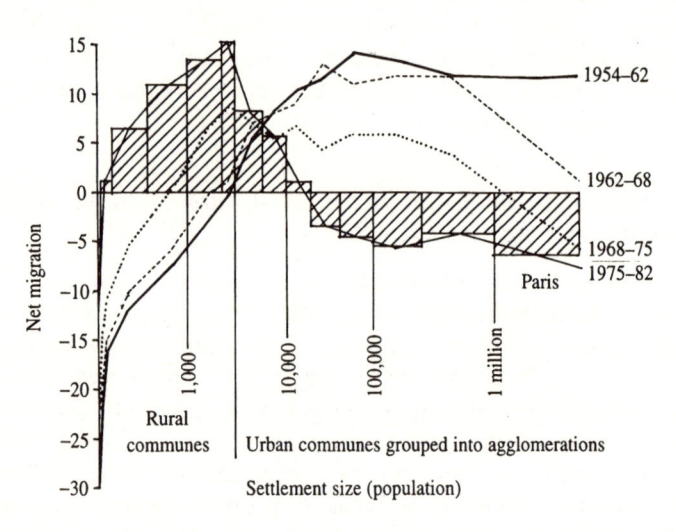

Fig. 9.3 France: annual net migration rates per thousand population, 1954–82, by settlement size category

development of the change towards counterurbanisation with peak migration gains in small and medium-sized towns and major losses in the larger cities. No set of data on this relationship is completely free from difficulties of interpretation arising from the areal definitions employed, but analyses at the functional region and planning region levels confirm the main features of the picture presented by these agglomeration/rural commune data; France became a country dominated by counterurbanisation as a result of Paris and the largest cities becoming areas of net migration loss, while residual rural depopulation became overshadowed by major net migration gains in the small and medium-sized towns and in the more populous rural communes.

Explanations of the transition from urbanisation to counterurbanisation

The preceding sections of this paper have done little more than describe the broadest features of change in one aspect of population redistribution (the net migration–settlement size relationship) and shown how the transition from one redistribution pattern to another has occurred in one particular country. The challenge now is to relate these trends in urbanisation and counterurbanisation to patterns of regional growth and decline, and to identify which processes have been most instrumental in bringing these trends and patterns about.

As a first step we can review the debate on the reasons why counterurbanisation came to replace urbanisation in the late 1960s and 1970s. There is a wide measure of agreement on certain points. For example, it is now accepted that the redistribution trends of the 1970s were not equivalent to the spatial extension of the suburbanisation trends of the 1950s and early 1960s. New processes involving the creation of job and housing opportunities well beyond the outer edges of the commuting zones of the major cities were involved. Similarly, although it is true that the absolute numbers of people leaving agricultural backgrounds diminished over the period, it is also the case that employment in agriculture continued to decline during the 1960s and 1970s, and that some of the rural regions which most spectacularly changed from being areas of net migration loss to become areas of net migration gain (e.g. Brittany) retained high levels of agricultural employment throughout. So the decline in rural depopulation only goes a small part of the way towards an explanation of the turnaround. Another point of agreement is that changes in transport and communications technology plus site constraints on old industrial premises facilitated the dispersal of economic activity. But this dispersal was not general; it was of very specific kinds of activities, even when it involved the relocation of functions normally associated with metropolitan cities such as parts of central government departments. Finally, it is often claimed that one of the effects of the increasing role of state-provided goods and services is a spatial standardisation of the general conditions of production and consumption; this too might be expected to facilitate dispersal. But here the agreement ends. There are three other main paths to explanation, to which I shall add a fourth.

The first emphasises the importance of individual preferences in determining migration flows. It claims that sometime around 1970 there was a marked urban to rural shift in place preferences: previously people had been attracted by the bright lights of the city; now they found the city environment to be stressful, dangerous and distasteful, and in pursuit of their 'village in the mind', they moved to rural and small-town locations. Undoubtedly, the late 1960s and early 1970s saw a questioning of the values of urban-industrial society, but there were few who were in a position to realise an intention to escape the 'rat race'. It is indeed the case that those retired people with sufficient wealth and mobility were more inclined during this period to move to rural areas, and there were some economically active people whose job situations permitted them to choose between jobs in different places. But most people were constrained by their work, wealth and family situations to remain in those places where there were jobs for them to do. So as an approach to the explanation of counterurbanisation this emphasis on individual preferences exaggerates the choices open to people; it is far too voluntaristic.

The second approach emphasises labour market processes within a neo-classical economic theory of migration. If the patterns of migration have changed then it is to changes in the geographical distributions of unemployment and wage rates that one must look for the causes. Unfortunately for this approach these changes simply did not occur. In general, the metropolitan cities retained their higher than average wages and lower than average unemployment, while the small-town, rural and peripheral areas retained their low wages and high unemployment. The neo-classical economic theory of migration is also at odds with the facts of contemporary migration when it comes to gross migration rates. The theory predicts that there will be an inverse relationship between gross in-migration rates and gross out-migration rates; in fact the rates were positively correlated (Fielding, 1971). The problem is not that the processes affecting population redistribution were non-economic, but that ideas drawn from neo-classical economics were far too simplistic to provide us with a proper understanding of them.

The third approach sees the replacement of urbanisation by counterurbanisation as a product of redistribution processes set in motion by public policy. It is pointed out that regional development efforts were intensified in several Western European countries during the 1960s and that special measures were taken to promote economic growth in rural and peripheral regions, often in conjunction with restrictions on the further development of the main metropolitan cities. The problem with this line of argument is that although the sums of money spent on urban and regional policies were considerable, they were insignificant in comparison with state expenditure on other items (such as defence procurement) which also had regional economic effects (often contrary to those of urban and regional policy). Furthermore, the calculations of the impacts of state-aided investments often ignored both the questions of what would have happened in the absence of policy, and the displacement effects of such investments. Finally, it is interesting to note that in cases where both agricultural and old industrial regions were in receipt of assistance (for example, Nord and Brittany in France), it was the agricultural regions which experienced the migration turnaround rather than the old industrial regions. In short, the type of area seems to have been more important than whether or not it was policy-supported.

To summarise, so far we have the following: urbanisation slowed down by a decrease in the exodus from agriculture, improved transport and communications, state service provision in non-metropolitan locations, increased retirement migration and some urban dropouts. Are these sufficient to explain the reversal in population redistribution trends? Surely not.

What is missing is a set of ideas that links the population redistribution processes to the major economic and social changes that were transforming the countries of Western Europe over the post-war period. It is my judgement that these ideas are to be found in the political economy literature, and more specifically in the concepts of spatial divisions of labour, and of regimes of accumulation. These links are spelled out elsewhere (Fielding, 1986b, 1987), but the gist of the argument is as follows:

- Population distribution and redistribution at an inter-labour market area level are largely determined by the geography of production, and by the manner in which that geography of production changes over time. Particularly important in this latter respect are the spatial changes which accompanied agricultural and industrial restructuring and the growth of producer and other 'basic' services (for example, tourism).
- In 1950 the dominant geography of production could be characterised as regional sectoral specialisation. This implied that population redistribution would result from the differential growth rates of the various branches of the national economy, through the impacts that these growth rates would have on employment changes in single-sector towns and regions. This explains the rural depopulation and out-migration from old industrial regions, and the rapid population growths of the metropolitan cities with their high concentrations of modern consumer goods industries (that is, urbanisation);
- Most of the countries of Western Europe experienced a long period of economic growth after 1950 lasting into the early to mid-1970s. This growth was 'Fordist' in the sense that it was to a considerable extent based upon the mass production of standardised goods for mass markets (cars, clothes, televisions, furniture, records, washing machines . . .). This Fordist production required, and brought forth, large labour markets and mass suburbanisation; in this way it too enhanced the urbanisation process;
- The key to Fordist growth, however, was the link between higher incomes, improved living standards and a buoyant demand for modern consumer goods. The problem was that, in the context of keener international competition, especially from Far Eastern producers, higher incomes meant lower profits. At the same time, economic growth had resulted in labour shortages, and these shortages were particularly severe in the major metropolitan cities. The responses to these problems included (a) a major concentration of ownership and control so that companies increasingly became multi-plant, multi-product and multi-national in character; (b) the export of capital for production in cheap labour sites in Third World countries; and (c) the importation of labour from the European periphery in the form of 'guestworkers'. A further response was to disinvest in the major cities and

128   *Differential Urbanization*

to seek out reserves of labour within national territories; these were to be found partly among women in existing industrial regions but also among young people and other low-wage workers in free-standing cities, and in small and medium-sized towns in rural and peripheral regions. The response resulted in a spatially dispersed pattern of branch plant and back-office developments during the 1960s and early 1970s.

- These changes helped to bring about a new spatial division of labour. This hierarchical division of labour was characterised by the spatial separation of tasks within the production process; notably this meant the separation of command from execution, and of white-collar employees from blue-collar, but also a considerable separation of technical and professional workers from other white-collar workers, and of the skilled manual workers from the semi-skilled and unskilled. Typically, the higher-paid workers were located in the main metropolitan cities (London, Paris, Milan, Frankfurt . . .), and in the 'prestige environments' nearby, while those with lower status and pay were confined to the industrial conurbations and to the rural and peripheral regions. One of the effects of this change was to produce a marked shift towards regional sectoral diversification, and hence a reduced sensitivity in many towns and regions to sectoral changes in the national economy. It also implied, however, the external control of most urban and regional economies, and a deindustrialisation of the major cities. The migration implications of the emergence of this new spatial division of labour included (a) a reduced need for manual worker mobility; (b) an enhanced need for 'service class' mobility (to manage and supply technical services to this spatially dispersed production); (c) a major reduction in the migration of less skilled people to the principal cities as these cities experienced disinvestment (especially in manufacturing industry); and (d) a continued out-migration from the principal cities sustained by the major industrial and service sector investments in free-standing cities and in small and medium-sized towns in rural and peripheral regions. The product of these migration effects was counterurbanisation.

Migration and social change in south-east England, 1971–81

Some of these speculations about the relationships between migration and the changing geography of production listed above can be checked against the special data provided by the Office of Population Censuses and Surveys' Longitudinal Study, 1971–81. These data refer to the situations of about half a million individuals in England and Wales whose census forms for 1971 were linked up with those for 1981 (Fielding and Savage, 1987). By this means one can trace through the changes which men and women have experienced and connect their migration behaviours to their social class origins and to occupational change. Some first results of this analysis are presented in Tables 9.1 and 9.2.

Several interesting results can be drawn from these tables:

- Despite a significant overall decline in the size of the labour market in south-east England there was an increase of 380 700 in the number of men

and women in the service class (that is, in professional, technical and managerial occupations). This came about from a high level of transfers from other classes in the south-east and from the recruitment of new entrants onto the labour market. South-east England was becoming markedly more 'middle class' during the 1970s (and at a higher rate than England and Wales as a whole).
- Despite the large in-migration into the south-east's service class by men and women who were already in the service class in 1971 (65 000), there was an even larger out-migration by the same kinds of people (92 300); if it were not for the many young men and women who entered the south-east's service class from education elsewhere in England and Wales, the net migration balance for this class would have been negative! This runs completely counter to neo-classical economic theory, but fits in rather well with the concept of the emergence of a new spatial division of labour.
- In general, out-migration slightly exceeded in-migration, but the out-migration rates for those in the labour market at both dates were much greater than the equivalent in-migration rates. This establishes the fact that counterurbanisation was not just the result of retirement migration.
- The other distinctive features of both the in-migration and out-migration streams are (a) the extraordinarily high migration rates for the service class; this is particularly so for those who were in that class at both dates, and for those in education in 1971, but it is also true for those who transferred in from another class, and (b) the equally remarkable low migration rates for the blue-collar section of the working class, and for those who were unemployed at both dates.[3]
- Finally, despite their low out-migration rates, the blue-collar working class lost out through their exchanges with other regions in England and Wales; this is because of their extraordinarily low in-migration rates. Once again this again is in line with what was expected on the basis of the ideas discussed in the earlier section.

To summarise, the information from the Longitudinal Study strengthens our confidence in the spatial divisions of labour approach; it does not prove that perspective to be correct, but with several of our expectations about migration based upon the approach being realised, it will have to be a good model that displaces it.

Conclusions: population redistribution in the 1980s

It was pointed out earlier that the counterurbanisation of the 1970s had largely disappeared by the early to mid-1980s, but that there is little sign so far of a simple return to the urbanisation process of the early post-war era. Rather, at a time of no clear relationship between net migration rate and settlement size, there appears to be a re-emergence of broader regional patterns of growth and decline. In Britain, for example, a regenerated London economy has put an end to net migration losses, and the capital city now joins the buoyant provincial cities, free-standing towns and smaller places in East Anglia and south-west England in experiencing net migration gains, and by this means reasserting the north–south

Table 9.1 Social class balance sheet for the south-east region of England 1971–81

	SC	PB	PWC	PBC	UE	T
In class X in SE in 1971, and in SE in 1981	13 160	4 283	20 787	25 341	1 855	65 426
In class X in 1971, but elsewhere in 1981 (including deaths)	2 829	753	3 367	4 085	360	11 394
Total in X in SE in 1971	15 989	5 036	24 154	29 426	2 215	76 820
Additions:						
Transfer in from other classes in SE	5 011	2 188	3 457	3 553	2 018	16 227
Entries from education in SE	3 498	428	6 362	4 990	1 799	17 077
Other entries in SE (e.g. married woman)	1 070	318	3 630	1 731	261	7 010
In-migrants from E and W of which:	2 389	159	1 338	818	317	5 021
in X in 1971	650	26	282	281	16	1 255
in LM other than X in 1971	375	83	181	146	113	898
in education in 1971	1 230	30	657	281	152	2 350
in others in 1971	134	20	218	110	36	518
Total additions	11 968	3 093	14 787	11 092	4 395	45 335

	SC	PB	PWC	PBC	UE	T
Subtractions:						
Transfer out to other classes in SE	2 660	1 510	5 246	5 827	984	16 227
Deaths	1 059	496	1 590	2 777	203	6 125
Retirement in SE	1 814	628	2 638	4 357	313	9 750
Other exits in SE	858	284	4 856	2 092	312	8 402
Out-migrants to E and W of which:	1 770	257	1 777	1 308	157	5 269
in X in 1981	923	86	418	487	26	1 940
in LM other than X in 1981	333	87	514	435	80	1 449
in retirement in 1981	292	45	256	247	16	856
in other in 1981	222	39	589	139	35	1 024
Total subtractions:	8 161	3 175	16 107	16 361	1 969	45 773
Net change 1971–81	+3 807	−82	−1 320	−5 269	+2 426	−438
Total in X in SE in 1981	19 796	4 954	22 834	24 157	4 641	76 382
In class X in SE in 1981, and in SE in 1971	17 407	4 795	21 496	23 339	4 324	71 361
In class X in SE in 1981, but elsewhere in 1971	2 389	159	1 338	818	317	5 021

Source: One per cent sample data from the OPCS Longitudinal Study
Note: The data include migration to and from the rest of England and Wales but not to and from the rest of the world. Untraced records are also excluded
SC = service class; PB = petite bourgeoisie; PWC = white-collar working class; PBC = blue-collar working class; UE = unemployed; T = total

Table 9.2 In-migration and out-migration rates for the south-east region, 1971–81

	SC	PB	PWC	PBC	UE	T
In-migrants by social class						
in 1981 of which:	133.5	31.8	57.0	30.5	92.5	65.4
in same class in 1971	36.3	5.2	12.0	10.5	4.7	16.4
in other class in 1971	21.0	16.6	7.7	5.4	33.0	11.7
in education in 1971	68.7	6.0	28.0	10.5	44.3	30.7
in others in 1971	7.5	4.0	9.3	4.1	10.5	6.8
Out-migrants by social class						
in 1971 of which:	98.9	51.5	75.6	48.8	45.8	68.8
in same class in 1981	51.6	17.2	17.8	18.2	7.6	25.3
in other class in 1981	18.6	17.4	21.9	16.2	23.3	18.9
in retirement in 1981	16.3	9.0	10.9	9.2	4.7	11.2
in other in 1981	12.4	7.8	25.1	5.2	10.2	13.4

Source and key: as for Table 9.1
Note: Rates per thousand population in each class (where population = (1971 + 1981)/2). Deaths excluded. Migration to and from the rest of England and Wales only

divide which had been so contradicted by London's demographic decline in the 1970s. South-east France, southern West Germany and north-central Italy seem to be similarly privileged as regions of growth within their national territories.

It was mentioned in the introduction that it was quite possible for several redistribution processes to contribute to a single spatial outcome. When discussing the transition from urbanisation to counterurbanisation, considerable emphasis was placed on the spatial dispersal of investment as a solution to the problems of Fordist accumulation. With hindsight it can be seen that this solution was partial and temporary. After 1973, Fordist forms of production began to give way in Western European countries to a variety of developments which are sometimes loosely grouped together under the title of 'flexible accumulation'. The list would include the fragmentation of the production process and the related demise of the mass collective worker, the orientation of investment towards new products for highly specialised markets, the development of new work relations such as those intended by the term 'japanisation', the use of the new computer technologies, notably CAD/CAM, the subcontracting of parts of production to small independent high-tech companies, and more generally, the growth of the small-firm sector. Like their branch plant predecessors, these new investments also tend to be located away from the major metropolitan cities, but unlike branch plants and back offices, the only non-metropolitan places that are deemed to be suitable are those that can attract and retain the highly qualified labour upon which these new activities depend. Thus during the late 1970s and 1980s we have witnessed employment growth and net migration gain in a number of prestige environments. These places tend to have two characteristics: first, they are often scenically attractive, and second, they usually have a social base and a past history appropriate for an emergent entrepreneurial culture and an information-based economic development.

It may be that some of the interpretations of population redistribution processes fostered by the spatial divisions of labour approach will need to be modified in the light of these developments. In particular, one of the strengths of that approach lies in the way it emphasises the economic and social connections between towns and regions, cemented by the flows of capital, goods and services, and migrants within a hierarchical spatial division of labour. But could it be that one is now witnessing a growing disjuncture, a regional form of 'social closure', in which certain areas are successfully linked into a global financial, cultural and scientific community, while others continue to suffer the dismantling of Fordist production, their inhabitants deterred from migration to the privileged regions by the high costs of entry (for example into the housing market), and by their inappropriate skills and lack of cultural capital?

Notes

1 In practice, the regression analyses were based on data for population density rather than settlement size. For a justification of this procedure see Fielding, (1982) note 1.
2 The problems which arise in the use of these data are discussed in Fielding (1986b, p. 236).
3 This produces the nice paradox that those who need to migrate the least, that is, the better off and more secure, migrate the most, while those who need to migrate the most, that is, the poor and those most at risk of unemployment, migrate the least.

References

Fielding, A. J. 1971: *Internal migration in England and Wales*. Centre for Environmental Studies University Working Paper 14.
Fielding, A. J. 1982: Counterurbanisation in Western Europe. *Progress in Planning* 17 (1), 1–52.
Fielding, A. J. 1986a: Counterurbanisation in Western Europe. In Findlay, A. and White, P. (eds), *Western European population change*. London: Croom Helm, 35–49.
Fielding, A. J. 1986b: Counterurbanisation. In Pacione, M. (ed.), *Population geography: Progress and prospect*. London: Croom Helm, 224–56.
Fielding, A. J. 1987: *Population redistribution in Western Europe: trends 1950–80 and the debate about counterurbanisation*. Paper in British Section Regional Science Association, Stirling.
Fielding, A. J. and Savage, M. 1987: *Social mobility and the changing class composition of south east England*. University of Sussex, Urban and Regional Studies, Working Paper 60.

---

10 A. G. Champion,
'The Reversal of the Migration Turnaround: Resumption of Traditional Trends?'

From: *International Regional Science Review* 11(3), 253–60 (1988)

Contributions of the Vining Group

Over the past ten years, Vining and his colleagues have made an important contribution to the debate about the breakdown of the long-established tendency of population concentration. They sought to demonstrate that population deconcentration in the United States during the early 1970s represented a 'clean break with the past' (Vining and Strauss, 1977). They pointed to the very widespread occurrence of a migration reversal for major metropolitan regions across the developed world (Vining and Kontuly, 1978a). They concluded that 'in the developed world at least, the century-long migration towards the high-density core regions is over' (Vining and Pallone, 1982, p. 339). After an even more extensive review of core–periphery trends in less developed countries, Vining (1986) reckoned that the contrast in recent trends between the developing and developed world can be accounted for in terms of the developmental model of spatial concentration and dispersal.

This work has not gone unchallenged, as for instance in the case of Gordon's (1979) criticism of the use made of the Hoover index of concentration by Vining and Strauss (1977), nor has Vining's team been slow to admit inaccuracies in their work arising from problems over data and geographical areas (Vining and Kontuly, 1978b, 1979). It can indeed be claimed with some justification that these studies are crude and broad-brushed, that they pay scant attention to the niceties and nuances of individual national settlement systems, and that, in proceeding from describing patterns to attributing processes, they may be guilty of generating more heat than light. Nevertheless, in their studies Vining and his colleagues demonstrated convincingly that in the 1970s population distribution in many developed countries was changing in a way very different from previous experience. In so doing, they raised a great many new questions about the factors affecting long-distance migration and lifted the focus from its early preoccupations with rural America (e.g. Beale, 1975; Morrison and Wheeler, 1976; McCarthy and Morrison, 1978; Brown and Wardwell, 1980) and residential preferences (e.g. Berry, 1976; Zuilches, 1980) to a wider concern with economic forces operating at the international scale (e.g. Fielding, 1982).

In almost every respect, the latest contribution (Cochrane and Vining, 1988) is characterized by the same set of weaknesses and strengths as these previous studies, but similarly in my view the benefits of this work far outweigh its shortcomings. Among the more technical aspects, it is good to see the emphasis which the authors place on migration in their examination of core–periphery shifts rather than overall population change despite the much greater difficulties encountered in data collection. At the same time, their secondary measure of regional growth rate differentials provides a very useful backup, since it enables the comparison of regional performance after allowing for variations in the national population growth rate caused by fluctuations in natural increase and international migration.

The weakest element of their approach remains the definitions of the core regions, which are adopted unchanged from Vining and Pallone (1982). In that study it was stated that:

> The core regions of a country are those regions which are economically and politically dominant: they contain the principal cities of the country and have traditionally experienced high rates of net inmigration from the other less urbanized, peripheral regions. The identification of these regions poses little difficulty and should not be controversial
>
> (Vining and Pallone, 1982, p. 340)

Yet, given the absence of any detailed explanation of the way in which national territories have been dissected, the impression is of a largely intuitive allocation procedure rather than one based on the application of consistent criteria related, for instance, to population density, economic structure, or some other clear indicator of 'core' status. No justification is presented, for example, for identifying three separate core areas in Sweden, a country of barely 15 million people, while for the United Kingdom, a country of over 50 million people, only one region is identified. Though extensive, that region excludes large sections of the Midlands like the Birmingham, Coventry, and Leicester urban regions which have traditionally been among the most dynamic elements of postwar Britain in both demographic and economic terms. Moreover, the study does not adequately address the scale problems involved, for example, in comparing trends for what are essentially metropolitan areas in some countries like Sweden with those for the four very broad subdivisions of Northeast, Midwest, South, and West in the United States at the other extreme. Lastly, while the core areas are generally meant to be overbounded in order to avoid the problem of local spillover effects, the reliance on official statistical areas rather than any more customized set of area definitions means that the core regions will be overbounded to greater or lesser extents in the various countries, making detailed comparisons among national experiences a dangerous exercise.

Fortunately, however, in practice the existence of these technical weaknesses does not appear to jeopardize the principal conclusions of Cochrane and Vining's study. In the first place, the authors are sensible enough to confine their comparisons among countries to the level of general tendencies towards concentration and dispersion, using the precise levels of migration and population change for core and periphery regions only for comparisons within each country. Second, they are lucky enough to find a very high degree of similarity in broad trends among countries, particularly within the three groups of advanced non-Communist countries. Whereas they observe a convergence in the growth rates of core and periphery in most countries of Northwest Europe and North America in the early 1980s, the countries in their third group – those on the periphery of West Europe (Finland, Norway, Sweden, Spain, and Italy) plus Japan – generally appear to have experienced divergence. This distinction is, however, more apparent than real and stems from a contrast in the traditional levels of differential growth in their core and periphery areas. The countries in this third group were, for a combination of geographical and historical reasons, recording much stronger growth in their core regions in the 1950s and 1960s and, even in the 1970s when their attractiveness was much diminished, the core regions did not go through a major and sustained migration reversal. Divergence in the 1980s is therefore associated with renewed core growth, which is precisely the same reason for the convergence recorded by the other two groups of countries. Despite its technical deficiencies, therefore, this study carries the clear message that, following a decade of a relative shift in population trends in favor of peripheral regions within countries of the more developed Western world, the early 1980s have been very widely associated with a further reversal involving a relative recovery of the core regions.

This finding can also be corroborated from the experience of the United Kingdom, a country which continues to trouble Vining and his colleagues owing to its traditional neglect of migration monitoring (Vining and Kontuly, 1978a, 1978b). In fact, as Cochrane and Vining (1988) note briefly, it is now possible to obtain statistics on migration flows at regional and county level from the National Health Service Central Register, as well as a breakdown into natural increase, net migration, and other changes on an annual basis (but in both cases, relatively little of these data are actually published by the official statistical agencies). Analyses based on these data reveal a marked reduction in the rate of population loss for Britain's principal cities since the mid-1970s, a major increase in the level of net migration from north (periphery) to south (core), and a particularly impressive recovery by London and the rest of South East England from the very substantial migration losses of the early 1970s (Britton, 1986; Stillwell, 1985; Stillwell and Boden, 1986; Champion, 1987). The southward drift of migrants, which had caused policymakers great concern in the 1950s and early 1960s, had fallen to around 10–20 000 people a year by the early 1970s but then accelerated to around 40–50 000 towards the end of the decade and touched 70 000 in the mid-1980s. Greater London's migration losses peaked at 120 000 in the year 1970-71, fell fairly steadily to half this level ten years later, and dropped even more rapidly to under 20 000 by 1983–84 (Champion and Congdon, 1988b; Champion, 1989).

This recognition of the forces of population concentration at work again is not completely new, or, in some quarters, unexpected. If this reversal of the metropolitan migration turnaround about which Vining and his colleagues wrote so confidently only a few years ago has come as something of a shock to them, this fact is well disguised in their most recent paper. Cochrane and Vining refer without comment to the first signs of a slowdown in nonmetropolitan growth emanating from the US Bureau of the Census from 1984 onwards, as well as to several of the studies that point to a resurgence of metropolitan populations and reconcentration in other countries. Some authorities, however, have from the first been skeptical about the so-called 'rural renaissance' and 'clean break' ideas, stressing the importance of the lateral extension of metropolitan areas through boundary spillover and the emergence of new metropolitan centers in areas currently defined as nonmetropolitan (e.g. Gordon, 1979). Moreover, proponents of the 'urban life-cycle' concept, particularly in Europe, have for some time been predicting (rather bravely in the prevailing climate of opinion) the onset of 're-urbanization' (e.g. Berg et al. 1987; Klaassen et al. 1981).

Towards explanation

Nevertheless, so many and plausible have been the reasons put forward over the last few years to account for counterurbanization and the rural migration turnaround that the discovery of a widespread reversion to population concentration appears extraordinary. In his literature review prior to examining counterurbanization trends in Australia, Hugo (1989) identified no less than nine lines of explanation, embracing economic recession, decentralization policy, rural resource development, urban diseconomies, state welfare payments, reduced distance friction, changing residential preferences, changing socio-demographic composition, and structural change in the economy. It would now appear necessary to check whether all these are the powerful forces for population deconcentration that their various proponents have maintained, to discover whether any of these have changed in their nature between the early 1970s and the early 1980s in such a dramatic way as to offset the overall balance towards deconcentration, and possibly to look for any new ingredient which could have such an effect.

One possible approach to the latest trends is to treat them merely as a short-term downward flexure in the rate of population deconcentration. Turning on its head the explanation of the counterurbanization of the 1970s as a temporary phenomenon resulting from a chance combination of factors which is unlikely to recur, this approach assumes that deconcentration is now the rule, yet recognizes that, just as with the urbanization trends of the past, the process is not uniform over time but fluctuates in sympathy with economic conditions and associated building cycles. Some evidence in support of this interpretation can be found in terms of variations in the pace of shorter-distance suburbanization and metropolitan decentralization noted in many countries in the postwar period and apparently related to income growth, availability and cost of house-mortgage credit, household formation rates, and road construction programs. Moreover, between the more buoyant days of the early 1970s and the depths of recession in the early 1980s, a major reduction in gross migration propensities occurred in many countries. To the extent that it was a general phenomenon affecting all migration flows equally, it would have the effect of slowing down the deconcentration process. In addition, given that, in the United Kingdom at least (Champion and Congdon, 1988a), this reduction appears to have had different effects on different population groups (e.g. a greater impact on family-age adults predominantly engaged in centrifugal moves than on young adults, who are more strongly represented among inmigrants to metropolitan areas), this phenomenon could even have brought about a temporary reversal in the counterurbanisation process.

An alternative, and more sophisticated, approach would set the deconcentration process within a wider context. As outlined in more detail by Champion and Illeris (1989), this perspective recognizes the existence of three distinct sets of factors influencing the distribution of economic activity and associated employment. First, there are a number of forces operating over the longer term in favor of deconcentration, such as the improvement of transport and communications, the more dispersed distribution of educational and other facilities, the increasing preferences for owner-occupied housing, and the growth of tourism and outdoor recreation. Second, some factors pull towards concentration into large cities and more urbanized regions, particularly the growth of business services and corporate headquarters and other activities requiring a high level of national and international accessibility and a large supply of highly qualified manpower. These two sets of centrifugal and centripetal forces are both likely to vary in strength over time, as described above for deconcentration, but in addition there exists a third group of factors which may have different geographical effects at different times depending on the prevailing circumstances; for instance, an increase in public-service provision at one time followed by a contraction or demographic changes which may mean a large bulge of school-leavers at one time and a boom in household formation and family rearing somewhat later. These groups of factors can be expected to produce considerable variations in rates of metropolitan growth and expansion over time, especially if they are interlinked to any significant extent and operating in tandem with each other.

A third approach would be to highlight the role of two separate processes in reshaping the space economy and to stress that the two tend to operate at rather different scales. These two are population deconcentration and regional restructuring. In their conclusions, Cochrane and Vining (1988) see the former merely as the outcome of the latter, whereby, 'Out-migration to peripheral regions at the expense of the core regions seems only to be ... a temporary phenomenon that occurs while an economy adjusts to the new spatial location requirements of post-industrial economic activity.' A more rigorous analysis by Frey (1987) observes these two processes operating alongside each other in a relatively independent manner and in fact concludes that population deconcentration has the greater power in explaining recent differences in population performance across metropolitan America. The contrast between these two conclusions may be largely due to the difference in scales between the two studies. Whereas regional restructuring is defined in terms of the macroregions and core–periphery comparisons used by Cochrane and Vining, counterurbanization operates predominantly at the intraregional scale, as it redistributes people down the urban hierarchy from larger to smaller settlements. In this context, population deconcentration produces differential regional growth only when people from larger cities are forced to look beyond the boundaries of their macroregions for the type of settlement in which they prefer to live. This necessity fluctuates over time according to the scale of pressures on more accessible smaller cities and rural areas, as is just now being demonstrated in southern Britain, where the South East region was able to absorb the majority of London's outmigrants in the early 1980s but has more recently begun to lose increasing numbers to adjacent regions, including the East and West Midlands (which are classified in the periphery by Cochrane and Vining).

If this interpretation is anywhere near correct, the latest tendencies towards reconcentration could be as much the result of a temporary combination of favourable conditions as many now deem the widespread trend towards counterurbanization and periphery growth to have been in the early 1970s. The growth of business services, the surge in young adults, and the lower level of new housebuilding for families are among the events which have coincided with the revival of the large metropolis, yet not all of these are unidirectional or likely to proceed at the same pace in the future.

The reality is that population decentralization is a long-established and deep-seated process primarily related to consumption, which can be expected to continue to shape the extent and structure of metropolitan systems and to affect the position and role of the principal centers within them. Superimposed on this general tendency are rather less predictable changes in the scale and location of investment in export-oriented production activities, which are characterized by a sophisticated spatial division of labor between macroregions. The more routine, mass-production components appear to be volatile, and often ephemeral, elements of a particular regional landscape, whereas the control and development functions are much more robust in size and location. Even the latter, however, are not immobile, as evidenced by their rapid growth in certain prestige regions outside the main core regions in Western Europe, such as southern Germany and southeast France. In the foreseeable future, it may be that a continued growth in the demand for highly qualified personnel, coupled with the onset of labor supply restrictions resulting from the 'baby bust' of the 1970s, will again strengthen the role of individual residential preferences in the labor market and lead to a renewed upsurge in population deconcentration during the course of the next decade. The generation-based cycles, now firmly implanted in the demography of advanced Western countries, provide a powerful framework within which the accompanying (but relatively autonomous) oscillations in the economy must work themselves out.

References

Beale, C. L. 1975: *The revival of population growth in nonmetropolitan America*. Washington, DC: Economic Research Service, US Department of Agriculture.

Berry, B. J. L. 1976: The counterurbanization process: urban America since 1970. In B. J. L. Berry, (ed.), *Urbanization and counterurbanization*, Beverly Hills, Calif.: Sage, 111–43 (*see* this volume, Chapter 1).

Berg, L. van den, Burns, L. S. and Klaassen, L. H. 1987: *Spatial cycles*. Aldershot: Gower.

Britton, M. 1986: Recent population changes in perspective. *Population Trends* 44, 33–41.

Brown, D. and Wardwell, J. M. (eds) 1980: *New directions in urban–rural migration: The population turnaround in rural America*. New York: Academic Press.

Champion, A. G. 1987: Recent changes in the pace of population deconcentration in Britain. *Geoforum* 18 (4), 379–401.

Champion, A. G. 1989: The United Kingdom. In Champion, A. G. (ed.), *Counterurbanisation*. London: Arnold

Champion, A. G. and Congdon, P. D. 1988a: An analysis of the recovery in London's population change rate. *Built Environment*.

Champion, A. G. and Congdon, P. D. 1988b: Recent trends in London's population. *Population Trends* 53, 7–17.

Champion, A. G. and Illeris, S. 1989: European population redistribution in the 1980s. In Hansen J. C. and Hebbert M. J. (eds), *Unfamiliar territory: The reshaping of European geography*. Aldershot: Gower.

Cochrane, S. G. and Vining, D. R. 1988: Recent trends in migration between core and peripheral regions in developed and advanced developing countries. *International Regional Science Review* 11 (3), 215–45.

Fielding, A. J. 1982: Counterurbanisation in Western Europe. *Progress in Planning* 17 (1), 1–52.

Frey, W. H. 1987: Migration and depopulation of the metropolis: regional restructuring or rural renaissance? *American Sociological Review* 52 (2), 240–57.

Gordon, P. 1979: Deconcentration without a 'clean break' *Environment and Planning A* 11 (3), 281–90 (*see* this volume, Chapter 4).

Hugo, G. 1989: Counterurbanisation in Australia. In Champion, A. G. (ed.), *Counterurbanisation*. London: Arnold.

Klaassen, L. H., Molle, W. T. M. and Paelinck, J. H. P. 1981: *Dynamics of urban development*. Aldershot: Gower.

McCarthy, K. F. and Morrison, P. 1978: *The changing demographic and economic structure of nonmetropolitan areas in the 1970s*. Santa Monica, Calif.: Rand Corporation.

Morrison, P. A. and Wheeler, J. P. 1976: Rural renaissance in America. *Population Bulletin* 31 (3), 1–27.

Stillwell, J. H. C. 1985: Migration between metropolitan and nonmetropolitan regions in the United Kingdom. In White P. E. and van der Knaap B. (eds), *Contemporary studies in migration*. Norwich: GeoBooks.

Stillwell, J. H. C. and Boden, P. 1986: *International migration in the United Kingdom: Characteristics and trends*. Leeds: School of Geography, University of Leeds, Working Paper 470.

Vining, D. R. Jr. 1986: Population redistribution towards core areas of less developed countries, 1950-1980. *International Regional Science Review* 10 (1), 1–46 (*see* this volume, Chapter 12).

Vining, D. R. and Kontuly, T. 1978a: Population dispersal from major metropolitan regions: an international comparison. *International Regional Science Review* 3 (1), 49–73 (*see* this volume, Chapter 6)

Vining, D. R. and Kontuly, T. 1978b: Population dispersal from major metropolitan regions: Great Britain is no exception. *International Regional Science Review* 3 (2), 182.

Vining, D. R. and Kontuly, T. 1979: Population dispersal from major metropolitan regions: a correction concerning New Zealand. *International Regional Science Review* 4 (2), 181–82.

Vining, D. R. and Pallone, R. 1982: Migration between core and peripheral regions: an international comparison. *Geoforum* 13 (4), 339–410.

Vining, D. R. and Strauss, A. 1977: A demonstration that the current deconcentration of population in the US is a clean break with the past. *Environment and Planning A* 9 (7), 751–8 (*see* this volume, Chapter 3)

Zuilches, J. J. 1980: Residential preferences in migration theory. In Brown, D. and Wardwell, J. M. (eds), *New directions in urban–rural migration: The population turnaround in rural America*. New York: Academic Press, 163–88.

PART TWO
Major Trends in Migration in the Less Developed World since the 1970s

SECTION ONE
INTERNATIONAL TRENDS

11 H. W. Richardson,
'Polarization reversal in developing countries'

From: *Papers of the Regional Science Association* 45, 67–85 (1980)

Polarization reversal and the spatial development process

In a recent fugitive, though reasonably well-circulated, paper (Richardson, 1977) I coined the term *polarization reversal* (hereafter PR). Although this clumsy, inelegant phrase is merely a new name for a well-known, if underresearched, spatial phenomenon, it seems to have fired the imagination of some observers of spatial planning in developing countries for it has already aroused some comment and analysis (Hwang, 1979; Linn, 1978; Lo and Salih, 1979; Renaud, 1977). It seems appropriate to begin a dialog with these commentators and to elaborate the significance and meaning of the concept. There are no research results presented here; the paper is intended as a stimulus to research.

PR may be defined as the turning point when spatial polarization trends in the national economy give way to a process of spatial dispersion out of the core region into other regions of the system. This definition requires elaboration by placing PR in the broader context of a descriptive theory of national spatial development. The urban-industrial process of national development begins in one or two regions only, primarily because of the scarcity of investment resources. The choice of regions is determined by initial location advantages (resource endowments, or a key immobile resource such as a port) or because it was the first area opened up from outside (and this eventually converts into the greater market size of the excolonial primate city). This initial start becomes a cumulative causation process explained by increasing returns to scale and the consequent polarization of labor and any surplus capital from other regions. The core–periphery relationship is thus established (see p. 153), where the core region consisting of the primate city and its hinterland dominates the rest of the space economy, called the periphery. This periphery is dominated by the core and dependent on it, and its rate of development is controlled and distorted so as to further the core's economic interests.

At a more advanced stage of development, a spatial transformation begins to occur within the core region. The population and agglomeration of economic activities in the primate city become so large that a monocentric spatial structure becomes inefficient and costly. Congestion costs and rising land values induce some economic activities to decentralize to satellite centers within the core region. These centers may intercept new migrants who are attracted by job opportunities expanding at a more rapid rate than in the primate city. But this intraregional decentralization does not count as PR, because the core region (including the primate city) continues to grow at a faster rate than the rest of the country.

Subsequently, however, conditions emerge that make dispersion into other regions of the system efficient. These conditions are probably associated with the generation of agglomeration economies and other scale economies at *selected* locations in the periphery, and these reflect the diffusion of technical knowledge from the core, rising population and incomes, expanding markets (locally, in the core region, and abroad), the exploitation of local resources, lower input costs, improvements in communications, the build-up of infrastructure and other factors making economic expansion at these locations profitable. The dispersion process may be accelerated by obstacles to continued rapid expansion in the core region, such as soaring land and labor costs, increasing congestion (even with the spatial structure of a polycentric metropolitan region), pressure on housing and infrastructure, and an above-average rate of increase in living costs. These obstacles accelerate the industrial decentralization process and induce an increasing number of migrants to choose urban destinations outside the core region. This process of interregional dispersion is the main feature of PR.

However, the dispersion takes place very unevenly, with most of the growth outside the core region occurring at a limited set of relatively large urban centers. In a sense, the national concentration within the core region is replicated by regional concentration in major regional centers. This spatial concentration reflects the critical role played by agglomeration economies in attracting both economic activity and population. At a later phase in the PR process, the intraregional decentralization observed earlier in the core region is repeated within the developing regions.[1] As a result, stable regional urban hierarchies emerge in each of the affected regions. Finally, the decentralization forces in all regions (but especially in the core region) may become so strong that the major cities begin to lose population absolutely. This decline in metropolitan populations has already been observed in the North East and Midwest in the United States, in some parts of Western Europe, and is possibly imminent in Japan. What lies beyond this phase of population redistribution is unknown, because it is outside observable experience.

From this description, it is clear that spatial development evolves in the form of different phases of agglomeration and dispersion. PR develops not when continuous polarization ceases, but only when intraregional decentralization is accompanied by *inter*regional dispersion. It is this extension of the dispersion process outside the core region into other regions of the system that marks the beginnings of PR. Even then, dispersion is a misleading term since the need to generate agglomeration economies as attractors for factors of production implies spatial concentration within these regions; hence, the term 'concentrated dispersion' is sometimes used to describe the process.

The above description of spatial development, as a context for defining PR, represents a set of stylized facts summarizing what has been observed in the advanced countries. Although a few developing countries may have reached the phase when PR begins (see below, pp. 151–52), none of them has passed through the spatial sequence traced in the more developed countries. This raises several nagging questions. Is the spatial development process summarized above a deterministic, predictive model or is it a 'conditional' model that can be thrown off course by the absence of certain preconditions or by the intervention of new forces? Are there differences between the developed and developing countries that limit the transferability of the model? What is the periodicity of the different phases of spatial development, is the periodicity similar in all cases, and what does this imply for the length of time between the takeoff in rapid urbanization/industrialization and onset of PR? Is the process wholly spontaneous (that is, the product of market forces), or has it been influenced by public policy? Assuming that intervention can have some impact, and that PR is desirable if achieved at moderate efficiency cost, when should policymakers intervene, and what form should appropriate intervention take? Although some comments on the questions are included in this paper, satisfactory answers require detailed empirical research, some based on cross-sectional analysis, and some in the form of country case studies.

The timing of intervention

The 'timing of intervention' argument implies that spatial dispersion policies will have their maximum effectiveness if they are implemented close to the PR turning point. An inference from this argument is that premature intervention – that is, in the period when polarization forces are strong – may be ineffective, will certainly be costly in terms of resources (for example, the opportunity cost of the infrastructure investments made in intermediate-size cities), and may hamper economic development to an unjustifiable extent even in countries where equity and other nonefficient objectives are important.

This argument has been challenged directly by Linn (1978). First, he argues that it may be difficult to unravel the effect of policy from spontaneous dispersion forces. But this problem remains even if policies are introduced in a rapid polarization phase, since policies presumably then mitigate polarization but to an unknown extent. This requires a comparison of actual with 'expected' spatial trends, a comparison which cannot be avoided regardless of whether there is polarization or dispersion. Second, he suggests that policy intervention close to the PR phase may be unnecessary, since dispersion will then occur naturally. Appraisal of this argument depends upon the rate of PR. If it is very slow, the role of intervention may be crucial in accelerating PR rather than a pointless frill. Third, he argues that the focus of a PR strategy on secondary cities may create problems of adjustment by accelerating the growth of already rapidly growing cities.

This argument is related to Linn's most important point: the implications of the concept of absorptive capacity for spatial intervention. He points out that population pressure (measured by a rapid rate of urban population growth) is only one side of the coin, the demand side. The other side (the supply side) can be represented by the 'absorptive capacity' of the city, the ability of the city to adjust to population growth without creating excessive strains. Absorptive capacity may be limited by any of three constraints: jobs, housing supply, and the provision of public services. If the absorptive capacity of a city is high relative to its existing population, it is appropriate to steer migrants to such a city. Similarly, if policy instruments can increase absorptive capacity at relatively low cost in a particular city, it becomes a candidate for receiving migrants. These principles hold regardless of where these cities stand in relation to PR. For example, the absorptive capacity criterion *might* suggest that further growth of the primate city should be stimulated. But, in view of the pressures on public services and housing in most primate cities, application of this criterion will usually be consistent with some kind of decentralization strategy. The implication of the absorptive capacity approach in the PR context is that policymakers are relieved from focusing primarily on the development phase when PR takes hold, since in each phase of spatial development the ratio of population pressure to absorptive capacity will vary from city to city. There will always be somewhere where intervention is relatively efficient. The attraction of the analysis is its reconciliation of interurban and intraurban policies, with interurban measures focusing on population distribution (that is, relative population pressure) while intraurban measures are primarily directed to increasing the absorptive capacity of cities. In this way, the dichotomy between urban size distribution policies and urban management is avoided.

A problem with this analysis is that almost all cities in developing countries suffer from severe lags in infrastructure and services, so that there are no obvious candidates for stimulation, only a choice from among the least deficient. The main weakness, however, is the simplistic assumptions behind the absorptive capacity approach. Because migration and job creation are very closely related, attempts to increase urban absorptive capacity through job creation have complex feedbacks on population pressure by inducing new migrants, usually with a substantial multiplier effect (such as the Todaro effect; Todaro, 1969). The elasticity of supply of housing is very high as a result of squatting and self-help housing, and there is no evidence that urban expansion in developing countries is constrained by housing pressures. Similarly, congestion in the use of public services, while a major diseconomy, is unlikely to dissuade inmigration, because the gap in public service standards per capita between urban and rural areas is so wide and because migrants do not receive their fair share of public services in any event.[2]

If absorptive capacity is difficult to specify, spatial policymakers are left with the very general, but familiar, prescription that it is insufficient to guide migration without taking action to increase public service investments and new jobs in the selected destination cities. This, in turn, raises questions about the opportunity cost of these investments and job creations, bringing us back to the PR issue, namely, when do the social returns from dispersion rise toward parity with the social returns from polarization? Unless these comparative returns are close to each other, the efficiency costs of a decentralization strategy may be too high. The urban absorptive approach does not free the spatial planner from the fetters of resource constraints or from the fact that productivity varies with location.

Some other observations on the timing problem are in order. The risks of premature intervention are sensitive to the type of development strategy adopted. For example, the risks are much lower in societies with a rural development emphasis where sustaining basic minimum needs is considered more important than standard economic efficiency criteria. On the other hand, the risks may be very high when the dominant macroeconomic objective is economic growth. Moreover, much depends on the type of intervention, with the risks of early intervention being much higher when it involves the use of resources with a high opportunity cost. Many types of intervention involve public infrastructure investment in facilities that have a long physical life and that are not malleable; if these facilities are underutilized in their early years their cost-effectiveness may be very low.[3]

A related problem is the irreversibility of projects that are very large in scale, even though they have a long gestation period. Thus, changes in government may not permit substantial revision of spatial policies in the light of experience, because the actions of one government commit those of the next, a commitment only avoidable at a huge economic loss. If large infrastructure investments are initiated at the wrong place at the wrong time (or, more strictly in the context of this argument, at the right place but at the wrong time), such losses will be incurred even if the mistake is subsequently realized. If this happens, the sole remaining hope is that the scale of investment is sufficiently large to achieve results, in spite of the wrong timing.

One other aspect of timing deserves emphasis. Even if it is too early to intervene on a large scale to promote PR, it is not too early to be very hesitant about committing investments that reinforce polarization. Such investments will not usually fall under the heading of spatial policies at all. Instead, they are components of sectoral policies, undertaken to achieve sectoral objectives, that incidentally tend to make the primate city or core region more attractive for production and consumption (see below). Also, they may be so large in scale – for example, a subway system for the primate city – that their implementation absorbs a share of the national development budget so big that other projects, including those with a decentralizing impact, have to be abandoned or at least squeezed. Of course, it may be impossible to stop all of these primate city projects, and some of them may be fully justifiable on national development or other grounds. The main policy implication is the need to scrutinize very carefully anticipated spatial impacts of such investments prior to any decision to go ahead with them.

Another type of intervention that may be insensitive to timing arises where there is a variety of simultaneous migration streams, in particular where the dominant rural–metropolitan stream is accompanied by rural–small town, circular, rural–rural, or even urban–rural migration. In these latter cases a modest degree of intraregional investment in transportation, services, and job creation strategies may help to internalize a higher proportion of migration flows within the region and contain outmigration to the core region. These investments, provided that they do not absorb too many scarce capital resources, may be highly productive in serving other goals (interregional equity especially, and perhaps even efficiency).

Where to intervene?

An important issue in a PR strategy is where to intervene. The implication of the stylized description of the model of spatial development at the beginning of the paper is that the large regional cities provide the most obvious focus. The justification is that interregional decentralization of economic activities and/or site selections outside the core region by new or expanding firms are key aspects of PR, and that their promotion requires the generation of strong agglomeration economies. Unless firms are large enough to internalize them (a rare occurrence in developing countries), these economies can be generated only in sizable urban centers.[4]

Other parts of the settlement hierarchy have a role in the PR process, but typically at a later stage. Of course, care should be taken to ensure that promotion of regional metropolises does not lead to excessive *intra*regional polarization with the result that many smaller centers in the region start to lose population.[5] This means that some investments will be needed in these lower ranks to maintain and improve services and to stimulate off-farm employment opportunities. Action to improve the rural settlement pattern *per se* does not have much of an impact on PR except when it results in a sharp drop in outmigration, which is not the typical case.

Implicit polarization policies

A common and valid argument is that polarization toward the primate city is not necessarily solely determined by spontaneous market forces, but is reinforced by nonspatial policies of a macro or sectoral nature. If this is the case, one obvious means of preparing the way for PR in the late polarization phase is to abolish all the implicit centralizing policies that do not serve priority objectives. Such a strategy may be very low cost, particularly when there are external diseconomies associated with continued expansion of the primate city.

There are many examples of implicit polarization policies, though of course not all of them are found in any one country. These include: import-substitution industrialization strategies combined with higher tariffs on industrial products than on raw materials and foodstuffs which shift the internal terms in favor of the urban-industrial sector; similarly, price controls on food subsidize the urban dweller at the expense of the farmer; rapid transit investments in the primate city fend off incipient congestion costs; the growth permitted in public sector activities without attempts to decentralize administrative functions; discriminatory freight rates and utility charges; risk-averse lending by financial institutions based in the primate city; 'open door' policies favoring multinationals with their strong preference for a core location; pricing policies (e.g. on fuel) that make peripheral consumers pay the full-cost freight charges; the absence of pollution fees and congestion taxes; food export tax that may induce an exodus from small farms into the cities; and large-scale water supply or electric power schemes to accommodate primate city demand without recouping the full cost through user charges. These are merely a few from a long potential list of examples.

However, abolition of implicit spatial policies fostering polarization is not easy. Some of these policies may be the pet instruments of politicians. Population distribution goals may be of secondary importance compared with other goals, even sectoral goals. The simultaneous abandonment of a wide range of these policies may be judged too risky a disturbance of the status quo, threatening both the credibility and stability of government.

A related difficulty is how policymakers should deal with so-called metropolitan problems, such as traffic congestion, pollution, housing and public service deficiencies and unemployment. Policies to relieve these pressures, sometimes called 'ameliorative' or 'accommodationist' strategies are desirable from the point of view of improving the welfare of the primate city's residents, and perhaps are conducive to increased economic efficiency, but have the unfortunate side effect that they may make the city more attractive to outsiders (especially individuals and households but also firms), thereby reinforcing primacy. Allowing economic and social conditions in the primate city to deteriorate (the Havana solution) has sometimes been recommended but is often difficult politically. Also, the main sufferers from a 'run down' strategy may be the urban poor, so that any spatial equity gains (more resources for other areas) may be outweighed by adverse income distribution effects.

A more satisfactory approach, though the answers await policy analysis research, is to devise metropolitan ameliorative policies that relieve congestion and other, related problems without making the city more attractive to migrants. Much depends on the variables to which migrants respond. If migrants respond to job opportunities more than to housing and public service levels (a reasonable assumption in developing as opposed to developed countries), it may be possible to improve social service provision, including more emphasis on the delivery of such services to deprived sections of the metropolitan population, without inducing inmigration provided that this is combined with a policy of severe restrictions on employment growth.[6] In any event, if policymakers wish to anticipate PR rather than to nudge it along once under way, such restrictions are almost obligatory. Decentralization of employment policy is a key element in PR strategies.

Empirical studies

Although there have been studies on the relative strength of spatial polarization and dispersion forces in developing countries,[7] it is hardly surprising that there have been few empirical analyses of PR since there has hitherto been little serious work on developing indicators of PR. However, the little work that has been done is worth a mention as a background to future research.

The first study was by Renaud (1977) on South Korea. He argued that PR is under way in South Korea, citing as evidence the convergence of regional growth indicators, the declining share of GDP absorbed by Seoul gross regional product (GRP), a narrowing in interregional and interurban indicators, the diversion of migration flows away from Seoul toward other urban areas, and a weakening of the positive relationship between income and city size. More recently, Hwang (1979) has referred to a different, though overlapping, set of variables in the Korean case: physical constraints on development and high land prices in Seoul; a decline in the capital's locational advantages for manufacturing (reflected in a fall in Seoul's share of national manufacturing output) combined with the development of large-scale industry in the south of the country (such as at Pohang, Woolsan, Masan, and Yeosu), partly policy-induced by the provision of moderately priced infrastructure and tax benefits; a fall in the population and GRP growth rates of Seoul; and, again, a convergence in interregional income disparities.

Lo and Salih (1979) have considered the case of Japan, chosen not as a former developing country but because their choice of PR indicators was available. They argue for a relationship between PR and regional income convergence (the Williamson hypothesis (1965)), a decline in urban primacy (as measured by the El-Shakhs index (1972), and a shift toward equality in the personal income distribution (Kuznets' law (1955)) as reflected in the Gini coefficient. A turning point is identified in each of these three indicators in the early to mid-1960s.

Lo and Salih (1979) identify four preconditions for PR: economy-wide full employment; agglomeration diseconomies in manufacturing in the core region; the cementing of interregional linkages in a way that facilitates the diffusion of development and spread effects; and the development of a complex organizational structure in business that permits the proliferation of branch plants and easy intra- and interfirm communications between the core and the periphery. There are problems with each of these preconditions in the developing country context. Large cities, in particular, have been subject to severe labor absorption difficulties, as a result of high inmigration rates and of relatively low rates of employment creation, especially in import-substituting manufacturing. The emergence of agglomeration diseconomies in the core region may not induce much industrial decentralization if *net* agglomeration economies are still much larger than the puny agglomeration economies generated in the small cities of the periphery. The hierarchical diffusion model does not always work, and a high degree of interurban connectivity may instead stimulate continued upward polarization toward the core. Income levels and the structure of demand in peripheral regions may not justify the creation of branch plants, and the large-scale industrial base may continue to be composed of monolithic, monopolistic plants clinging to core locations to be close to government and the metropolitan amenities that cater for their managers. Although most of these points are conditional, the prospects for PR remain bleak in most developing countries, if these are indeed the key preconditions.

However, it is not clear that all these preconditions (or at least these specific preconditions) are necessary. They appear to rely heavily on observations of

Japanese experience. For instance, it is improbable that full employment is a precondition of PR in developing countries, since firms are rarely induced to decentralize as a result of labor shortages in the primate city.

In a similar study, Gedik (1978) showed that high levels of economic development (measured by real GNP per capita) were followed by a narrowing in interregional per capita income disparities and a decline in the rate of population concentration (measured by the annual number of net migrants to the metropolitan regions),[8] with both indices declining after the early 1960s; *see also* Mera (1978). She suggested that the dominant cause of PR in Japan was the high rate of economic development, indicating a PR process of the type observed in other advanced countries, such as the United States, Sweden, and West Germany. In other words, it is largely a spontaneous process rather than the result of government intervention. More specifically, the main economic reason was a decline in the economic incentive to migrate to metropolitan areas. This is explained by several factors: a narrowing in interregional income disparities,[9] the decentralization of job opportunities, and – especially after the oil crisis of 1973–74 – a reduction in the formerly fast rate of growth in the demand for labor in metropolitan areas. An important noneconomic factor was a change in values in favor of environmental quality, increasing the numbers preferring a lifestyle outside the major metropolises. A growth pole strategy implemented after the Comprehensive National Development Plan of 1962 was generally unsuccessful, with public investment lagging behind planned levels and following rather than leading population deconcentration, a focus on capital-intensive industries, and the failure to raise per capita income levels. However, the emergence of PR in Japan was facilitated by the presence of a set of favorable preconditions: an equitable interpersonal income distribution and relatively narrow interregional disparities, both reflecting the nation's tax and subsidy system, a stable political system and shared goals, a highly educated labor force, advanced technological knowledge, and high levels of savings and investment. Needless to say, these preconditions also are rarely present in developing countries.

Linn (1978) tested some measures of PR in Colombia, and concluded that on the whole Colombia did not pass the test. Over the period 1951–73, the largest four cities and especially the capital city, Bogotá, grew faster than all other urban size groups. Also, over the same period, the degree of primacy increased. However, when the size class analysis is abandoned in order to examine growth rates in the thirty largest cities individually, then some of them grew faster than Bogotá. The rapidly growing intermediate size cities (<350 000 in 1973) included Bucaramanga, Cúcuta, Buenaventura, Tuluá, Valledupar, and Villavicencio, none of these located in the core region. Migration rates (interprovincial because of the lack of urban migration data) indicated a continued polarization towards the big four metropolitan areas (Bogotá, D E; Medellín in the department of Antioquia; Cali in Valle; and Barranquilla in Atlantico), though there are some other centers of attraction such as Valledupar (in Cesar), Uribia (La Guajira) and Villavicencio (Meta).

Although GRP per capita was highest in Bogotá and Barranquilla, and relatively high in Medellín and Cali, none of the big four experienced fast GRP per capita growth rates and gross investment per capita levels were low outside Bogotá (ranked second). A definite slowing down in urbanization marks the two intercensal periods 1964–73 and 1951–64, with a drop to 3.7 per cent per annum from 5.4 per cent, but, perhaps paradoxically, this appears to have been associated with a slight increase in polarization. Polarization toward Bogotá was more marked than in the other major metropolises, and in Cali's case there has been a marked diffusion of development into the surrounding region. Yet there is dynamism in parts of the periphery, with rapid growth experienced by some of the relatively small towns located in agricultural regions. However, this 'selective process of diffusion' (Linn 1978, p. 55) cannot qualify as PR, for Bogotá's position in the national urban hierarchy is stronger than ever.

Moreover, the polarization in favor of Bogotá is in danger of being reinforced by policy impacts. In particular, the proposed PIN (Plan de Integracion Nacional) emphasizes improvement in the trunk road network linking the three major cities and bounding the so-called Golden Triangle. However, this transportation investment strategy strengthens the three main cities at the expense of the rest of the country, and increases Bogotá's competitive advantage relative to Medillín and Cali. An alternative sequencing of the road network (for example, linking the Pacific and Atlantic coast by improving the Cali–Medellín and Medellín–coast legs) would assist rather than counteract PR.

Relationships with other spatial concepts

Because PR is concerned with the key forces of agglomeration and dispersion that lie at the heart of spatial planning, it is not surprising that there are some relationships between PR and other spatial concepts, not to mention some non-spatial development concepts. For example, as suggested above in the Japanese case in particular (*see* pp. 149–52; *see also* Mera, 1978), one possible indicator of PR is a reduction in a coefficient measuring the dispersion of interregional per capita incomes, and this implies, of course, a connection between PR and the Williamson hypothesis (the inverted-U path followed by the regional dispersion coefficient as development advances and matures) However, Williamson (1965) himself hardly mentioned spatial dispersion trends, and it is obvious that the upper turning point in regional income dispersion antedates the onset of PR. In particular, continued spatial polarization (especially in terms of population concentration, though this also implies the spatial concentration of economic activity) is compatible with interregional income convergence, primarily because rural–urban migration increases rural per capita incomes, via first-round arithmetic effects and second-round effects on rural productivity and technical change, and because of sizable urban–rural remittances (Stark, 1978). In addition, though with less assurance, urban income growth and welfare may be dampened by the impacts of inmigration on metropolitan congestion. If the Williamson turning point antedates PR, the critical question for policy is by how much and whether there is any regularity across countries in the PR lag. This question has not been examined even in the advanced countries.

A similar point can be made with respect to indices of primacy. The correspondence between a decline in the primacy index and the beginning of PR is far from direct. A decline in the 4-city index, for instance, may be unaccompanied by dispersion further down the hierarchy, while a fall in a high Pareto coefficient of the city size distribution might merely reflect the growth of middle-sized cities *within the core region*, which fails to count as PR on the definition given above. On the other hand, in probably rare circumstances an increase in the 4-city index might be compatible with PR if there were above-average urban growth further down the hierarchy. In general, however, a decline in primacy may be a prelude to PR. But PR does not inevitably follow a decline in the primacy indices, and when it does the question of the length of the lag is again unsettled.

PR can obviously be related to the third and fourth stages of the core–periphery model (preindustrial localized settlement pattern; core–periphery; spatial dispersion into selected parts of the periphery; and the development of a spatially integrated system of interdependent regions) when the core–periphery relation of dominance breaks down. Although the core–periphery model was intended as a paradigm rather than as a deterministic predictive model, Friedmann (1966, 1973) did argue that the core–periphery relationship would eventually be undermined. The critical changes are: evolution of sociocultural patterns into an urbanized, though highly heterogeneous, society; the dispersion of economic activity; a gradual change in the settlement pattern from primacy toward lognormality; and decentralization of power into a polycentric decision-making system. The middle two of these changes are merely descriptive, but the first and last highlight the critical role of social and political preconditions for successful PR. There is no reason why these preconditions should develop. The sociocultural transformation may be aborted by the inability to destroy the strait-jacket of dualism, particularly spatial dualism. In addition, the combination of spatial dualism and potentially unstable centralized political regimes may prevent the evolution of a decentralized political structure. The prediction that the core–periphery relation will break down may reflect a liberal's faith in progress more than a hard assessment of the facts. The most optimistic interpretation is that social and political changes are less a precondition for PR than a consequence of PR.

One constructive reaction to this pessimism is to find a way for regions on the periphery to break free from core dominance via self-help and an internally driven mode of development. If this is possible, PR can be initiated through changes in the development process in the periphery rather than in the conventional model by transformations in the core region (for example, rising congestion costs and other agglomeration diseconomies, evolution of a multiplant industrial structure, decline in the attraction of the primate city for migrants).[10] The 'selective spatial closure' concept (Stöhr and Tödtling, 1979) is the most recent and best articulated example of this type of strategy.[11]

The underlying idea is that spatial integration between the core and the periphery tends to increase initial disparities (in other terminology, backwash dominates spread). Although these disparities might be reduced by a strong central redistributive mechanism, the periphery still suffers from limited access to decision-making. 'Selective spatial closure' attempts to minimize the backwash effects. Several preconditions are needed: explicit inclusion of social and political processes in spatial development policy; treating distance frictions as positive rather than negative by increasing the potential for disaggregated decision systems rather than regarding these frictions as an obstacle to spatial equilibrium; more emphasis on small-scale man–environment relations; and a shift in power from sectoral (vertical) units to territorial (horizontal) units. Strategies to promote selective spatial closure may emphasize the supply side (adoption of appropriate technology, local ownership and exploitation of natural resources, local control over human resources) or the demand side (accommodating regional differences in preferences, and resisting the diffusion of external innovations). Moreover, other measures include unidirectional transport cost subsidies favoring the periphery, building transport infrastructure between and within less-developed regions rather than giving priority to linking the core and the periphery, and subsidies to provide basic services to compensate for the failure to exploit scale economies.

These ideas are suggestions that have rarely been tested. There is a clear-cut conflict between selective spatial closure and PR. Since PR presumes an earlier spatial integration between the core and the periphery, probably implying initial polarization and subsequent dispersion, the implication is that the precondition of spatial integration would be preempted by a strategy of selective spatial closure. On a long-term view, however, selective spatial closure might lead to PR in the sense of more rapid growth of population and expansion of economic activity in the periphery than in the core. But the process of PR would be quite different from that predicted by the standard spatial development model.

In the interregional context, there is clearly some connection between changes in spread and backwash, as perceived by Myrdal (1957) and Hirschman (1958), and PR. But the connection is not explicitly a close one, since the spread–backwash model focuses on changes over *time* and in non-spatial variables such as economic structure and per capita incomes (Salvatore, 1972), whereas PR is, above all, a spatial dispersion process.

However, the spatial aspects of spread and backwash also merit attention. In the typical national space economy of a developing country, there are both spread and backwash effects around the primate city. These effects decay over distance in a way that may be represented by a gravity model.[12] But the important point is that spread decays much faster than backwash. The high value of the spread coefficient will decline during the process of development, but it remains much larger than the backwash coefficient. In other words, the growth of the primate city and continuation of the national development process will eventually be associated with favorable impacts of its surrounding hinterland, usually taking the form of decentralization of economic activity and population within the core region. The backwash effects, on the other hand, are typically felt over the whole country, including the most isolated areas of the periphery.[13] The most important of these effects are the flows of migrants and profits that are not reinvested in the rural sector but are transmitted to the core. These backwash effects persist, and may even be intensified, as the economy and the core region expands. In any event, the value of the backwash coefficient remains unchanged.

This interpretation of the spatial dynamics of the spread–backwash model suggests a policy strategy to initiate, or accelerate, the onset of PR. If spread decays more rapidly over space than backwash, the solution must be to create additional sources of spread. This may be achieved via the promotion of regional metropolises in several regions of the country. Spread effects in the form of radial diffusion will be created in the hinterland around each of these regional centers. This will ensure that a much higher proportion of national space will be accessible to spread effects. Also, since these regional metropolises will compete with the primate city for migrants and savings, the massive backwash effects from the periphery to the core will be reduced, in effect, by sharing them out among a larger number of centers. As a result, the balance between backwash and spread effects will be much more equal when there are several large cities than in the primate city dominant case. This argument strengthens the case for policies that focus on promoting *large* cities outside the core region as a strategy to aid PR.

Growth pole strategies obviously suggest themselves as a means to initiate PR. So much has been written about growth poles that it would divert attention from the PR process itself to make more than a few brief comments. Since growth center policies focus on urban nodes of many different sizes, and since they sometimes involve the creation of new settlements from scratch (or quantum jumps in the size of existing small settlements), they are less sharply directed to promoting PR than the regional metropolis approach. The latter focuses on the generation of agglomeration economies on such a scale as to encourage the decentralization of economic activities located in the core. Growth poles, on the other hand, are often developed as a focus for new industries exploiting local resources (e.g. ores, petroleum). Moreover, in developing countries the frequency with which growth poles turn out as industrial enclaves with no linkages with the regions where they are located but with strong links with the international economy (through multinational ownership) is not an optimistic sign for their value as an instrument to help PR.

'Counterurbanization' is defined by Berry (1978, p. 42) as a 'process of population deconcentration; it implies a movement from a state of more concentration to a state of less concentration.' This definition suggests a similarity with PR, but there are significant differences. Although Berry refers to different types of counterurbanization occurring in different parts of the world (including the Chinese model), most of his analysis focuses on the United States case. Here he emphasizes the slowing down in both population growth and economic growth and the expansion of population outside metropolitan areas. PR in the developing country context is not associated with a secular downward trend in economic growth; on the contrary, it is a consequence of economic growth and its symptoms are most visible in those developing countries that have grown most rapidly. The key initiating factor in PR is much more the interregional decentralization of economic activity with population shifting in response, whereas counterurbanization stresses the changes in people's tastes about where they want to live. Most important of all, PR is

156 Differential Urbanization

associated with population deconcentration only in the core region itself or at the scale of the national space economy, at least initially. In other regions, PR almost always involves more spatial concentration with intraregional polarization toward regional cities continuing over a long period of time.

Research implications

The justification for this paper is to clarify some of the conceptual issues and policy problems associated with PR as a prior step in undertaking serious research. It is believed that the research questions posed by the PR debate are as challenging as most facing regional scientists. Only a few of a wide range of research issues will be mentioned here.

Simulation models of PR

One interesting research question is the simulation of the spatial concentration and subsequent dispersion process using some kind of dynamic nonlinear model. Such a model of PR would focus on its critical determinants, might enable predictions to be made of when PR occurs, and might permit simulation experiments of the impact of policy variables.

Spatial dispersion measures

The second technical issue, and perhaps the one closest to the traditional work undertaken by regional scientists, is the design of relevant measures of spatial dispersion. The coefficients developed will have to be feasible within the constraint imposed by available data sets in developing countries. Although the familiar Hoover index that has been used in a PR-related context is easily applicable because data are available in the majority of developing countries (certainly in those close to the PR turning point), it is a crude and unsatisfactory measure. Changes in the Hoover index over time are barely suggestive of the complex spatial dynamics associated with PR

PR indicators

The distinction between spatial dispersion measures and PR indicators is that the former are intended as relatively objective measures of the distribution of economic activity or population over national geographic space while the latter represent signals to policymakers that policies to initiate or accelerate PR might be appropriate. These indicators may be spatial or nonspatial. Several possibilities have been suggested in this paper and in recent studies, such as per capita income and other general development indicators, Gini coefficients and other measures of the income distribution, the urbanization level, measures of industrialization, changes in the Pareto (city size distribution) coefficient or in the degree of primacy, interregional per capita income convergence, differential growth rates of urban size classes (especially a faster growth in the secondary cities than in the primate city), relative changes in the direction or spread of migration flows, interregional decentralization of industry, tightening of the core region's labor market, changes in industrial structure in favor of multiplant firms, completion of the interregional transportation and communications network, and emergence of serious congestion costs and agglomeration diseconomies in the core region.

However, these suggestions are all *ad hoc*, and there has as yet been no attempt to standardize the particular indicators that might presage the onset of PR in different countries, or even to ask whether a standardized set is possible. One problem, of course, is the limited number of developing countries where PR is under way and the dangers of extrapolating from the experience of developed countries by devising and using a set of indicators based on the PR process there. Another important consideration is that the indicators should be built up from data that are readily available in developing countries.

Case studies

Even if a general set of PR indicators can be developed, there is a need for case studies of specific countries. The pace and form of PR are likely to differ from country to country depending upon the existing settlement, geography, development 'style,' and culture. Also, in cases where PR has been influenced by policy, methodologies for unraveling the policy impact from spontaneous effects must be applied to particular cases.

Cross-sectional analyses

Regardless of the importance of case studies, insights into the PR process can be gained from cross-sectional (and intertemporal) comparisons across countries. Since spatial polarization processes are so prevalent in developing countries and have many common features, it is reasonable to presume that the subsequent PR process will be similar in many respects. An important outcome of cross-country comparisons may be a deeper understanding of some of the more detailed aspects of PR that might aid the design of specific policies relating to the location and type of intervention. These studies would also assist generalization (if generalization is possible) on some of the critical timing questions: When does PR tend to occur during the industrialization process? How long does it take for PR to become established? When is the optimal time for intervention? On this last point, the most critical of all, international comparisons may shed light on the three main policy choices in respect to timing: to intervene at an early stage, where the role of policy is to counteract polarization; to intervene somewhat later in anticipation of PR (that is, close to the PR turning point); or only to intervene when signs of spatial dispersion have emerged, so that the role of policy is to guide and reinforce observed market trends.

Migration and relocation

Changes in the degree of spatial concentration are the product of the interaction between public investment decisions (especially with respect to investment in urban infrastructure and in interregional transportation and communication networks) and the private decisions of individuals, households, and firms to choose where to live or work, including the decision not to move (locational inertia). The latter have the dominant role in PR, especially if it is believed that public infrastructure tends to follow demand rather than to lead it. Thus, understanding changes in the determinants of migration and in the location decisions of firms is a vital ingredient in theoretical explorations of PR. Although much is known about both these issues in general, they have never been studied in a PR context. Emphasis needs to be placed on the dynamics of migration and relocation processes, especially of situations where there are changes in direction and dramatic shifts in rates. For example, how and why is a dominant rural–primate city migrant pattern replaced by strong *intra*regional rural–urban migration streams to intermediate-size cities? And what factors lead to the cumulative entry of new firms into a particular locality, or to the cumulative exodus of plants from the core of the primate city? Placing these processes in a PR context gives them a new slant that is suggestive for future lines of research.

Other research questions

There are many more general research questions on the nature of the PR process, some of which have been touched upon in this paper, but all of which require more detailed research. A list of such questions will suffice to indicate the potential scope. How should PR be defined and measured? What are the preconditions and/or bottlenecks? Is PR easier to introduce in countries with a particular settlement pattern (for example, a system of regional metropolises rather than a heavily primate economy)? Is PR a necessary condition for transformation into a developed economy? Is the spatial ambit of the model always the national economy? What types of policy (strategies and instruments) might help to stimulate PR? Is PR predictable in terms of tending to occur close to a given time after the onset of industrialization? How relevant is the experience of the developed countries in helping to understand the PR process in developing countries?

These are merely a few of the interesting issues raised by study of the PR process.

Notes

1. This process has been described as 'spatial cycles' (Berg *et al.*, 1979). The cycle of agglomeration, then dispersion, first occurs in the core regions and then is repeated in other regions of the system. *See also* Hansen (1978) for a summary of some recent research sponsored by IIASA that touches upon spatial decentralization trends in developed countries.
2. Linn's ideas on absorptive capacity appear to have been derived from earlier observations by Bertrand Renaud. In the original formulation (*ca* 1975) absorptive capacity was restricted to labor absorption and not discussed in the PR context. Infrastructure and services entered the analysis only as constraints on the rate at which cities (especially secondary cities) could absorb new firms and hence create employment. Housing and residential services, on the other hand, are merely a

derived demand from the income generated by employment opportunities. Personal communication from B. Renaud, 4 September 1979.

3 Unfortunately, even projects with demonstrably high net benefits may not be undertaken in developing countries if the costs stream is heavily skewed to the early years relative to the benefits stream. This is because of capital shortage (the limited revenues and borrowing capacity of national governments, the embryonic character of local capital markets, and the capriciousness of access to foreign capital and loans from international agencies).

4 *See* p. 154 for a rationale for promoting large cities that relies on spatial hypotheses about backwash and spread.

5 Some observers have argued that the multiplier effects in high-order centers of income expansion in low-order centers are greater than the multiplier effects in small towns of expansion in the regional centers. In other words, multiplier repercussions tend to be stronger up rather than down the urban hierarchy. *See* Moseley (1973).

6 These restrictions do not necessarily apply to small-scale enterprises (SSEs). Measures to promote SSEs (especially financial aid and technical assistance) may stimulate income and employment without inducing migration.

7 For example, in a study of Brazilian urbanization, Tolosa (1975) showed that decentralization trends were evident in six out of the nine largest cities. However, since this decentralization was intraregional, it is doubtful that it qualifies as PR even in a spatially large country such as Brazil.

8 A related measure is the degree of convergence in the growth rates of prefectures, metropolitan areas, and municipalities (Glickman, 1977; Kawashima, 1977). Another measure used by other observers in the Japanese case (Sakashita, 1978; Vining, 1974) is the Hoover index, which measures concentration by the difference between the observed value of an area's population and the hypothetical value if the population density of each region were equal. However, the Hoover index is very sensitive to the degree of spatial disaggregation.

9 This resulted from a convergence in wage rates among sectors brought about by a tightening of the labor market in both the urban and the rural sectors, reinforced by tax subsidies to rural local governments.

10 In the fashionable terminology of today, this implies 'development from below' rather than 'development from above.' The agropolitan development approach offers another 'development from below' strategy, though with specific emphasis on rural areas (Friedmann and Douglass, 1978). It is arguable that the end result of a successful agropolitan strategy would be a halt in the polarization process rather than PR. It is difficult to envisage how the modern activities of the core region could be decentralized to agropolitan districts.

11 The underlying idea behind selective spatial closure can be traced to Hirschman (1958, p. 199): 'If only we could in some respects treat a region as though it were a country and in some others treat a country as though it were a region, we could indeed get the best of both worlds and be able to create situations particularly favorable to development. Their advantage consisted largely in their greater exposure to the trickling-down effects and in their ability to call for help from the larger unit to which they belong. Their disadvantage seemed to lie principally in their exposure to polarization effects, in their inability to develop production for exports along lines of comparative advantage, and in the absence of certain potentially development-promoting policy instruments that usually come with sovereignty. A nation attempting to develop its own backward regions should therefore provide certain "equivalents of sovereignty" for these regions.'

12 Hierarchical diffusion mechanisms will be weak in a primate developing country with a poorly formed national urban hierarchy. See Pedersen (1970).
13 Exceptions may exist in a very large country, where there is no connectivity at all between the core region and sparsely populated, unexploited territory on the nation's fringes.

References

Berg, L. van der, Drewett, R., Klaassen, L. H., Rossi, A. and Vijverberg, C. H. T. 1979: *Urban Europe: A study of growth decline*. Oxford: Pergamon Press.

Berry, B. J. L. 1978: The counter-urbanization process: how general? In Hansen, M. (ed.), *Human settlement systems: International perspectives on structure, change and public policy*. Cambridge, Mass.: Ballinger, 25–49.

El-Shakhs, S. 1972: Development, primacy and systems of cities. *Journal of Developing Areas* 7, 11–36.

Friedmann, J. 1966: *Regional development policy: a case study of Venezuela*. Cambridge, Mass.: MIT Press.

Friedman, J. 1973: *Urbanization, planning and national development*. Beverly Hills, Calif.: Sage.

Friedmann, J. and Douglass, M. 1978: Agropolitan development: towards a new strategy for regional planning in Asia. In Lo, F. and Salih, K. (eds), *Growth poles and regional development policy: Asian experiences and alternative strategies*. London: Pergamon Press, 163–92.

Gedik, A. 1978: *Spatial distribution of population in postwar Japan (1945–75): and implications for developing countries*. DP 35. Institute of Socio-economic Planning, University of Tsukuba, Sakura, Japan.

Glickman, N. J. 1977: *Growth and change in the Japanese urban system: The experience of the 1970s*. Research Memorandum 77-39, International Institute for Applied Systems Analysis, Laxenburg, Austria.

Hansen, N. M. 1978: *Human settlement systems: International perspectives on structure, change and public policy*. Cambridge, Mass.: Ballinger.

Hirschman, A. O. 1958: *The strategy of economic development*. New Haven, Conn.: Yale University Press.

Hwang, M. C. 1979: A search for a development strategy for the capital region of Korea. In Rho, Y. H. and Hwang, M. C. (eds), *Metropolitan planning: Issues and policies*. Seoul: Korea Research for Human Settlements, 3–32.

Kawashima, T. 1977: *Changes in the spatial population structure of Japan*. Research Memorandum 77-25, International Institute for Applied Systems Analysis, Laxenburg, Austria.

Kuznets, S. 1955: Economic growth and income inequality. *American Economic Review* 45, 1–28.

Linn, J. F. 1978: *Urbanization trends, polarization reversal, and spatial policy in Colombia*. Westfälische Wilhelms-Universität Münster, Sonderforschungsbereich 26, Raumordnung und Raumwirtschaft, WP 12.

Lo, F. and Salih, K. 1979: Growth poles, agropolitan development and polarization reversal: the debate and search for alternatives. In Stöhr, W. and Taylor, D. R. F. (eds), *Development from above and below? A radical reappraisal of spatial planning in developing countries*. New York: John Wiley.

Mera, K. 1978: Population concentration and regional income disparities: a comparative analysis of Japan and Korea. In Hansen, N. M. (ed.), *Human settlement systems*. Cambridge, Mass.: Ballinger, 155–75.

Moseley, M. J. 1973: The impact of growth centers in rural regions I and II. *Regional Studies* 7, 57–94.
Myrdal, G. 1957: *Economic theory and underdeveloped regions*. London: Duckworth.
Pedersen, P. O. 1970: Innovation diffusion within and between national urban systems. *Geographical Analysis* 2, 203–54.
Renaud, B. 1977: Economic structure, growth and urbanization in Korea. Paper prepared for the Multi-Disciplinary Conference on South Korean Industrialization, Honolulu, Hawaii, June.
Richardson, H. W. 1977: *City size and national spatial strategies in developing countries*. Washington, DC: World Bank Staff WP 252.
Sakashita, N. 1978: Urban growth analysis in postwar Japan: fact findings on the distribution of urban population. Paper presented at the World Regional Development and Planning Conference, Institute of Socio-economic Planning, University of Tsukuba, Sakura, Japan, August.
Salvatore, D. 1972: The operation of the market mechanism and regional inequality. *Kyklos* 25, 518–36.
Stark, O. 1978: *Economic–demographic interactions in agricultural development: The case of rural-to-urban migration*. Rome: Food and Agriculture Organization.
Stöhr, W. and Tödtling, F. 1979: Spatial equity: some antitheses to current regional development doctrine. In Folmer, H. and J. Oosterhaven, (eds), *Spatial inequalities and regional development*. Boston: Nijhoff, 133–60.
Todaro, M. J. 1969: A model of labor migration and urban unemployment in less developed countries. *American Economic Review* 59, 138–48.
Tolosa, H. 1975: The macroeeconomics of Brazilian urbanization. *Brazilian Economic Studies* 1, 227–74.
Vining, D. R. Jr. 1974: The spatial distribution of human populations and its characteristic evolution over time: some recent evidence from Japan. *Papers of the Regional Science Association* 35, 157–78.
Williamson, J. G. 1965: Regional inequality and the process of national development: a description of the patterns. *Economic Development and Cultural Change* 13, 3–45.

12 D. R. Vining Jr,
'Population Redistribution towards Core Areas of Less Developed Countries, 1950–80'

From: *International Regional Science Review* 10, 1–45 (1986)

Introduction

The cessation of migration, in net terms, towards densely populated core regions appears to be a universal pattern for the wealthy countries of the West, including Japan (Vining and Kontuly, 1978; Vining *et al.*, 1981; Vining and Pallone, 1982; Vining, 1982a; Francart, 1983; Robert and Randolph, 1983; Illeris, 1984; Aydalot, 1984a, 1984b; Fielding, 1982; Mera, 1986; Myklebost, 1984; Moseley, 1984; Bedford, n.d.; Keeble *et al.*, 1982). In this article, estimates of net migration towards core regions are presented for a sample of 44 Third World countries for the same period over which the decline in net migration to the core regions of the West has been recorded, namely, the 1970s. As we shall see, no comparable decline occurred – contrary to predictions by some close observers of Third World economies (Critchfield, 1979, 1981; Brown, 1976a, 1976b) and despite claims that the largest of these cities are of sizes well beyond what is either economically efficient or ecologically viable (Bairoch, 1982; Sachs, 1980; Vacca, 1973). In fact, in a number of countries, there has been a tendency in the 1970s towards an increase in the rate of net movement of persons to the dominant core areas. The purpose of this article is to summarise these patterns and to present their implications for the standard economic model of regional growth and concentration (Alonso, 1968, 1971; Mera, 1973).[1]

Data and methods

Direct measurements of interregional migration rates in the countries of the developing world are scarce. Here, a surrogate measure of net migration is employed, namely, the difference between the core region's population growth rate and that of the nation as a whole:

$$M_c^{0,T} = 1/T \left[\ln (P_{CT}/P_{C0}) - \ln (P_T/P_0) \right] \tag{1}$$

where $M_c^{0,T}$ is the annual net migration rate for the core region over the period 0 to T; P_{C0}, P_{CT} are population of the core region at time 0 and time T, respectively; P_0, P_T are national population at time 0 and time T, respectively; and T is measured in years.

M_c will equal the true rate of net migration to the core region from the rest of the nation if and only if the core region's rate of natural increase equals that of the nation (Shryock *et al.*, 1975, pp. 625–7).[2] The assumption of equal rates of natural increase in core and nation is a plausible one (*see*, for example, the evidence on rates of natural increase in urban and rural areas in United Nations (1980)). If there is a discrepancy, however, M_c will probably underestimate the true rate of migration to the core region, since, in periods of declining rates of natural increase, the highly urbanised regions can be expected to lead the rest of the nation and have below-average rates of natural increase (O'Connell, 1981), at least in the initial stages of the transition to a lower rate of population increase. This effect may be attenuated by the selective net migration to the core region of younger adults in the fertile age groups (Alonso, 1973, p. 101).

An underestimated net migration rate for the core regions will cause a bias against the hypothesis that net migration rates for the core regions of the developing world are still high; therefore, any evidence found here in favour of the hypothesis will be all the stronger. In any event, rates of population increase have not fallen appreciably in the developing world over the period studied here so that the assumption of approximate equality in rates of natural increase between core and nation is probably still a valid one. Furthermore, changes in rates of net migration to the core regions will still be well reflected by the measure given in eqn (1), even though the actual levels of net migration may be somewhat different.

Estimates of net migration based on the actual rates of natural increase of the core regions themselves were obtained for two countries, Argentina and the Philippines. In the latter case, the estimates of net migration based on the core region's rate of natural increase and on the national rate of population increase are quite close, which is to say that the Philippine core area (the Manila region) has a rate of natural increase quite close to that of the nation as a whole (*see* Llosa, 1982). In the case of Argentina, however, when estimated using eqn (1) the rate of net migration to the Buenos Aires region is negative for the most recent intercensal period, 1970–80 (−2.2 per thousand per year), whereas the more accurate estimate using the rate of natural increase for the core region itself shows positive net migration over the same period (2.6 per thousand per year (Bertinotti de Petrei, 1980, p. 151)). However, the difference between the two estimates is slight, both being close to zero. Moreover, the decline in net migration to the Buenos Aires region between the periods 1960–70 and 1970–80 is detected by both measures (from 3.3 to −2.2 per thousand per year by the one measure and from 9.5 to 2.6 per thousand per year by the other). The rather large discrepancy in the period 1960–70 shown by these estimates is due to the much lower rate of natural increase in the Buenos Aires region than in the rest of the country in that period. Argentina is one of the few countries in the sample studied here to have experienced a decline in its rate of population increase (though, interestingly, not between 1960–70 and 1970–80), and it is likely that this decline has been particularly pronounced and rapid in the core region, which caused rates of natural increase to be lower there than in the nation as a whole, particularly in the period 1960–70 (11.2 versus 16.1 per thousand per year in the period 1960–70, 14.1 versus 17.9 in the period 1970–80 (Bertinotti de Petrei, 1980, pp. 150, 154)).

The core regions of each country are here defined as the regions containing and surrounding the country's most important and dominant city (in a few cases, cities), which is generally but not always the capital city. The work reported here may be regarded as a preliminary step towards a full analysis of regional net migration trends in Third World countries, which would include regions other than the core regions. The much richer data from the developed world on interregional net migration rates suggest, however, that the distribution of regional net migration rates is largely driven, in a statistical sense, by the rates of net migration to a few dominant, highly urbanised regions. In the developed countries, the decline in the rate of net migration to the core regions is generally reflected in increased rates of net migration to *all* of the peripheral regions (Vining and Pallone, 1982). Thus, a large part of the information about the history of the distribution of regional net migration rates probably is contained in the history of the core region's net migration rate. In particular, when the latter falls, as it eventually will, regional net migration rates will approach zero *en masse*, and the interregional population redistribution attendant upon economic development will cease. In short, a remarkably complete picture of a developing country's evolving population geography can be gotten by simply looking at the rate of net migration to its most important, dominant, and densely populated region, though to demonstrate this contention conclusively would require a much more extensive analysis of the data on regional population trends.

The identities and constituent administrative subdivisions of the core regions of the 44 countries studied in this paper are given in Appendix 1 of the original paper. The boundaries of these regions, subject to minor changes, are fixed over the entire study period. The core regions as defined here overbound the metropolises contained within them to a considerable degree. Overbounding should not alter the time trend in the rate of net migration to the core region, but it does produce an underestimate of the rate of net migration to a region experiencing strong net inmigration, and the greater the degree of overbounding the greater the underestimate. It is necessary to overbound, however, since population decentralisation within metropolitan regions has been occurring at a significant rate in non-Western cities, just as it has in Western cities (*see*, for example, the data on the Manila metropolitan area in Stinner and Bacol-Montilla, 1981), so that today the original administrative delimitations of the capital cities of a number of countries underbound their actual physical extent to a considerable degree. In fact, underbounding may be responsible for some unfortunate premature announcements of nation-wide population deconcentration in certain LDCs. For example, the literature contains several independent assertions and/or intimations of a retardation in the rate of net migration to Seoul (*see* UNCRD, 1983, p. 8; Mera, 1978; Higgins, 1978; Economic Planning Board, 1975, pp. 27–8). Certainly, the *city* of Seoul has experienced a declining growth rate relative to that of South Korea as a whole, but the Seoul *metropolis*, as a physical entity, has grown well beyond its administrative bounds. Seoul city plus the surrounding province of Gyeongi continues to experience very high rates of net inmigration, and the other regions, apart from the Southeast region containing South Korea's second city, Pusan, continue to experience very high rates of net outmigration, as has been shown elsewhere with direct estimates of interregional migration flows in South Korea (Vining and Pallone, 1982). Rates of net migration in the late 1970s for both the net out- and the net inmigration regions of South Korea are not significantly different from their average postwar levels.

El-Shakhs (1982; 1983, p. 7) speculates that decentralisation *within* the core region presages an overall interregional convergence of growth rates. It cannot be denied that metropolitan decentralisation precedes interregional deconcentration. The question is whether deconcentration is simply a diffusion of decentralisation in ever-widening circles around the core or occurs more or less independently. The data that are available on the developed world suggest that the convergence of growth rates comes rather rapidly, once a certain level of economic development is reached and a build-up of infrastructure and modern facilities in the non-core regions has been achieved. Hence, convergence is not the result of a gradual spread of the core outwards, as El-Shakhs may be suggesting; it is rather a fairly sudden bursting loose of the modern economy to cover the entire national space.

Estimates of net migration to the core areas of the Third World

Rates of net migration to the core regions of 44 developing countries over the post-war period 1950–80 have been estimated. These countries may be divided into six distinct categories based on their patterns of core-ward migration.

Category I. Five advanced developing countries in which significant declines in the rate of net inmigration to their core regions have taken place in the 1970s: *Argentina, Greece, Ireland, Spain,* and *Venezuela*. In *Argentina*, the *Buenos Aires* region (i.e. Buenos Aires province plus the federal capital) has for the first time in the history of the country experienced a decline in its share of the national population.[3] (A similar phenomenon has occurred in neighbouring *Uruguay*, where the *Montevideo* region had a slightly lower share of the national population in 1975 than in 1963. Given the paucity of censuses in Uruguay, however, it is not possible to pinpoint in which decade this decline began.) Furthermore, as Wilkie (1981) shows and as was predicted in the previous section of this paper, the regions with above average growth rates are spread throughout Argentina's periphery, including Patagonia, and are not confined, as El-Shakhs seems to be predicting, to just the semi-periphery surrounding the *Buenos Aires* region.

In *Greece*, the greater *Athens* region experienced a rate of net inmigration in the period 1971–81 which was less than one-third of that of the 1961–71 period. This break in what was actually a rising trend of net inmigration between the 1950s and 1960s seems to have been unanticipated in the literature on Greek regional demographic trends (e.g. Bennison, 1972; Georgakis and Tsiafetas, 1982). As in Argentina, virtually all of the peripheral regions have experienced an increase in their rates of net inmigration, and not just those regions near the *Athens* conurbation. This is, again, inconsistent with the suggestion that falling rates of net outmigration will gradually spread from the region immediately surrounding the core (Wilkie's 'core fringe') to the more remote regions (Wilkie's 'periphery'). If such a diffusion does take place, it does so very rapidly (i.e. within the space of a typical intercensal period of one decade), once a retardation in the core's growth rate relative to that of the nation occurs.

In *Ireland*, the *Dublin* region (*see* Hutt and Lutz, 1979 for a discussion of how best to delimit this region) is growing at a substantially lower rate, relative to the country as a whole, than it did in the 1950s and 1960s (*see* Horner and Daultrey, 1980 and Keane, 1984 for further discussion of the trends in the 1970s). This development, as in Greece, appears to have been unanticipated by students of Ireland's modern population geography (e.g. Bannon, 1975; Horner, 1974).

In *Spain*, the two dominant metropolitan regions, those surrounding and containing the cities of *Madrid* and *Barcelona*, have both experienced a decline in the rate of net inmigration, which, though still positive, is 50 per cent lower in the 1970s than in the 1960s. This development was, again, not anticipated in the literature on internal migration patterns in Spain (e.g. Bertrand, 1978) and does not seem to be widely recognised or appreciated today. The mirror image of this decline in net migration to Spain's dominant metropolitan regions is an increase in net migration to *all* of Spain's traditional peripheral and predominantly rural regions (Vining and Pallone, 1982).

Finally, in *Venezuela*, the rate of net inmigration to the Central region, consisting of *Caracas* and its immediate hinterland, was both high and stable (around 10 per thousand per year) until the 1970s when it fell to 2 per thousand per year (*see also* Eastwood, 1983, whose data, drawn from the *preliminary* 1981 census, shows virtual stability in the Central region's share of the Venezuelan population between 1971 and 1981).

Rondinelli (1982) claims that the bias of central government investment towards the core regions in the Third World is the dominant proximate cause of the ongoing concentration of population and assets in these regions. The estimates presented here indicate that at some level of national wealth, either this cause of concentration becomes inoperative or the central government actually acts to deter, through its investment policies, the further concentration of population, capital, and economic activity in the capital region. Japan is a particularly well-documented example of the switch in regional investment policy by central governments at a certain level of economic development (*see* Gall, 1983; Itakura, 1982; Mera, 1986), and Jones (1982) and Greenwood (1984) present evidence of a similar switch in Venezuela.

Category II. *Peru, Chile*, and *Egypt* have also experienced sharp declines in the net migration of persons to their core regions, i.e. the regions containing and surrounding *Lima* (*see also* Richardson, 1984), *Santiago*, and *Cairo*. Such declines are unusual in countries at their level of development and are best explained, at least in the cases of Peru and Chile, by the protracted stagnation in the economies of these two countries during a significant part of the 1970s and early 1980s. Egypt, on the other hand, has enjoyed quite rapid rates of economic growth over this same period. Egypt may be the sole example of what is referred to in the next section as premature deconcentration. Unlike other large cities in the Third World, Cairo may have reached a scale and density that make it significantly less attractive to both migrants and firms (Bodgener, 1984) (although its absolute size is exceeded by a number of Third World cities). Farrell (1983) has described some of the diseconomies of scale in the Cairo metropolis. Also, because of the compact distribution of Egypt's population in the lower Nile delta (4 per cent of Egypt's land area), accessibility even in its peripheral parts may be satisfactory despite Egypt's low level of economic development. Critchfield's thesis, which will be discussed at length in the next section, may also help to explain the Egyptian data. He argues that rural development has been sufficiently successful in certain poor, but agriculturally proficient, Third World countries so as to slow the migration of peasants to the large cities of these countries.

Interestingly, Ibrahim's exhaustive survey of internal migration trends in Egypt (Ibrahim, 1983) makes no mention of the rapid decrease in the rate of population redistribution to the Cairo area.[4] Morcos (1980), by contrast, discusses this emerging trend in the patterns of spatial redistribution in Egypt, attributing it in part to a more rapid decline in the rate of natural increase in the Cairo region than in Egypt as a whole. Egypt is one of the few countries in the sample used here to have experienced a substantial decline in its rate of population increase in the latest intercensual period, though postcensual estimates suggest that this decline was short-lived and that the rate of population increase is now at its highest level ever (Bureau of the Census, 1983, p. 84). The 1986 census in Egypt should tell us to what degree the break in the trend of Cairo's rate of growth revealed by the 1976 census was special to the period 1966–76, and to what degree it represents a sustained downturn in Cairo's rate of growth relative to that of the rest of Egypt. The most recent estimates of the Cairo region's population suggest that there is now net outmigration from this region; however, the method used to make these estimates is not revealed.

The rates of population redistribution towards the core regions are not actually low in these three countries, if we disregard the 1981 population estimates for the Cairo region; they are simply considerably lower than they have been. Indeed, the rates remain high compared to the prevailing rates of redistribution to the core regions of the developed world before the drop in these latter rates in the 1970s (*see* Vining and Pallone, 1982).

Category III. Twenty-three of the 44 countries studied here show either a stable or an increasing rate of population redistribution towards their core areas. The 23 countries are discussed below one at a time.

In *Brazil*, the rate of net migration to its most developed state, *São Paulo* (Haller, 1982), has been steadily increasing over the entire post-war period with no signs of retardation in the 1970s. There is evidence, however, of some population dispersal away from the greater São Paulo metropolitan area itself to smaller towns and cities within 150 kilometres of the metropolitan area (Townroe and Keen, 1984, p. 51). Indeed, Storper (1984, p. 152) presents data showing a levelling off in the concentration of manufacturing employment in São Paulo State as a whole (*see* his Table 7, last line, last two columns; also Vining, 1985 and Storper, 1985 for further discussion). Unfortunately, Storper does not explain how it is that São Paulo State is attracting population at its fastest rate ever, while manufacturing employment is no longer concentrating there. Is the tertiary sector growing so rapidly now as to compensate for the relative stagnation in São Paulo's manufacturing base? Can the tertiary sector in Brazil be so spatially disconnected from the manufacturing sector? Are the census data simply in error? Or is it Storper's data that are in error? These are some of the questions that remain unresolved.

In *Colombia*, the rate of net migration to the Bogotá region is high (over 10 per thousand per year) and increasing, though the available data do not extend very far into the 1970s. Williams and Griffin (1978, p. 28) also report an accelerating rate of net outmigration from Colombia's rural areas.

In the *Dominican Republic*, the rate of net migration to its capital, *Santa Domingo*, is both extraordinarily high (approximately 30 per thousand per year, which is among the highest rates of net migration to core regions recorded here) and stable, showing no signs of significant decline.

In *Ecuador*, the two major attractor zones, the regions containing and surrounding *Quito* and *Guayaquil*, have experienced a high and stable rate of net inmigration (approximately 10 per thousand per year) over the period 1950–82. These two zones now contain over 42 per cent of the population, whereas in 1950 they contained only 30 per cent. Munro (1981, p. 784) reports government restrictions on interprovincial migration, designed to 'stem the flow of peasant families into the squatter settlements of Quito and Guayaquil', but these restrictions have apparently not been effective.

In *Mexico*, net migration to the Mexico City region is both high and stable (around 10 per thousand per year), with only the slightest signs of decline. Pommier (1982) provides further discussion of Mexico City's growth trends. The trends themselves, and the lack of any retardation in them, have become a major policy concern (at least before the financial and economic crisis overwhelmed all other concerns in Mexico), for the new government is committed to economic, demographic, and political decentralisation (*see* Acosta-Romero, 1982; Cantu-Segovia *et al.*, 1982).

In *Panama*, the rate of net migration to the Panama City region is high (over 10 per thousand per year) and stable.

In *Algeria*, the rate of net migration to the Algiers region is high (over 10 per thousand per year) and increasing.

In *Libya*, the rate of net migration to the regions containing and surrounding its two major points of population concentration, *Tripoli* and *Benghazi*, is high (15 per thousand per year) and increasing, though the available data do not extend very far into the 1970s. For a more detailed analysis of trends of population redistribution in Libya, *see* Rashed and Nawas (1977) and Attir (1983).

In *Tunisia*, the rate of net migration to the north-eastern region, which contains and surrounds Tunis, is likewise increasing.

In *Indonesia*, the rate of net migration to the region containing and surrounding the capital city, *Jakarta*, is high and increasing (over 15 per thousand per year). Jakarta City itself, or DKI Jakarta, experienced a decline in its rate of net inmigration, but this appears to be simply due to the spread of Jakarta beyond its administratively defined boundaries. Jakarta plus the surrounding districts of Bogor, Bekasi, Tanggerang, and Serang have experienced a significant increase in their aggregate rate of net migration between the periods 1961–71 and 1971–81. The Jakarta government's efforts to stem the flow of migrants to the capital, insofar as they have succeeded, appear to have done so only by diverting some of that flow to the immediately adjoining districts (*see* Hugo, 1979 for a brief discussion of and further references on the regulations adopted in 1970 to discourage migration to Jakarta). Persistent reports of large, government-sponsored population movements from Java to the so-called outer islands, which would tend *ceteris paribus* to reduce the flow of population to the Jakarta area, have yet to be verified by data (*see*, however, Arndt, 1984a, 1984b) and indeed probably cannot be verified until the 1991 census, as the biggest push is said to have occurred over the 1979–84 five-year plan (Caufield, 1984; Crossette, 1984).

In *Malaysia*, the rate of net migration to the *Kuala Lumpur* region is both high (over 10 per thousand per year) and increasing.

In the *Philippines*, the rate of net migration to the *Manila* region is both stable and high (around 10 per thousand per year), but with some sign of decline in the 1975–80 period. This last development may be due to a faltering pace of economic development in the Philippines in the late 1970s. Indeed, the other countries might well exhibit a similar phenomenon but one which would go undetected here because they, unlike the Philippines, held no censuses in the mid-1970s.

In *South Korea*, the rate of net migration to the *Seoul* region is both very high (consistently over 20 per thousand per year) and stable. The data are inconsistent with reports elsewhere of a slowdown in Seoul's rate of growth relative to the nation (e.g. 'The success of the Republic of Korea in slowing the growth of Seoul is unique among the developing countries...' (ESCAP, 1983, p. 22); *see also* UNCRD, 1983; Mera, 1978; Higgins, 1978; Economic Planning Board, 1975, pp. 27–28; Mahar, 1984, p. 317; Richardson, 1983, p. 35 and additional references in Richardson, 1980, p. 74). This inconsistency is probably due to the underbounding of Seoul in those studies. A plausible hypothesis is that the growth of intermediate-sized cities in South Korea is to a significant extent confined to Seoul's immediate hinterland (Nelson, 1983, pp. 291–2). Indeed, discussions of intermediate-sized cities are often plagued by a failure to pay adequate attention to the spatial arrangement of these cities relative to the core as well as to the distribution of their growth rates as a function of their distance from the core (Rondinelli, 1982 and 1983 and Smith *et al.*, 1983 are exceptions in this regard).

In *Taiwan*, the rate of net migration to the Taipei region is very high (around 20 per thousand per year) and remains above its post-war average. For a further discussion of this rising trend in the rate of net migration to the Taipei region, *see* Wang (1977) and Vining and Pallone (1982).

In *Thailand*, the rate of net migration to the *Bangkok* region is both high (over 10 per thousand per year) and increasing. For further discussion of this region's growth, *see* Sternstein (1984).

In *Bangladesh*, the rate of net migration to the *Dhaka* region has increased quite markedly in the 1970s. For a description of pre-1970 internal migration patterns in Bangladesh, *see* Khan (1982) and Krishnan and Rowe (1978).

In *India*, the rate of net migration to its three national cities (*Calcutta*, *Bombay*, and *Delhi*) is both high (around 10 per thousand per year) and increasing. This observation, in fact, may be generally made about India's largest cities (*see* Crook and Dyson, 1982, p. 153).

In *Pakistan*, the rate of net migration of the *Karachi* region is both high and increasing. For a discussion of pre-1970 internal migration patterns in Pakistan, *see* Helbock (1975a, 1975b), and, for a specific discussion of Karachi, *see* Scholz (1983).

In *Portugal*, the rate of net migration to the *Lisbon* region is over 10 per thousand per year and remains close to its post-war average, though it is lower in the 1970s than in the 1960s.

In *Turkey*, the rate of net migration to its three largest cities, *Istanbul*, *Ankara*, and *Izmir*, is quite high (upwards of 20 per thousand per year, in the aggregate) and stable. Government policy to curtail the rapid concentration of population and economic activity in Turkey's largest cities (Levine, 1980) appears to have failed. The rate of redistribution to Turkey's largest cities, particularly Ankara, did fall quite markedly in the period 1975–80, but until the 1980–85 data are available and show a persistence in this trend, this decline must be considered temporary and probably due to the massive political violence and social unrest that Turkey, and particularly its largest cities, experienced during the period 1975–80.

Ethiopia has never held a census, and those population data that are available are of dubious quality. Nonetheless, these data show the same extraordinarily rapid redistribution of population towards the capital region over the 1960s and 1970s as is observed elsewhere in Africa (as will be discussed below).

In *Sierra Leone*, the rate of net migration to the *Freetown* region is both high (over 20 per thousand per year) and increasing, though here, as with Colombia and Libya, the available data do not carry us very far into the 1970s.

In *Haiti*, the rate of redistribution towards the *Port-au-Prince* area was below but not much below its post-war average during the 1970s and in any event remains close to 10 per thousand per year.

Category IV. There are seven countries in sub-Saharan Africa for which census data are available on regional populations at the end of the 1970s or beginning of the 1980s: *Botswana, Gambia, Guinea-Bissau, Kenya, Malawi, Tanzania,* and *Zambia*. In all seven of these countries, the rates of net migration to their capital cities are high (more detailed descriptions for some of these countries may be found in Banyikwa, 1982; Christiansen, 1984; Hayuma, 1983; Nag, 1981). In Botswana, Gambia, Guinea-Bissau, Tanzania, and Zambia, they are or have been the highest of any recorded in the world. In most African countries, however, the capital cities are small both in absolute size and as a proportion of the national population. It is doubtful that the rates of net inmigration to these cities can be sustained as these cities expand in population and come to hold a significant proportion of the national population. Indeed, Lusaka's rate of net migration fell from 73 per thousand per year in the 1960s to 31 per thousand per year in the 1970s. The latter rate, however, is still among the highest ever recorded in the Third World. Thus, though rates of core-ward migration are falling in some countries of Africa, such declines are probably inevitable, given the extraordinarily high levels they started at and given the small population bases upon which they were calculated. Moreover, they remain, even if declining, in the upper range of the distribution of core-ward migration rates in the Third World. For this reason, these countries are classified here as a separate category and are not included in either Category II (developing countries with low per capita incomes in which declines in core-ward migration rates have taken place) or Category III (developing countries with either stable or accelerating rates of core-ward net migration). In short, the African countries which had censuses at the end of the 1970s and which had more than two censuses show declining rates of core-ward net migration, but it is doubtful that these declines will continue to the point where the rates of net migration are actually low (below, say, 10 per thousand). A more likely development is that they will level off at the modal rate of core-ward migration found in the developing world (between 10 and 20 per thousand per year), as represented by the Category III countries here, which are over 50 per cent of the study sample.

Jordan also published a census at the end of the 1970s. The capital region surrounding and containing Amman, in spite of its containing over 50 per cent of Jordan's population, continues to grow at a significantly higher rate than the rest of the country. Unfortunately, since only two censuses are available for the post-war period, no trends in this growth rate can be established.

Category V. There are two countries in the sample in which the rate of redistribution to the core region is negligible: *Cuba* and *Sri Lanka*. In the latter, this rate has actually been negative over most of the post-war period, and the core's share of the national population has therefore been declining. At the end of World War II, the core region that emerged during the colonial period and is the site of its modern capital, *Colombo* (Kirk, 1981; Moore, 1984), contained 21.3 per cent of the total population of the country. By 1981, this share had fallen to 20.8 per cent. Outside of the Communist nations of the Third World, such as Cuba (to be discussed shortly), North Korea, Vietnam, and Cambodia, no other country in the developing world besides Sri Lanka has had the experience of continuous deconcentration away from the core region.

Sri Lanka is materially a poor country, so it cannot be included with the countries of Category I above, i.e. those countries which have passed the threshold of development whereupon a retardation in the rate of population redistribution towards the core regions can be expected. Sri Lanka is well below that threshold above which a significant decline in the rate of redistribution to a nation's core region takes place (which is estimated here to be, based on the experience of the Category I countries, very approximately $3000 per capita GDP in 1975 US dollars). Nor can it be included with the countries of Category II, i.e. those countries well below this threshold but whose core regions experienced a significant decline in this rate during the 1970s. Sri Lanka's core region has been growing at a slower rate than the rest of the country over most of the post-war period. Thus, explanations of the absence of population concentration in Sri Lanka must be sought elsewhere than in the simple stage model of economic development and population concentration, whereby the latter is first an increasing and then a decreasing function of the former, and pauses in the pace of the latter are brought on by slowdowns in the pace of the former. Sri Lanka is on a different path of spatial development from most other countries studied here, as it is of social development in general (Sen, 1981; Isenman, 1980).

Cuba, like Sri Lanka, has not experienced the massive redistribution of population towards its core region that has been typical of the vast majority of developing countries (*see also* Landstreet and Mundigo, 1983). The relative growth of the *Havana* region was only slightly above that of the nation as a whole during the 1950s and 1960s; during the 1970s, Havana's share of the Cuban population actually fell so today it is almost exactly equal to what it was estimated to be in the 1943 census.

The other Communist countries in the developing world (e.g. Vietnam, *see* Lang and Kolb, 1980; Wiegersma, 1983, p. 101; China, *see* Jowett, 1984; Rowe, 1984) likewise exhibit patterns of internal population redistribution quite different from the well-known pattern of rapid redistribution in favour of a few dominant urban regions, which is observed in those countries of the Third World where the Communist Party does not have administrative and political control. Vining and Pallone (1982) and Vining (1982a) have shown that the Soviet bloc countries of Eastern Europe exhibit rates of population redistribution to their core regions that are generally below those of Western Europe at a comparable stage of economic development (with the possible exception of the Soviet Union itself; *see* Houston, 1979; Rowland, 1983), particularly after taking into account the very low rates of natural increase in these core regions. The exceptionalism of the Communist countries has been ignored by students of spatial development and policy (e.g., Gordon, 1982; Rondinelli, 1982), who have given the impression that the expansion of core regions at the expense of peripheral regions is a universal phenomenon of economic development. This view is erroneous, as may be seen here in the data of Cuba and Sri Lanka. Alternative paths of spatial development are clearly feasible, though it is quite possible that they entail an economic development of an entirely different character, e.g. one that emphasises a predominantly agrarian and solar-based mode of economic subsistence, a model of development which found its most extreme and, therefore, clearest expression in the recent history of Cambodia (Ponchaud, 1978; Kiljunen, 1984; Shawcross, 1984).

Category VI. *Israel* and *Syria* are simply anomalies, probably for geopolitical reasons. In Syria, there has been a low and falling (though not insignificant) rate of redistribution towards the *Damascus* region. It is possible that this decline is due to the highly vulnerable position of Damascus, which is only 40 kilometres from the borders and army of Israel. Israeli occupancy of the Golan Heights 'provides that army with observation over, and control of, the plain as far east as Damascus' (Elliot and Lee, 1980, pp. 640–1). Therefore, the economy of Damascus may not be able to attract capital and population from the rest of the country because of the risks perceived by the periphery in investing in a city that is so vulnerable to destruction and/or isolation. Possibly it may be Syrian state policy to deter the further concentration of capital and population in a forward capital (Spate, 1942), where they would be vulnerable to sudden destruction. Finally, Damascus itself may have become, to a degree, peripheralised in the Syrian space economy as access to its historic port, Beirut, became increasingly problematic in the post-war period (*see* Makarius, 1883, as quoted in Diab and Wahlin, 1983, p. 115: 'Beirut is the Flower of Syria and the center of her sciences, the seaport of Syria and the outlet for her produce'), and as the economic positions of such northern cities as Homs, Latakia, and Aleppos, which now have an easier and surer access to the Mediterranean, correspondingly strengthened. Despite these hypotheses, little is understood of the organisation of Syria's space economy in the post-war period, particularly in the 1970s when the rate of redistribution to Damascus fell to a low, by Third World standards, 2 per thousand per year. Syria has not even approached that level of economic development that is normally associated with a slow-down of the pace of redistribution towards the core regions.

The entire history of *Israel* is marked by the absence of population redistribution towards the two major urban regions along its coast surrounding and containing *Tel Aviv* and *Haifa* (Soen, 1977). Over this same period, Israel has had a rapidly growing industrial economy and, for this reason, cannot be classified along with Cuba and Sri Lanka in the Category V countries, in which the capital regions have also not attracted population from the peripheral regions but in which industrialisation has had only limited success. The exceptionality of Israel is almost surely due to the fact that the very survival of this state and of the majority of the people living within it depend upon its maintaining secure borders, preventing the infiltration of, among others, hostile former inhabitants and their descendants, and preventing the Arabisation of peripheral areas, such as the Galilee, by the natural increase of Israeli Arabs. These goals can only be accomplished by ensuring that the peripheral regions do not become depopulated of Jews and that the Jewish population does not drift away from the periphery towards a few large, already demographically extremely dominant coastal centres. Thus, Israel has pursued a policy of settling its Jewish immigrants in the peripheral regions (Gradus, 1983), which has more than compensated for a moderate net *internal* flow of the Jewish population to the coast and away from the borders. Clearly, the survival of the Israeli state and of the Jews living in it will always be preferred by that state and people to the achievement of a somewhat greater economic efficiency and product through a higher concentration of population in a few large centres. Plaut (1983) rather glosses over this point in his standard economist's critique of Israeli spatial policy.

Discussion

As a country industrialises, the densely populated core regions generally expand in population and other assets at the expense of the sparsely populated, non-metropolitan regions. To account for this phenomenon, economists and geographers have posited the existence of agglomeration and scale economies. 'For the same kind of reason for which increasing returns lead to monopoly in terms of microeconomics,' writes Nicholas Kaldor, 'industrial development tends to get polarised in certain "growth points" or in "success areas", which become areas of vast immigration from surrounding centres or more distant areas' (Kaldor, 1975, p. 356). As Weber put it 80 years ago in his celebrated statistical study of cities, 'Production increases with increasing density, and more particularly with increasing concentration' (Weber, 1899, p. 417). The general principle is this: that capital, including human capital, can earn a higher return, the higher the density of capital already present in an area, and that this sets off a movement of capital and population to the high-density core regions in a self-reinforcing cycle, as described by Ullman (1958), Jefferson (1939), Murphey (1978), Myrdal (1957), Clark (1964–5), Mera (1973), and Alonso (1971).

At the beginning of the 1970s, few urban economists or regional scientists foresaw any limits to the economic advantages of large urban size and population concentration, i.e. limits that might emerge to retard the growth of large urban regions relative to more sparsely populated areas (e.g. Alonso, 1971; Mera, 1973). Subsequent developments during the 1970s, however, convinced most urban theorists that beyond a certain level of economic development, the economic advantages of further population concentration begin to disappear and even to become disadvantages (*see*, for example, Long, 1981; Wheaton and Shishido, 1981; Mera, 1977). In particular, a broad trend towards radically lower rates of redistribution towards core regions or even outright deconcentration emerged in the 1970s throughout the developed world (Vining and Kontuly, 1978; Vining *et al.*, 1981; Vining and Pallone, 1982; Vining, 1982a).

The link between the phenomenon of deconcentration (or cessation of concentration) and the level of economic development has become explicit in the literature, though most regional scientists will concede that the nature of this link is not well understood. For example, John Long writes,

> In general, deconcentration is much more than just a coincidence of individual factors such as increased retirement, growth of land grant colleges, flight from urban crime, or a return-to-nature movement. Deconcentration is a consequence of advanced economic development, and it is now occurring in other advanced economies – like Sweden, Great Britain, Germany and Japan. This association with economic development has a major influence on the prospects for continued deconcentration.
>
> (Long, 1981, p. 87)

Wheaton and Shishido are even more emphatic and precise in their discussion of this link:

> Up to $2000 [1970 US dollars] per capita [approximately the level of per capita income in Argentina, Uruguay, Greece, Venezuela, Spain, and Ireland during the 1970s, but 50 percent higher than that of Turkey and South Korea and from three to five times that of Bangladesh, India, Indonesia, Sri Lanka, and Pakistan], greater capital usage increases scale and agglomeration economies quite rapidly. ... Urban concentration *must* increase with the level of development, until the latter approaches $2000 [$2800 1975 U.S. dollars]. After that, spatial decentralisation sets in.
>
> (Wheaton and Shishido, 1981, p. 29, emphasis added).[5]

Mera (1977) likewise emphasises the high level of development necessary for the trend towards concentration in a country's core regions to begin to slacken and eventually exhaust, if not reverse, itself.

Somewhat surprisingly, declines in labour and capital productivity in the core regions of the developed world, relative to labour and capital productivity in the peripheral regions, cannot be invoked to explain the shifts in internal migration patterns in the developed world in the 1970s. Keeble *et al.* (1982) show for Western Europe that productivity differentials between core and periphery there widened rather than diminished in the 1970s, as did per capita income differentials (Commission of the European Communities, 1981, p. 54;

Mayer, 1984, p. 18). Also, Hulten and Schwab (1984) show that labour productivity grew at approximately the same rate in both the core and the peripheral regions in the United States, though it should be noted that these results are not consistent with those reported by Casetti (1982). Finally, Carlino (1982) finds that the optimal-size US metropolitan area (that size for which labour productivity is highest) is still around 3 000 000 persons, which is much higher than the city size of maximum population growth.[6] In other words, no radical change in the relationship between city size and productivity has been found to accompany and thereby to explain the observed changes in the city size–growth rate relationship. A system-wide change in the economy seems to occur at a certain level of economic development that enables deconcentration, or at least the cessation of concentration, to take place (*see* Vining, 1982b); diseconomies of *absolute* size or density in the core regions explain very little of this phenomenon. What the nature of these system-wide changes is remains unclear and largely unresearched.

How advanced the level of economic development can be before an exhaustion of the concentration trend takes place is illustrated by the case of Japan. It was not until 1975, when Japan's per capita GDP was approximately $5000 (1975 US dollars) (Summers and Heston, 1984), that the rate of net migration towards the great Pacific megalopolis between Tokyo and Osaka approached zero (Vining and Pallone, 1982). By contrast, no such slackening in core-ward migration has taken place in the 1970s in Taiwan and South Korea, which have the most advanced economies in Asia outside of Japan (*see* Vining and Pallone, 1982). Nor has there been a slackening in core-ward migration in such rapidly industrialising countries as Brazil, Mexico, Thailand, or Malaysia. Indeed, the preponderance of the evidence from those countries studied here whose per capita incomes in the 1970s did exceed the $2800 (1975 US dollars) inflection point found by Wheaton and Shishido (1981) for Greece, Ireland, Spain, Argentina, and Venezuela suggests that only a slackening in the *pace* of population redistribution towards the core can be expected when this level of development is reached, not an actual deconcentration or what Wheaton and Shishido call decentralisation.[7] The Athens, Dublin, Madrid, Barcelona, and Caracas regions continue to gain population through net inmigration; what happens at roughly $3000 per capita income is that this rate is significantly reduced from a previously very high level. Wheaton and Shishido's analysis, however, concerns interurban decentralisation from large to small cities, whereas the analysis here is of interregional deconcentration, and the former doubtless precedes the latter.

Some recent reports from close observers of the Third World (e.g. Critchfield, 1979, 1981) have challenged the standard economic model of spatial development whereby countries experience a kind of lock-step spatial evolution as their economies industrialise – with core regions growing inexorably larger at the expense of peripheral regions until a high level of economic development is reached, at which point the exhaustion of this process sets in rather quickly. These reports do not deny that some countries will move and are moving along this path successfully. They are the rapidly industrialising nations of East Asia, namely, South Korea, Taiwan, Malaysia, and Thailand (the city states of Singapore and Hong Kong, where the terms concentration and deconcentration are largely without meaning, are ignored here), the remaining less than thoroughly industrialised countries of Europe, namely, Turkey, Spain, Portugal, Greece, and Ireland, the rapidly industrialising countries of Latin America, such as Brazil, Mexico, Argentina, and perhaps Chile, and the oil-rich countries of the Middle East. In sub-Saharan Africa, probably only South Africa would belong to this group. These nations, in the aggregate, however, constitute a rather small proportion of the developing world's population and do not even contain a majority of its largest cities (cities, say, of over 3 million persons). There is a second group of countries, according to these reports, containing a much larger population in the aggregate and the majority of the Third World's large cities, where, it is argued, the dynamic sector in the economy is the rural, agricultural sector rather than the urban, industrial sector.

Countries belonging to this second group include India, Indonesia, Sri Lanka, Egypt, Pakistan, Bangladesh, and China – old, very densely populated countries with very large populations (together they contain over 60 per cent of the Third World population), countries so large in population that many find it difficult to imagine industrialisation ever attaining sufficient scale to lift any significant proportion of their populations to income levels now reached by even still relatively poor countries, such as South Korea or Greece. What these countries do possess, however, are large numbers of highly skilled agriculturalists who are able and willing to adopt new technologies (such as new seeds, new irrigation techniques, and chemical fertilisers) and who could raise productivities in the rural sector very rapidly if given access to these technologies. According to scattered reports now reaching the West, this is precisely what began to happen in these countries during the 1970s and is going on at an increasing rate in the 1980s (Simon, 1981). Coupled with rapidly falling birth rates, rapidly rising production in the rural areas could lift living standards in these areas to the point where a reversal in migration between rural areas and the great cities is theoretically plausible. Some observers of the Third World claim it is not only theoretically plausible but already occurring in South and East Asia and scattered places elsewhere.

'Agricultural prosperity,' writes William Tucker, 'quickly changes people's habits. In many areas, the hopeless migration of starving villagers to city slums has reversed. Many Third World megalopolises are now losing population as villagers find they can return to the land and make a living' (Tucker, 1981. p. 18). Critchfield, whose book, *Villages*, Tucker reviews, is only somewhat more circumspect:

> Nobody had made a statistical study, but my guess (based on interviews as an AID consultant in 30-odd villages across Java) was that Jakarta was probably the first great Asian city to experience reverse migration, with more peasants going out than coming in
>
> (Critchfield, 1981, p. 173).

Critchfield goes on to make the more general claim:

> great swollen cities are disgorging some of their inhabitants. In a turnabout of the long march out of the countryside and into the cities, armies of ex-peasants are leav-

ing the Asian cities to return to their villages, which can once more feed them. The vast exodus from village to city is ending, probably has ended already. Cities like Bombay and Jakarta appear to be experiencing reverse migration, though it may take some time until we have the figures to prove it.

(Critchfield, 1981, p. 322).

Critchfield's model is basically the following: (1) a rise in agricultural productivity increases the resources available to rural areas and (2) a drop in the birth rate in these areas (which seems particularly strong in the generally agnostic, Buddhist, or peripherally Moslem countries of East and South Asia) relieves some of the population pressure on these resources. In combination, these two events cause a sufficient rise in the standard of living among villagers to attract former peasants away from the industrially stagnant cities. In other words, a decreased push from the villages, due to a lower rate of natural increase, and an increased pull in the rural areas, due to a rising relative wage and higher rate of job growth in the rural sector, act together to reverse the direction of migration between village and city. Though the modernisation and resultant reorganisation of agriculture might also reduce the number of persons able to directly make their living from agriculture (Critchfield, however, virtually alone among students of the Green Revolution, denies that this is happening), the higher production attendant upon this modernisation could create more broadly a demand for labour in the rural towns and peripheral cities where the industries producing the inputs to this more capital intensive but still necessarily spatially dispersed agriculture would be located. Without a rapidly growing industrial sector in the large cities of the core region to attract population, the agrarian towns and peripheral cities could theoretically retain the surplus population released by the modernisation of agriculture (if it is indeed released) to a degree not observed in the West or in rapidly industrialising countries such as South Korea and Taiwan where the industrial sector in the large core cities is much more dynamic and rapidly growing than the rural sector, despite a thorough modernisation of the latter.

A perhaps more controversial prediction of Critchfield's model is that it will achieve a better fit in societies where resistance to birth control is low (or, which is essentially the same thing, where the relative status of women is high) and where there is a long tradition of settled agriculture whose peasantry is receptive to innovation (i.e. the countries of East and South Asia) and will achieve an increasingly poor fit in countries with higher resistance to birth control (i.e. Middle Eastern and Catholic countries) and more shallow traditions of settled agriculture (i.e. the countries of sub-Saharan Africa).

There is nothing inherently implausible in Critchfield's model, as summarised above. The question is whether anything resembling its predictions with respect to migration between core and peripheral regions, and the relative population growth rates of the two, can be detected on any significant scale in the Third World. Population statistics are the most reliable of any statistics produced in the Third World, if not elsewhere. An obvious first step is to employ these statistics to see if any trace of the phenomenon suggested by Critchfield's necessarily more limited observations can be detected in these statistics. In particular, one can ask if a 'premature' slackening in the growth of the primate cities of any significant number of countries has occurred during the 1970s and, more particularly, if it is occurring along the cultural and geographical gradient predicted by Critchfield. More generally, one can ask if the standard model of population deconcentration now prevalent in regional science and urban economics has the generality which seems to be claimed for it or if there are indeterminate 'regions' of the model where if the assumed path of economic development is not taken or, perhaps, is abandoned, then the path of spatial development is likewise indeterminate.

The large number of censuses taken around 1980 could provide a unique and exciting window on this phenomenon, if it has occurred. It is possible, however, that the phenomenon of 'premature' deconcentration occurred too late in the 1970s for the 1980 censuses to pick it up, though Critchfield's own observations suggest deconcentration beginning in the mid-1970s rather than in the late 1970s. His first publication reporting this phenomenon (Critchfield, 1979) was based on observations from the mid-1970s onwards. In any event, the information to be gained from the data presented here on the growth over the 1970s of large cities in the Third World can provide a baseline for future studies, even if the phenomenon of premature deconcentration did occur too late for detection by the 1980–81 round of censuses.

The data, which are nearly exhaustive of the important countries of the non-communist Third World, provide very little support for Critchfield's thesis. Pakistan, Bangladesh, India, Indonesia, the Philippines, and others all show either increasing or stable levels of net flows towards their largest cities and core regions (*also see* Batubara, 1982). Pakistan and Bangladesh show a particularly rapid increase in the rate of these net flows. The acceleration has not been as rapid in India, but there is no doubt that an accelerating trend is present there as well. Crook and Dyson (1982) provide some interesting data from the 1981 Indian census on recent trends in the growth of India's largest cities. Of India's 10 largest metropolitan areas, only three had significantly lower rates of growth in the 1971–81 period than in the preceding inter-censual period, 1961–71: Bombay, Madras, and Hyderabad. In the case of Bombay the addition of nearby, closely associated smaller agglomerations (e.g. Thane) would give this agglomeration a higher growth rate in the 1970s than in the 1960s. India's population growth rate overall, meanwhile, has remained stable. Only Sri Lanka in South Asia shows an absence of core-ward migration, but this is a country which has never in the post-war period conformed to the basic agglomeration model of the regional science school, so that it cannot be said to support Critchfield's claim that the 1970s marked a turning point in the migration of persons towards the largest cities of those countries of the Third World with long histories of settled agriculture but without the potential for large-scale industrialisation. Egypt alone, where the Cairo region is growing at a rate much below past projections, is consistent with Critchfield's thesis. Indeed, Critchfield has recently acknowledged that the data *so far* do not, in general, support his thesis (Critchfield, 1984).

Conclusion

The basic model of regional population redistribution now current in the regional science literature predicts that population concentrates in a country's most urbanised and dominant regions until a rather high level of economic development is reached (following Wheaton and Shishido, 1981, it was estimated here to be a little below $3000 in 1975 US dollars), whereupon a diminishment in this tendency takes place. Countries with per capita incomes much below $3000 rarely show declining rates of net inmigration to their core regions. Two of the three countries in the sample that do – Peru and Chile (Egypt is the third) – experienced quite significant declines in economic growth, and the standard model predicts just such an association between slower economic growth and slower coreward migration. Egypt, on the other hand, did not experience a decline in its rate of economic growth and may well be an instance of Critchfield's thesis. Two other countries – Cuba and Sri Lanka – are true exceptions. In neither has there been significant population redistribution towards the core regions over the entire post-war period, indicating that other paths of spatial development are possible. These two countries are at least consistent with the spirit of Critchfield's thesis. Both are self-consciously agrarian societies, with highly literate and educated populations and relatively low birth rates but with small industrial bases and low, faltering rates of industrialisation (in the case of Sri Lanka, *see* Hooley, 1983). Neither, however, seems to be the leading edge of future spatial developments in Third World countries, and their small sizes give them virtually no weight in the overall trend in large city growth rates in the Third World. The basic pattern first observed in Western countries, i.e. of a close association between economic development and population concentration, appears to be being followed in an entirely predictable fashion by non-Western countries.

Notes

1 Demographers, in studying the distribution of human populations, have largely concentrated on urbanization, i.e. the fraction of population living in urban areas, rather than, as is done here, the distribution of population across regions. Urbanization, however, has little meaning apart from that of modernization itself, i.e. the shift of population from the rural, agricultural sector to the urban, industrial and service sector. In fact, urbanization need not entail any regional or spatial shifts of population at all, if, for example, agricultural villages become urban (non agricultural) places *in situ*. Thus, the fact that urbanization accompanies economic development is not remarkable – the two are simply different words for the same phenomenon. A relationship between the regional or spatial concentration of population and economic development, on the other hand, is a non-tautological one and of greater interest – though some claim that regional concentration itself is due to nothing more than the uneven distribution of urbanization, i.e. to the fact that certain regions have a higher proportion of their population in urban areas at the advent of industrialization and that because these areas attract industrial investment and, therefore, rural migrants in proportion to their respective sizes, they cause their regions to grow faster than others. No demonstration of this interesting theorem has been offered, however, with the possible exception of Sweden (*see* the remarks in

Ahnstrom, 1982, p. 70). One reason for this lack of empirical demonstration is that it is difficult to distinguish between what is urban and what is not, in a modern or modernizing economy (for further discussion, *see* ESCAP, 1984, p. 21). The criteria of density and population size are arbitrary, and the economic definition returns us to the original problem: how to measure urbanization in such a way as not to make it tautologically identical to economic development.

2. The rate of net immigration from abroad to the core region should also be equal to the rate of net immigration to the nation as a whole for this relation to hold. This component of growth is ignored here, since it is insignificant as a component of growth in most of the countries in the sample. The exceptions include Cuba, where net immigration is negative and disproportionately concentrated in the Havana region, effectively cancelling out the positive net internal migration to this region (*see* Landstreet, 1976, p. 156, and the annual issues of *Anuario Estadístico de Cuba*), an important detail missed in Slater's (1982) otherwise comprehensive review of spatial policy in Cuba; Israel, where net immigration to the periphery has more than compensated for net internal migration to the core; Greece, where net immigration to and from Greece is probably disproportionately to and from its peripheral regions; and possibly Egypt, where net immigration is most probably also negative and concentrated in the capital city, Cairo, though there are no data to demonstrate this.

   Also, the natural increase in the core region due to net inmigration is counted as net inmigration to the core region rather than natural increase. In the rare case of net outmigration from the core region, the natural increase in the peripheral regions due to net outmigration from the core region is counted as net outmigration from the core region during the period in which it occurs.

3. There is a simple relationship between the change in a region's share of the national population and its growth rate relative to that of the nation (*see* United Nations, 1980, p. 34). *See* Williams (1983) for a clear and exhaustive treatment of the basic mathematics of spatial population redistribution.

4. Like the Ministry of Information (1983) and Leishman (1985), Farrell mistakenly characterizes Cairo's rate of growth as unchecked and its population as 14 million in 1983 (the latter cannot be much more than 11 million, if the 1976 census count of 9.2 million, which includes the population of a very large region surrounding Cairo, is accepted and if an extrapolation from this population is made using the region's 1966–76 average annual growth rate of 2.6 per cent) – Grove (1982, p. 441) and Pipes (1984) likewise report Cairo's population as being between 12 and 15 million.

5. The US GDP price deflator rose by 37 per cent over the period, 1970–75 (International Monetary Fund, 1981, p. 35). This gives a rough estimate of what the 1975 dollar is worth in 1970 dollars, an estimate which should be of sufficient precision for the purposes of this paper.

6. The overall impression given by the recent literature on regional (as well as personal) income differentials and the spatial distribution of economic activity is one of considerable confusion. Apparently, no consensus has emerged as to whether regional per capita incomes do converge at higher national income levels, as suggested in Williamson's (1965) seminal article, or as to whether regional income shares and the spatial distribution of economic activity stabilize after a certain point in the development of an economy (*see* Krebs 1982, for a review of the recent literature, also Saith 1983, for a critique of the convergence hypothesis with respect to the personal income distribution). The data on regional population shares, by contrast, unambiguously show a stabilization in these shares at the higher income levels (and in a few countries, like the USA, actual reductions in the dominant region's

share of population). One concludes either that demographic and economic trends become decoupled at some point in the economic development of a country or that the data on regional population shares are simply the more reliable indicators of the regional distribution of economic activity in a country.

7 It would appear that in countries which have been very rapidly propelled to or beyond Wheaton and Shishido's threshold of roughly $3000, such as Libya (for estimates of the latter's real gross domestic product, see Summers et al., 1980), there are substantial lags in the response of core-ward migration to level of economic development. In the case of Portugal, whose per capita real GDP also exceeded $3000 by 1980, there has been some decline in the rate of net migration to the core region, and further declines may be expected as its economy continues to grow.

References

Acosta-Romero, M. 1982: Mexican federalism: conception and reality. *Public Administration Review* 42, 399–404.

Ahnstrom, L. 1982: The concentration of a compound – the deconcentration of its parts: the economically active population of the Stockholm region, 1950–1957. *Geografiska Annaler* 64B, 69–75.

Alonso, W. 1968: Urban and regional imbalances. *Economic Development and Cultural Change* 17, 1–14.

Alonso, W. 1971: The economics of urban size. *Papers of the Regional Science Association* 26, 67–83.

Alonso, W. 1973: *National interregional demographic accounts: A prototype.* Monograph no. 17. Berkeley: Institute of Urban and Regional Development, University of California.

Arndt, H. 1984a: Little land and many people. *Far Eastern Economic Review* 1, 40–1.

Arndt, H. 1984b: Transmigration in Indonesia. Geneva: International Labour Office, Population and Labour Policies Programme (WEP 2-21/WPP. 146, UNFPA Project No. GLO/79/P83).

Attir, M. 1983: Libya's pattern of urbanisation. *Ekistics* 50, 157–62.

Aydalot, P. 1984a: Questions for regional economy. *Tijdschrift voor Economische en Sociale Geografie* 75, 4–13.

Aydalot, P. 1984b. Note sur les migrations interrégionales en France 1975–1982. Paris: Université de Paris I (Panthéon-Sorbonne), Cahier nr. 40, Dossiers du Centre Économie, Espace, Environnement.

Bairoch, P. 1982: Employment and large cities: problems and outlook. *International Labour Review* 121, 519–33.

Bannon, M. 1975: Ireland. In Jones R. (ed.), *Essays on world urbanisation.* London: George Phillip, 204–11.

Banyikwa, W. 1982: Binnewanderung in Tansania. *Geographische Berichte* 27, 113–24.

Batubara, C. 1982: SE Asia's urban problems. *Australian Planner* 20, 133.

Bedford, R. n.d.: Repopulation of the countryside. In Bedford, R. and Sturman, A. (eds), *Canterbury at the cross-roads: Issues for the eighties.* Geographical Society Miscellaneous Series no. 8. Christchurch: University of Canterbury, New Zealand, 277–306.

Bennison, D. 1972: Urban growth and migration in Greece since 1951. *Journal of the Durham Geographical Society* 14, 150–70.

Bertinotti de Petrei, N. 1980: Las migraciones en Argentina en el período 1970–80. *Estudios* (Cordoba, Argentina) 15 (3), 148–60.

Bertrand, J. 1978: Note sur les migrations intérieures récents en Espagne. *Géographie et Recherche* (Université de Dijon) 7, 3–19.
Bodgener, J. 1984: Cairo: making a start on urban renewal. *Middle East Economic Digest* 28 (16), 10–11, 13, 15.
Brown, L. 1976a: The urban prospect: re-examining the basic assumptions. *Population and Development Review* 2, 267–78.
Brown, L. 1976b. The limits to growth of Third World cities. *Futurist* 310, 307–15.
Bureau of the Census. 1983: *World population 1983: Recent demographic estimates for the countries and regions of the world.* Washington, DC: Government Printing Office.
Cantu-Segovia, E., Medina-Aguiar, J. and Basave-Benitez, A. 1982: The challenge of managing Mexico: the priorities of the 1982–1988 administration. *Public Administration Review* 42, 405–9.
Carlino, G. 1982: From centralisation to deconcentration: economic activity spreads out. *Business Review* (Federal Reserve Bank of Philadelphia), May–June, 15–25.
Casetti, E. 1982: Technological progress, capital deepening, and manufacturing productivity growth in the USA: a regional analysis. *Environment and Planning* A 14, 1577–85.
Caufield, C. 1984: Indonesia's great exodus. *New Scientist*, 17 May, 25–7.
Christiansen, R. 1984: The pattern of internal migration in response to structural change in the economy of Malawi 1969–1977. *Development and Change* 15, 195–218.
Clark, C. 1964–65: The location of industries and population. *Town Planning Review* 35, 195–218.
Commission of the European Communities. 1981: *The regions of Europe.* Brussels.
Critchfield, R. 1979: Revolution of the village. *Human Behaviour* 8, 18–27.
Critchfield, R. 1981: *Villages.* New York: Anchor Press/Doubleday.
Critchfield, R. 1984: Reassessing Asian villagers' return from the city. *Asian Wall Street Journal*, 5 May, 12.
Crook, N. and Dyson, T. 1982: Urbanisation in India: results of the 1981 census. *Population and Development Review* 8, 145–56.
Crossette, B. 1984: In Indonesia, a wilderness is now home. *New York Times*, 25 December, 28.
Diab, H. and Wahlin, L. 1983: The geography of education in Syria in 1882. With a translation of 'Education in Syria' by Shahin Makarius, 1883. *Geografiska Annaler* 65B, 105–28.
Eastwood, D, 1983: Venezuela: the 1980 census shows continued rapid population growth. *Geography* 68, 345–47.
ESCAP (Economic and Social Commission for Asia and the Pacific, Population Division, United Nations): 1983. *Population distribution and development policies in the ESCAP region.* Population Research Leads no. 13. Bangkok.
ESCAP. 1984: *Urbanisation in Thailand and its implications for the family planning programme.* Population Research Leads no. 17. Bangkok.
Economic Planning Board. 1975: *Statistical handbook of Korea 1975.* Seoul.
Elliot, S. and Lee, I. 1980: Guns and geography. *Geographical Magazine* 52, 640–6.
El-Shakhs, S. 1982: National and regional issues and policies in facing the challenges of the urban future. In Hauser, P., Gardner, R., Laquain, A. and El-Shakhs, S. (eds), *Population and the urban future.* Albany: State University of New York Press, 103–80.
El-Shakhs, S. 1983: Polarisation reversal and the future of super cities: policy and planning implications. Paper presented to the North-eastern Regional Science Association Meetings, New York.

Farrell, W. 1983: Cairo is crumbling under the burden of its people. *New York Times*, 29 May.
Fielding, A. 1982: Counterurbanisation in Western Europe. *Progress in Planning* 17, 1–52.
Francart, G. 1983: Le rééquilibrage démographique de la France. *Économie et Statistique* 153, 35–46, 85–9.
Gall, N. 1983: The rise and decline of industrial Japan. *Commentary* 76, 27–34.
Georgakis, G. and Tsiafetas, G. 1982: Stochastic projection of the population distribution in Greece. *European Demographic Information Bulletin* 13, 120–33.
Gordon, P. 1982: Spatial planning for the LDCs. *Socio-economic Planning Sciences* 16, 193–4.
Gradus, Y. 1983: The role of politics in regional inequality: the Israeli case. *Annals of the Association of American Geographers* 73, 388–403.
Greenwood, J. 1984: Regional planning in Venezuela: recent directions. *Third World Planning Review* 6, 239–54.
Grove, A. 1982; Egypt has too much water. *Geographical Magazine* 54, 437–41.
Haller, A. 1982: A socio-economic regionalisation of Brazil. *Geographical Review* 72, 450–64.
Hayuma, A. 1983: The growth of population and employment in Dar es Salaam city region, Tanzania. *Ekistics* 50, 255–9.
Helbock, R. 1975a: Differential urban growth and distance considerations in domestic migration flows in Pakistan. *Pakistan Development Review* 14, 53–84.
Helbock, R. 1975b: Urban population growth in Pakistan: 1961–1972. *Pakistan Development Review* 14, 315–33.
Higgins, B. 1978: *Review of Economic growth and structure in the Republic of Korea*, by P. Kuznets. *Journal of Economic Literature* 16, 609–11.
Hooley, R. 1983: Recent economic trends in Sri Lanka. *Journal of Southeast Asian Studies* 14, 262–65.
Horner, A. 1974: Future population change in the Dublin region. *Irish Geography* 7, 120–6.
Horner, A. and Daultrey, S. 1980: Recent population changes in the Republic of Ireland. *Area* 12, 129–35.
Houston, C. 1979: Administrative control of migration to Moscow, 1959–1975. *Canadian Geographer* 23, 32–44.
Huff, D. and Lutz, J. 1979: Ireland's urban system. *Economic Geography* 55, 196–212.
Hugo, G. 1979: Indonesia: migration to and from Jakarta. In Pryor, R. (ed.), *Migration and development in Southeast Asia: A demographic perspective*. Kuala Lumpur: Oxford University Press, 192–203.
Hulten, D. and Schwab, R. 1984: Regional productivity growth in US manufacturing: 1951–1978. *American Economic Review* 74, 152–62.
Ibrahim, S. 1983: *Internal migration in Egypt: A critical review*. Monograph Series no. 5. Cairo: Supreme Council for Population and Family Planning, research office, Population and Family Planning Board.
Illeris, S, 1984: Danish regional development during economic crisis. *Geografisk Tidsskrift* 84, 53-62.
International Monetary Fund. 1981: *International financial statistics, supplement on price statistics*. Supplement Series no. 2. Washington, DC.
Isenman, P. 1980: Basic needs: the case of Sri Lanka. *World Development* 8, 237–58.
Itakura, K. 1982: On the structure of economy and society in Tohoku. *Science Reports of Tohoku University, 7th Series (Geography)* 32, 71–87.
Jefferson, M. 1939: The law of the primate city. *Geographical Review* 37, 461–85.

Jones, R. 1982: Regional income inequalities and government investment in Venezuela. *Journal of Developing Areas* 16, 373–90.
Jowett, A. 1984: China: land of the thousand million. *Geography* 69, 252–7.
Kaldor, N. 1975: What is wrong with economic theory? *Quarterly Journal of Economics* 89, 347–57.
Keane, M. 1984: Accessibility and urban growth rates for the Irish urban system. *Economic and Social Review* 15, 125–39.
Keeble, D., Owens, P. and Thompson, C. 1982: EEC regional disparities and trends in the 1970s. *Built Environment* 7, 154–61.
Khan, A. 1982: Rural–urban migration and urbanisation in Bangladesh. *Geographical Review* 72, 379–94.
Kiljunen, K. (ed). 1984: *Kampuchea: Decade of the genocide.* London: Zed Press.
Kirk, W. 1981: Cores and peripheries: the problems of regional inequality in the development of Southern Asia. *Geography* 66, 188–210.
Krebs, G. 1982: Regional inequalities during the process of national economic development: a critical approach. *Geoforum* 13, 71–81.
Krishnan, P. and Rowe, C. 1978: Internal migration in Bangladesh. *Rural Demography* 5, 1–21.
Landstreet, B. 1976: Cuban population issues in historical and comparative perspective. Ph.D. dissertation, Cornell University.
Landstreet, B. and Mundigo, A. 1983: Migraciones internas y cambios en las tendencias de urbanización en Cuba. *Demografia y Económica* 17, 409–47.
Lang, M. and Kolb, B. 1980: Locational components of urban and regional public policy in post-war Vietnam: the case of Ho Chi Minh City (Saigon). *GeoJournal* 4, 13–18.
Leishman, K. 1985: Egypt: the future of the past. *Atlantic* 255 (1), 21–7.
Levine, N. 1980: Antiurbanization an implicit development policy in Turkey. *Journal of Developing Areas* 14, 513–38.
Llosa, R. 1982: Social, economic and demographic factors relating to interregional migration in the Philippines: 1970–1980. *IIPS Newsletter* (International Institute for Population Studies, Bombay) 23 (4), 1–10.
Long, J. 1981: *Population deconcentration in the United States.* Washington, DC: US Bureau of the Census.
Mahar, D. 1984: Population redistribution within LDCs. *Finance and Development and Cultural Change* 21, 309–24.
Makarius, S. 1883: Education in Syria. In Diab, H. and Wahlin, L. 1983: The geography of education in Syria in 1882. *Geografiska Annaler* 65B, 105–28.
Mayer, J. 1984: Regional employment development: the evolution of theory and practice. *International Labour Review* 123, 17–34.
Mera, K. 1973: On the urban agglomeration and economic efficiency. *Economic Development and Cultural Change* 21, 309–24.
Mera, K. 1977: The changing pattern of population distribution in Japan and its implications for developing countries. *Habitat International* 2, 455–79.
Mera, K. 1978: Population concentration and regional income disparities: a comparative analysis of Japan and Korea. In Hansen, N. (ed.), *Human settlement systems.* Cambridge, Mass.: Ballinger, 155–75.
Mera, K. 1986: Population stabilisation and national spatial policy of public investment: the Japanese experience. *International Regional Science Review* 10, 147–65
Ministry of Information (Egypt). 1983: *Egypt: Facts and figures.* Cairo.
Moore, M. 1984: Categorising space: urban–rural or core–periphery in Sri Lanka. *Journal of Development Studies* 3 (20), 102–22.

Morcos, W. 1980: Trends and patterns of urbanisation in Egypt during the 1966–1976 intercensual period. *Population Studies (Egypt)* 52, 1–17.
Moseley, M. 1984: The revival of rural areas in advanced economies: a review of some causes and consequences. *GeoJournal* 15, 447–56.
Munro, D. 1981: Problems on the way to Ecuador's democracy. *Geographical Magazine* 53, 782–86.
Murphey, R. 1978: *Patterns on the earth: An introduction to geography*, 4th edition. Chicago: Rand McNally.
Myklebost, H. 1984: The evidence for urban turnaround in Norway. *Geoforum* 15, 167–76.
Myrdal, C. 1957: *Rich lands and poor.* New York: Harper.
Nag, P. 1981: Population redistribution: aspects of Zambian national development. *Geographical Review of India (Calcutta)* 43, 41–9.
Nelson, J. 1983: Population redistribution policies and migrants' choices. In Morrison, P. (ed.), *Population movement: Their forms and functions in urbanisation and development*. Liège: International Union for the Scientific Study of Population, 281–312.
O'Connell, M. 1981: Regional fertility patterns in the United States: convergence or divergence. *International Regional Science Review* 6, 1–14.
Pipes, D. 1984: Under construction. *New Republic*, 17 December, 43.
Plaut, S. 1983: The economics of population dispersal. *Urban Studies* 20, 353–7.
Pommier, P. 1982: The place of Mexico City in the nation's growth: employment trends and policies. *International Labour Review* 121, 345–60.
Ponchaud, F. 1978: *Cambodia: Year zero.* New York: Holt, Rinehart, & Winston.
Rashed, M. and Nawas, F. 1977: Patterns of population change in Libya: growing and declining places, 1964–1973. *Libyan Journal of Agriculture* 6, 93–100.
Richardson, H. 1980: Polarization reversal in developing countries. *Papers of the Regional Science Association* 45, 67–85 (*see* this volume, Chapter 11).
Richardson, H. 1983: Population distribution policies. *Population Bulletin of the United Nations* 15, 35–49.
Richardson, H. 1984: Planning strategies and policies for metropolitan Lima. *Third World Planning Review* 6, 123–38.
Robert, S. and Randolph, W. 1983: Beyond decentralisation: the evolution of population distribution in England and Wales, 1961–1981. *Geoforum* 14, 75–102.
Rondinelli, D. 1982: Intermediate cities in developing countries. *Third World Planning Review* 4, 357–86.
Rondinelli, D. 1983: Dynamics of growth of secondary cities in developing countries. *Geographical Review* 73, 42–57.
Rowe, W. 1984: Urban policy of China. *Problems of Communism* 33 (6), 75–80.
Rowland, R. 1983: The growth of large cities in the USSR: policies and trends, 1959–1979. *Urban Geography* 4, 258–79.
Sachs, I. 1980: Cities and resources. *Social Science Information* 19, 673–84.
Saith, A. 1983: Development and distribution: a critique of the cross-country U-hypothesis. *Journal of Development Economics* 13, 367–82.
Scholz, F. 1983: Urbanisation in the Third World: the case of Pakistan. *Applied Geography and Development (Tübingen, FDR)* 21, 7–34.
Sen, A. 1981: Public action and the quality of life in developing countries. *Oxford Bulletin of Economics and Statistics* 43, 287–319.
Shawcross, W. 1984: Cambodia's burial. *New York Review of Books*, 10 May, 16–20.
Shryock, H., Siegel, J. and associates. 1975: *The methods and materials of demography*, 2, third printing (revised), US Bureau of the Census. Washington, DC: GPO.

Simon, J. 1981: *The ultimate resources*. Princeton, NJ: Princeton University Press.
Slater D. 1982: State and territory in postrevolutionary Cuba: some critical reflections on the development of spatial policy. *International Journal of Urban and Regional Research* 6, 1–34.
Smith, W., Huh, W. and Demko, G. 1983: Population concentration in an urban system: Korea 1949–1980. *Urban Geography* 4, 63–79.
Soen, D. 1977: Israel's population dispersal plans and their implementation, 1949–1974: failure or success? *GeoJournal* 1, 21–26.
Spate, O. 1942: Factors in the development of capital cities. *Geographical Review* 32, 622–31.
Sternstein, L. 1984: The growth of the population of the world's pre-eminent 'primate city': Bangkok at its bicentenary. *Journal of Southeast Asian Studies* 15, 43–68.
Stinner, M. and Bacol-Montilla, M. 1981: Population deconcentration in metropolitan Manila in the twentieth century. *Journal of Developing Areas* 16, 3–16.
Storper, M. 1984: Who benefits from industrial decentralisation? Social power in the labour market, income distribution and spatial policy in Brazil. *Regional Studies* 18, 143–64.
Storper, M. 1985: Response to Professor Vining's 'Query'. *Regional Studies* 19, 164.
Summers, R. and Heston, A. 1984: Improved international comparisons of real product and its composition: 1950–1980. *Review of Income and Wealth* 30, 207–62.
Summers, R., Kravis, I. and Heston, A. 1980: International comparison of real product and its composition: 1950–1977. *Review of Income and Wealth* 26, 19–66.
Townroe, R. and Keen, D. 1984: Polarisation reversal in the state of São Paulo, Brazil. *Regional Studies* 18, 45–54 (*see* this volume, Chapter 13).
Tucker, W. 1981: A better life in the world's villages. *Wall Street Journal*, 31 July, 18.
Ullman, E. 1958: Regional development and the geography of concentration. *Papers and Proceedings of the Regional Science Association* 4, 179–202.
UNCRD (United Nations Centre for Regional Development [NAGOYA]).1983: *UNCRD Newsletter* 15, Jan.
United Nations. 1980: *Patterns of urban and rural population growth*. Population Studies, no. 68. United Nations, New York: Department of International Economic and Social Affairs.
Vacca, R. 1973: *The coming dark age*. Garden City, NY: Doubleday.
Vining, D. 1982a: Migration between the core and the periphery. *Scientific American* 247, 44–53.
Vining, D. 1982b: On a catastrophe model of regional dynamics. *Annals of the Association of American Geographers* 72, 554–55.
Vining, D. 1985: Industrial decentralisation in Brazil: a qeuery. *Regional Studies* 19, 163–4.
Vining, D. and Kontuly, T. 1978: Population dispersal from major metropolitan regions: an international comparison. *International Regional Science Review* 3, 49–73 (*see* this volume, Chapter 6).
Vining, D. and Pallone, R. 1982: Migration between core and peripheral regions: a description and tentative explanation of the patterns in 22 countries. *Geoforum* 13, 339–410.
Vining, D., Pallone, R. and Plane, D. 1981: Recent migration patterns in the developed world: a clarification of some differences between our and IIASA's findings. *Environment and Planning A* 13, 243–50.
Wang, I. 1977: The recent pattern of internal migration in Taiwan. *China Geographer* 6, 37–47.

Weber, A. 1899: *The growth of cities in the nineteenth century: A study in statistics.* Cornell Reprints in Urban Studies. Ithaca: Cornell University Press, 1963.

Wheaton, W. C. and Shishido, H. 1981: Urban concentration, agglomeration economies, and the level of economic development. *Economic Development and Cultural Change* 30, 17–30.

Wiegersma, N. 1983: Regional differences in socialist transformation in Vietnam. *Economic Forum* 14, 95–109.

Wilkie, R. 1981: The populations of Mexico and Argentina in 1980: preliminary data and some comparisons. In Wilkie, J. W. and Haber S. (eds), *Statistical abstract of Latin America.* Los Angeles: UCLA Latin American Centre Publications, University of California, 21, 637–54.

Williams, L. 1983: The urbanisation process: toward a paradigm of population redistribution. *Urban Geography* 4, 122–37.

Williams, L. and Griffin, E. 1978: Rural and small-town depopulation in Colombia. *Geographical Review* 68, 13–30.

Williamson, J. 1965: Regional inequality and the process of national development: a description of the patterns. *Economic Development and Cultural Change* 13 (Part 2), 3–84.

SECTION TWO
POLARIZATION REVERSAL TRENDS

13   P. M. Townroe and D. Keen,
'Polarisation Reversal in the State of São Paulo, Brazil'

From: *Regional Studies* 18, 45–54 (1984)

Introduction

'Polarisation reversal' is the term given to the turning point in the spatial pattern of growth and development in a nation when continuing relative concentration ceases and urban deconcentration or spatial decentralisation commences. The term was first used in 1977 (Richardson, 1977), although the phenomenon has been of interest for many years. It is linked to the ideas of 'counterurbanisation' and regional dispersal remarked upon and studied in the United States and other developed countries (e.g. Berry and Gillard, 1977; Vining and Kontuly, 1978; Hall and Hay, 1980). However, the particular usage of the term 'polarisation reversal' (PR) hitherto has been in the context of less developed nations. The focus of this paper is on those LDCs with fast growth rates of population which are reaching a transitional stage in their pattern of economic development. Our study of São Paulo State, Brazil, finds PR in the last decade. This may well be the first documented example of polarisation reversal in an LDC.

The paper briefly discusses why a PR turning point may be anticipated as the economy of a nation grows and the urban system develops. We then review the relatively few studies which have attempted to measure the PR turning point. Our survey of these studies finds that polarisation reversal is an ambiguous term. There is no agreed definition. Out of our examination of possible indices of PR and of alternative definitions of the core and the periphery emerges a new and narrower definition of PR. This definition and the earlier measures are then considered in the context of the State of São Paulo in Brazil. We find that the State of São Paulo underwent increasing polarisation of urban population from 1950 through to 1970, but that a PR turning point is evident in the 1970s. The paper provides a demonstration of polarisation reversal but does not seek to empirically explain the behaviour of the actors involved.

The onset of polarisation reversal

For initial simplicity, we may think of an economy which is effectively a single city region, with a large primate city and a surrounding scattering of smaller cities stretching away into a rural hinterland. We may then use as a working definition of PR that provided by Richardson: the point at which the growth rate of the secondary cities located outside the core comes to exceed that of the primate metropolitan centre (Richardson, 1977). This definition of PR allows a continuing urban concentration in absolute terms where the population of the primate city is greater than that of the secondary cities combined; but it has the virtue of distinguishing the *process* of polarisation from the *state* of primacy. This distinction is a cornerstone of this paper and is the prime reason for conducting the discussion of the urbanisation process in terms of PR rather than (more traditionally) solely in terms of relative degrees of primacy. Further definitions of PR will be discussed in the next section of the paper.

One way to think about the reasons behind relative urban deconcentration in an economy is to look for the dual of those factors encouraging changes towards greater urban concentration or polarisation (Friedmann, 1975). However, the interest here is in the turning point between the two sets of pressures. This turning point may be brought about by increases or decreases in the intensity of certain social and economic forces initially promoting concentration. At a given level, the interaction between these forces and the forces promoting deconcentration is such as to move the configuration of changes in the urban system from one direction to another. This is PR. Alternatively, new forces enter the system. The underlying issue is that of changes in the relative advantages of different locations for the performance of economic activity. Indeed, there is a close link here to the discussions on the role and evolution of secondary city growth centres (e.g. Hansen, 1980; Gaile, 1980).

The forces working on the pattern of urban development in a nation include the growth of material output and an associated increase in the complexity of the national economy. Demographic transition, degrees of social and economic inequality, patterns of rural development, and the institutions and social pathways for the diffusion of information and innovations may all work to increase or decrease concentration in the distribution of urban population. One group of factors works within the dominant city, the other within the hinterland.

Within the dominant metropolis, continuing growth may bring about increases in congestion, in crime and pollution, and in infrastructure deficiencies. Or the marginal costs of infrastructure and services may rise. Producer decentralisation is thus encouraged; the inmigration of consumers and workers is discouraged, thus bringing on PR (Ternent, 1976). These negative externalities reinforce the pressure of rising land values in the central city, values which tend to rise faster than prices in general. And as the economy develops, land-intensive activities (e.g. in the service sector) will tend to encourage an outward movement of land-extensive activities (many sectors in manufacturing).

The pressures in the labour market work somewhat differently. If metropolitan growth results in residential environmental deterioration and in higher taxes and longer journeys to work, then would-be inmigrants are deterred and existing residents become prospective outmigrants, unless compensation is available in the form of higher rewards from employment. Employers may be able to pay these higher rewards if they themselves are benefiting from metropolitan externalities and economies of agglomeration; or if they are in a dominant market position to pass forward the higher costs to the consumer. As the city grows and the economy develops towards European or North American levels, the office sectors, particularly government, may begin to outbid manufacturing for labour. Employers not able to pay the higher rewards are encouraged to move to the suburbs or to a secondary city. In labour-abundant countries these pressures are reduced if a steady flow of poor immigrants with low expectation hold down the bidding price. This is where issues of income distribution may become important, because the monetary/non-monetary trade-off in the migration decision of the poor inmigrant will take a very different form than that of the affluent skilled or professional prospective outmigrant.

The advantages, for a resident or for an existing company or for new investment, of remaining in the city as it grows in size may cease to rise in an absolute sense. Absolute advantages for inmigrants to the city may also stabilise as the city increases in population. However, for a resident or an inmigrant or a firm to react to a slowdown or decline in advantages, the *possibility* of an alternative location must exist. This is where we bring in a second set of factors influencing PR. The possibility of an alternative location for the resident will normally mean the guarantee or prospect of employment in a secondary city, access to public services of some national standard, and not more than a small fall in the real value of his money income. Similarly, for the company, outward movement from the metropolis will normally be regarded as possible only if the centres for deconcentration have relevant infrastructures, services and communications networks. For the resident, the migrant, the existing company or the new investment, location or relocation to a secondary city therefore rests on prerequisites before a calculation of balance of advantage is undertaken.

Prerequisites are not absolutes in these forms of locational behaviour, but they form a base upon which locational choices are made by trading off the *relative* advantages of the central metropolis against those of one or more of the secondary cities. If the information is available, a company for example will compare relative land values, transport costs and wage levels just as the resident will compare relative house prices, journey-to-work costs and income levels. Changes in the *aggregate* pattern of choices will induce PR.

There are also direct policy influences at work. Policies which protect or promote a designated sector of industry implicity protect or promote those urban centres in which the industry is located. If the industry is spatially concentrated the subsidy or tariff will have a spatially biased result. Location policies for state-owned enterprise can have a similar impact on urban development patterns. Those policies which seek to promote metropolitan decentralisation and/or lagging region development will normally have PR, or an acceleration of the reversal once started, as a prime strategic objective.

A further policy change which may bring PR closer is any improvement in the fiscal standing of municipal authorities in the secondary cities. At early states of development, the major industrial metropolis typically receives a net fiscal transfer per head and per unit of output from the rest of the economy. Political clout and rising public service costs may maintain this differential as the city grows. If local tax revenues are dependent upon property values or on industrial output and powers to borrow are limited, the public authorities in a secondary city have little ability to foster local industry or attract in new investment. They are dependent upon a centralised approach to decentralisation (Friedmann, 1975). And yet, the least primate urban systems around the world generally occur in countries with a federal, decentralised system of government (Henderson, 1982).

The political concerns with PR come from several directions simultaneously. They come from a desire to promote development in lagging regions and the rural hinterland; or from worries about economic and social costs of a city or region which appears to be getting too large or too dominant or which is growing too fast. The concerns are with the balance of private and social costs and benefits and with issues of equity as well as efficiency. The quality of economic understanding underlying these concerns has frequently been questioned (e.g. Alonso, 1968; Richardson, 1977; Renaud, 1981; Townroe, 1979). Nevertheless the politician will be interested in the timing of PR; and then whether and in what ways PR should be encouraged and fostered. Answers to these questions in any one nation leads us to consider alternative definitions and measures of PR.

The measurement of polarisation reversal

Some problems of definition

In offering a definition of PR based upon relative urban growth rates of population, Richardson (1977) viewed the onset of PR as the commencement of a persistent tendency for a few secondary cities to grow faster than the primate metropolis. This is but one of the many measures of PR used in the literature however. For example, Richardson also suggests the spatial concentration of industry as a possible PR measure. Renaud (1977) analysed population growth and migration in his study of PR in Korea; but he also considered the changing spatial concentration of gross regional product and of regional per capita incomes. Linn (1979) uses gross regional product, income, and investment statistics in his examination of PR in Colombia. A wide range of indices is listed in Table 13.1.

The different measures employed in the studies of urban concentration introduce considerable ambiguity to the use of the term polarisation reversal. For example, does the phenomenon refer to the concurrent reversal in the trends of spatial concentration of population, industry, and income, or may one turning point be out of phase with the others? At what point does PR begin if industrial deconcentration leads deconcentration of population?

Although the spatial patterns of industry, income, and other regional characteristics are of interest in their own right, this paper uses population as the sole measure of PR. This decision is based on operational considerations as well as on the strong policy concerns with the population size of primate cities. Population estimates are often available on an annual basis, while they are comprehensive over space, and are reported with a high level of spatial disaggregation. Employment and income data frequently do not have these advantages. Furthermore, differences in production technology and productivity across cities could cloud analysis of employment trends, while a movement of population can offer direct evidence of changing economic advantages over space.

The use of population as the measure of PR does not eliminate operational difficulties. Population growth trends are particularly sensitive to definitions of the geographic boundaries of the areas under analysis. The minimum boundary of the core region for a small country would include both the central city and its suburbs, and secondary centres within the dominant commuting shed. Underbounding of this area could lead to relatively rapid growth of the suburbs being interpreted as PR. The problem of changes in any geographical definitions of the metropolitan area that occur over time may best be handled by assuming one definition of metropolitan area boundaries for all time periods. This definition would be based on the boundaries that correspond to the most recent population estimates.

Further problems emerge once we move away from our small primate country example. If our primary concern is with the processes of spatial development which lead to the acceleration and then deceleration of the growth of large metropolitan areas, and we have a large country with several major cities, then there are two alternatives for measurement. One is to follow the approach used by Barat and Geiger (1973) in Brazil and to study each major centre with its hinterland independently of the other regions. The results will make possible a generalisation as to whether or not PR has taken place within city regions. The other is to regard the various dominant cities as collectively forming the core, and then make comparisons with growth in the periphery summed across the city regions. However, the experience of industrial and economic growth is so different between the major Brazilian cities that a collective approach would mask important variations. The collective approach would not cover the complete nation (e.g. Amazonia and the rural western states). We follow the former approach in this paper, focusing on a single large city region .

A further analytical issue is whether to study the total population of areas or regions or only the urban population. This is critical in order to distinguish between a study of urbanisation, or 'counterurbanisation' (Berry, 1978), and PR. Studying the growth of total population of a highly urbanised core region and a predominantly rural periphery is very close to just studying urban versus rural population growth (Falk, 1978). Analysis using urban population eliminates the danger of losing sight of vigorous growth of secondary cities in a periphery dominated by rural decline.

Polarization Reversal Trends 193

Table 13.1 Indices of relative geographical concentration

No.		
1.	Share of population in core (primacy ratio)[a]	$P_c / \Sigma P$
2.	Rate of change of share of population in core	$(P_{c,t+1} / \Sigma P_{t+1}) - (P_{c,t} / \Sigma P_t)$
3.	Share of population growth captured by core[b]	$(P_{c,t+1} - P_{c,t}) / (\Sigma P_{t+1} - \Sigma P_t)$
4.	Difference in average annual population[c] growth between core and periphery	growth rate P_c − growth rate P_p
5.	Difference in absolute population growth between core and periphery	$(P_{c,t+1} - P_{c,t}) - (P_{p,t+1} - P_{p,t})$
6.	Comparison of average annual population[d] growth rates between core and secondary city size classes	growth rate $P_{size\,j}$ − growth rate P_c
7.	Number of secondary cities growing[e] faster than the core	Number of cities for which growth rate P_j > growth rate P_c
8.	Differences in average annual growth[f] rates between the core and the total core plus three next largest cities	(growth rate P_c) − (growth rate $P_c + P_2 + P_3 + P_4$)
9.	Four-city primacy ratio[g]	$P_c / (P_c + P_2 + P_3 + P_4)$
10.	Ten-city primacy ratio[h]	$P_c / (P_c + P_2 ... + P_{10})$
11.	Urban deconcentration index (UD)[i]	UD = $[\sum_{i=1}^{n} (P_j / \Sigma P)^2] - 1$

[a] Used by Richardson (1977), Mera (1978) and Linn (1979)
[b] Used by Mera (1978) and Linn (1979)
[c] Used by Dillinger (1978)
[d] Modified from Linn (1979) and Falk (1978)
[e] Modified from Richardson (1980)
[f] Used by Dillinger (1978)
[g] Used by Owen and Witton (1973) and by Renaud (1981)
[h] Used by Linn (1979)
[i] Used by Henderson (1980) and Wheaton and Shishido (1981)

Notation:
Average annual growth rate = $(P_{t+n} / P_t)^{1/n} - 1$
$P_{c,t}$ = urban population in core in year t
$P_{p,t+n}$ = urban population in periphery in year $t + n$
$\Sigma P = P_c + P_p$
P_j = urban population of i th largest city
$P_{size\,j}$ = urban population of city size class j

Note that not all indices originally used *urban* populations

Indices of reversal

Once these preliminary definitional issues have been settled, there remains the question of how to measure the onset of polarisation reversal. Studies of urban concentration and deconcentration have used a number of different summary indices. Those from the major studies are listed in Table 13.1. There are three groups of indices shown in this table. Indices 1–5 all compare the core region with all the periphery cities as a group. These indices change as the relationship between the aggregate urban populations of the core and the periphery changes. Indices 6–10 compare the core with subsets of the cities in the periphery. Finally, the urban deconcentration index (UD index) compares all cities with one another. This index will change if one secondary city grows relative to another, even without a change in the aggregate urban population of the core and the periphery.

These indices can yield contradictory results. To illustrate possible inconsistencies, imagine a nation in which the core region is growing faster than the

194 Differential Urbanization

largest cities in the periphery and the cities in the periphery as a whole but slower than small cities in the periphery. It is not clear how indices 2–5 and index 10 would behave in this example. Indices 1, 8 and 9 would increase and suggest further polarisation. Index 6 would show that one size class of periphery cities was growing faster than the core region and index 7 would indeed show a persistent tendency for some secondary cities to grow faster than the core. The resulting increase in the UD index would also suggest PR. Thus, the answer to the PR question depends upon the index used as much as the underlying spatial trends.

The way in which each index relates to PR becomes clearer if we introduce a simple model of polarisation. The bell-shaped curve of geographical concentration (Alonso, 1980) shown in Figure 13.1, serves as the basis of this stylised model.[1] As shown by the solid curve, polarisation, defined as the concentration of the urban population in the metropolitan core, increases over time until reaching a peak. After reaching this peak, polarisation decreases over time. Alonso (1980) cautions that very little is known about the right-hand side of the bell. Polarisation may just reach a certain level and stabilise, or stabilise only when all the urban population is in the core region. Two additional concentration functions may be added to Alonso's original model in order to identify specific states of polarisation. The dashed curve in Figure 13.1 represents the rate of change of concentration and the dotted curve portrays the share of population growth captured by the core region over time. The rate of change of polarisation peaks at point A corresponding to the inflection point in the absolute concentration curve. At point B, the proportion of growth going to the core region equals the aggregate share of population in the core. At this point, change in polarisation is zero.

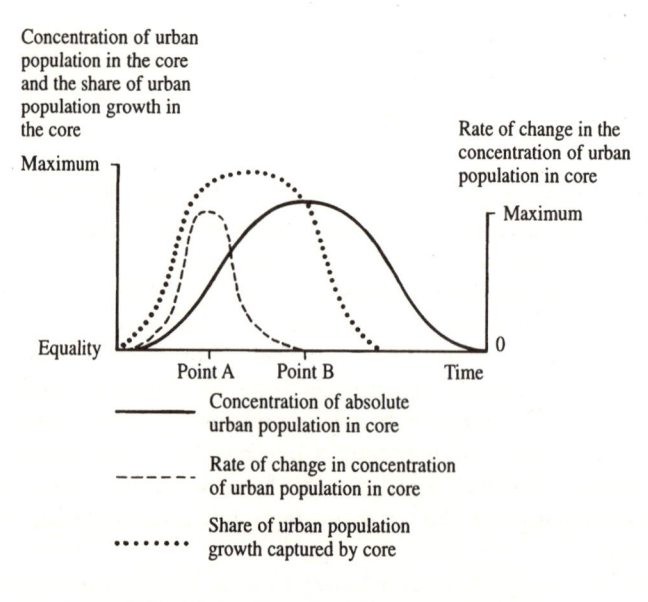

Fig. 13.1 A model of polarisation

From this model we can now suggest a definition of PR and some related terms, and discuss the appropriate indices for measuring PR. Assuming that these curves properly reflect the underlying model of other research workers who have considered urban concentration, polarisation reversal may be defined as point B in Figure 13.1 – *the point at which the concentration of the urban population in the core begins to decrease.* This is measured in index 1, the primacy ratio. It may also be defined by the point at which index 3, the share of growth captured by the core, equals the value of index 1, or by the point at which index 4, the difference in growth rates between the core and periphery, equals zero. If concentration stabilises only after point B, this would not be PR but may be defined as *polarisation stabilisation*. This can also be measured by indices 3 and 4. Point A in Figure 13.1 may be defined as the point of *reversal in the rate of polarisation*. Index 2 increased before this point and decreases after this point. We must caution, however, that the model in Figure 13.1 is a very stylised representation of population concentration. There may be many points of reversal in the rate of polarisation over time as the rate of concentration increases and decreases with economic growth and development. There is therefore the possibility of two or more points of polarisation reversal in the rate of polarisation and even reversal of PR.

The remaining indices cannot be used to define the PR turning point, but may be very useful to determining the underlying spatial changes related to PR. Comparison of growth rates of different city size classes over time, for example, will identify the types of cities bringing about PR.

Polarisation reversal in the state of São Paulo, Brazil

Development of the State of São Paulo

Use of a single state of a country as the area of analysis for PR would not be recommended for most countries since the state boundaries would not encompass the relevant city region. The well-developed city system and the size of the State of São Paulo in southeast Brazil make it an exception of this rule. The state is large. It has 25 million residents, about the same population as Canada, Yugoslavia or Colombia. It is also physically large, approximately the same size as the United Kingdom or West Germany. The State of São Paulo holds the largest industrial concentration in South America and has almost one-half of the industrial employment of Brazil. The state is also a rich agricultural region with coffee and sugar cane as major crops. Adjacent metropolitan areas are well distant from the dominant centre in the state. Rio de Janeiro is 430 kilometres distant, Belo Horizonte is 586 kilometres, and Curitiba is 408 kilometres away. Metropolitan São Paulo and the city system of the state can be regarded as one large city region.

With a population of 12.6 million in 1980, metropolitan São Paulo is the second largest LDC city in the world after Mexico City. The metropolitan area covers 6000 square kilometres, with the City of São Paulo as its centre, being the capital of the state. The 1980 Demographic Census shows 9 million people

196 Differential Urbanization

Table 13.2 Total and urban population for areas of the State of São Paulo, 1950–80 (millions)

Areas within the state[a]	Total population				Urban population[b]			
	1950	1960	1970	1980	1950	1960	1970	1980
Metropolitan São Paulo	2.7	4.8	8.1	12.6	2.3	4.5	7.9	12.3
City of São Paulo	2.2	3.8	6.2	9.0	2.1	3.7	6.2	8.9
Built-up fringe	0.3	0.7	1.4	2.6	0.2	0.6	1.4	2.5
Outer suburbs	0.2	0.3	0.5	1.0	0.1	0.1	0.4	0.9
Hinterland	6.5	8.2	9.6	12.4	1.7	3.0	4.8	7.6
Larger city inner region	0.9	1.4	2.2	3.6	0.6	1.1	2.0	3.4
Other inner region	1.1	1.3	1.6	2.2	0.3	0.4	0.7	1.2
Outer region	4.5	5.5	5.8	6.6	0.8	1.4	2.1	3.0
State of São Paulo	9.1	13.0	17.8	25.0	4.0	7.5	12.1	19.9
Brazil	51.9	70.0	93.2	119.0	13.0	22.4	37.7	58.1

Sources: Fundacão IBGE. *Censo Demográfico* 1950, 1970; *Censo Preliminar*, 1960, 1980; Fox (1975)
[a] Areas have constant boundaries corresponding to boundaries existing in 1950. Metropolitan São Paulo as defined by Davidovich and Lima (1975). 'Inner region' includes all areas within 150 kilometres of the metropolitan area plus all of the Paraiba Valley. 'Large city' refers to nine urban agglomerations
[b] Urban population is urban population as defined by the Brazilian Demographic Census for *municipios* with 20 000 or more inhabitants in 1970 or cities within urban agglomerations

living in the central city (as defined by 1950 boundaries) and 3.6 million residents living in the suburbs. Population trends are further detailed in Table 13.2.

The hinterland of the state extends 600 kilometres to the northwest of metropolitan São Paulo. Nine urban agglomerations within 150 kilometres of the metropolitan area account for 3.6 million of the state's population, as shown in Table 13.2. The area outside this inner region encompasses most of the rural population of the state. Cities in this region are distributed in a well-developed central place pattern. The hinterland cities are served by an extensive modern highway network radiating from metropolitan São Paulo. The entire state has undergone rapid improvements in infrastructure since 1960 (Keen, 1982).

Population growth rates for the cities of São Paulo State are shown in Table 13.3. Growth rates of inner-region cities have remained steady since 1950 while the growth of the metropolitan area has declined to the point where the urban populations of both large and the smaller cities in the inner region grew at a faster rate than the metropolitan area overall. The core region in the subsequent calculations has been taken as the metropolitan area. The boundaries of this area are drawn generously, with clear geographical barriers to the north and south and limited across boundary commuting to east and west. There were insufficient data on consistent boundaries over the 30-year period to test the consistency of the PR conclusion to this definition of the core region. The metropolitan area, as defined, is also the focus of related planning policies. Local political and professional concern has been expressed in the past on the relative rate of population growth of the metropolitan area, although this concern has not been translated into an active decentralisation policy.

Table 13.3 Average annual growth rates of total and urban population for areas of the State of São Paulo, 1950–80

Areas within the state[a]	Total population (%)			Urban population (%)[b]		
	1950–60	1960–70	1970–80	1950–60	1960–70	1970–80
Metropolitan São Paulo	6.1	5.4	4.4	6.8	5.8	4.6
City of São Paulo	5.7	5.0	3.7	6.1	5.1	3.8
Built-up fringe	9.3	7.6	6.4	10.8	8.1	6.2
Outer suburbs	4.4	6.1	6.7	9.2	9.7	9.6
Hinterland	2.4	1.6	2.6	5.8	4.8	4.8
Larger city inner region	4.6	4.6	5.1	5.9	5.5	5.8
Other inner region	2.1	2.1	3.2	5.2	4.6	5.3
Outer region	2.0	0.6	1.3	5.9	4.2	3.6
State of São Paulo	3.3	3.2	3.5	6.4	5.4	4.7
Brazil	3.0	2.9	2.5	5.6	5.4	4.4

Sources: As for Table 13.2
[a] As in Table 13.2
[b] As in Table 13.2

Indices of urban concentration

Analysis of the urban concentration indices shows increasing polarisation in the State of São Paulo from 1950 but polarisation reversal between 1970 and 1980. As shown in Table 13.4, the share of urban population in the metropolitan area declined from 62.4 per cent to 61.9 per cent between 1970 and 1980.[2] Index 2, the change in concentration, was negative for the last period, and the percentage of growth was greater for hinterland cities than the core. Index 4 was negative for 1970–80. These four indices show PR taking place around 1970. Population projections for the 1970s suggest that PR took place sometime between 1970 and 1980.[3]

The spatial trends underlying PR in the 1970s are further explained by index 6 in Table 13.5. Average annual growth rates of cities over 100 000 population between 1970 and 1980 were more than one percentage point higher than the growth rates for smaller cities. It was the growth of these large cities that brought about PR. Index 6 also demonstrates that PR did not arise out of a sudden increase

Table 13.4 Principal indices of polarisation reversal

No.	Index	1950	1960	1970	1980
1	Urban population in core (%)	58.0	60.2	62.4	61.9
		1950–60	1960–70	1970–80	
2	Changes in percentage of urban population in core	2.2	2.2	–0.5	
3	Urban population growth captured by the core (%)	62.8	65.5	61.0	
4	Difference in average annual percentage population growth between core and periphery (core–periphery)	+0.97	+0.97	–0.22	

Notes: *Core* is the São Paulo Metropolitan Area. Periphery is the cities over 20 000 urban population in 1970 outside the metropolitan area in São Paulo State

Differential Urbanization

Table 13.5 Descriptive polarisation reversal indices

No.	Index	1950–60	1960–70	1970–80
5	Difference in absolute population growth between core and periphery (in 000s)	+877	+1 603	+1 602
6	Comparison of average annual population growth rates[a]			
	Metropolitan São Paulo	6.8	5.8	4.6
	Secondary cities			
	20 000–50 000 n = 45	5.6	4.0	4.1
	50 000–100 000 n = 14	6.2	4.9	4.3
	100 000–250 000 n = 9	6.0	5.2	5.3
	250 000+ n = 2	5.5	5.3	5.4
7	Number of secondary cities growing faster than the core (n = 70)			
8	Difference in average annual percentage growth rates between the core and the total of core plus three next largest cities	+1.43	+0.43	−1.25

No.	Index	1950	1960	1970	1980
9	Four-city primacy ratio	0.846	0.862	0.868	0.853
10	Ten-city primacy ratio	0.761	0.778	0.786	0.770
11	Urban deconcentration index	2.92	2.72	2.53	2.57

[a] City size classes defined by the urban population of cities 1970. Cities do not move from one class to another over time.

in secondary city growth rates, since these rates are fairly constant by city size class between 1960–70 and 1970–80. Rather, PR resulted from a decline in the growth rate of metropolitan São Paulo. Whether this can be explained by reverse migration from the core to the secondary cities or by a redirection of migration patterns is not clear. Further analysis of the 1980 Demographic Census will shed more light on the factors behind PR in the State of São Paulo.

The results for index 7, the number of cities growing faster than the core, are consistent with those for index 6. Twenty-five cities experienced higher growth rates than the metropolitan area between 1970 and 1980, up from 15 in the previous period. An average of three out of four cities in the inner region, extending roughly 150 kilometres from metropolitan São Paulo, grew faster than the core from 1970 to 1980. One out of five cities located beyond 150 kilometres grew faster than metropolitan São Paulo. These high-growth outer-region cities tended to be larger cities. These patterns are consistent with a model of spatial change in which population growth diffuses both by distance away from the core and down the urban hierarchy within the periphery region.

The remaining indices in Table 13.5 are less useful to an understanding of the components of PR. Index 5 demonstrates that the measure of PR has to be based on relative, not absolute, growth. Indices 8 and 9 show that up to 1970, the metropolitan area grew faster than the next three largest cities, but after 1970, this pattern was reversed. A similar comparison of the core with the nine next largest cities, index 10, shows the same results: increasing concentration in the core from 1950 to 1970 and a decrease from 1970 to 1980. The urban deconcentration index compares all cities with one another with smaller values revealing greater inequality in population between cities. This index decreased between 1950 and 1970, but increased from 1970 to 1980. The two primacy indices returned to pre-1960 levels by 1980 but the UD index showed only a small change from 1970 to 1980, suggesting greater equality among the largest cities in the state but less equality between the largest and smallest cities.

Summary

This paper has removed some of the ambiguities surrounding the concept of polarisation reversal. After a brief review of factors that may work to bring about PR, we narrowed examination of PR to urban population only and defined the core and periphery regions. Starting with a list of potential PR indices culled from past studies of urban concentration, we defined the principal index of PR as the proportion of urban population in a city region that is in the core region. This definition and the definitions of polarisation stabilisation and reversal of the rate of polarisation are based on Alonso's model of urban concentration (Alonso, 1980).

Using these definitions we found that PR took place in the State of São Paulo between 1970 and 1980. This may be the first documented case of polarisation reversal in a less developed country.

The study is limited, however, to spatial changes in the distribution of the aggregate population and does not encompass components of this change such as migration trends. We also do not link PR in the State of São Paulo to other changes associated with economic development. There is no explanation of the behavioural changes involved in the reversal. Ideally, movements towards PR should be correlated with changing land values, demographic shifts, growth in infrastructure provision, the decentralisation of services and other trends. Unfortunately, sufficient time series of measures of these changes do not exist in São Paulo to permit the clear estimation of these linkages. This is a task for future detailed work. Thus, this paper is a first step towards additional research on population concentration in the State of São Paulo and other city regions of LDCs. The approach to modelling PR presented here will also be useful as a base for further discussion of what is meant by polarisation reversal and how it may be measured.

In conclusion, it should be noted that the onset of PR may weaken the case for active policy intervention to influence population distribution. Policies to decentralise residents and employers must rest on a more complex set of arguments couched in terms of both imperfectly operating markets (efficiency arguments) and socially undesirable outcomes from market processes (equity arguments). This argument set allows a judgement to be reached as to whether PR is to be regarded as a 'good' or a 'bad' thing. The case of São Paulo is examined within the context of a general review of the argument for policy intervention at the city region scale in Townroe (1983).

Notes

This paper was prepared while the authors were working for the Urban Development Department of the World Bank. The views and interpretations in the paper are those of the authors and should not be attributed to the World Bank, or to its affiliated organizations.

1 Alonso (1980), expressed concentration as a function of economic development. We have modified this relationship by replacing level of economic development by time with the assumption of a strong link between development and passage of time.
2 The data used for this analysis are urban population for each *municipio* (county) from the Brazil Demographic Census for 1950 and 1970 and from the Preliminary Census for 1960 and 1980. We have redefined urban population to include only *municipios* with over 20 000 urban population in 1970 or *municipios* that are part of urban agglomerations. Definitions of urban agglomerations are modified from the State Secretariat of Planning and Economic Affairs, *Organização Regional do Estado de São Paulo* (São Paulo: Secretaria de Económica e Planejamento, 1980), and Ablas and Azzoni (1978). This analysis uses constant boundaries of *municipios* based on 1950 definitions. A final modification was to adjust urban population estimates to correct for underclassification of urban population in some rapidly growing *municipios* (*see* A. Altman, *Componentes demograficos do crescimento urbano*: Regiao Metropolitana de São Paulo) Teresopolis, R. J.: Seminario Técnico sobre Dados, Medidas e Conseqüências das Migraçoes Internas, 1980).
3 The 10-year interval between demographic censuses in Brazil makes it difficult to identify the precise timing of PR in the State of São Paulo. It is not clear whether PR took place before or after 1970. The Fundação Sistema Educacional de Análise de Dados (SEADE) forecast city populations for 1980 based in part on statistics for the early 1970s. The SEADE figures grossly underestimated the actual 1980 population of secondary cities. This suggests that migration patterns changed after the early 1970s, bringing about PR.

References

Ablas, L. and Azzoni, C. 1978: *Requisitos locaçionais de indústrias*. São Paulo: FIPE.
Alonso, W. 1968: Urban and regional imbalances in economic development. *Economic Development and Cultural Change* 17, 1–14.
Alonso, W. 1980: Five bell shapes in development. *Papers of the Regional Science Association*. 45, 5–16.
Barat, J. and Geiger, P. P. 1973: Estructura económica das áreas metropolitanas Brasileiras. *Pesquisa e Planejamento Económico* 3, 673–713.
Berry, B. J. L. and Gillard, Q. 1977: *The changing shape of metropolitan America: Commuting patterns, urban fields and decentralization processes 1960–70*. Cambridge, Mass.: Ballinger.
Berry, B. J. L. 1978: The counterurbanization process: how general? In Hansen, N. M. (ed.), *Human settlement systems*. Cambridge, Mass: Ballinger, 25–49.
Davidovich, F. R. and Lima, O. B. 1975: Continuação do estudo de aglomeraçoes urbanas no Brasil. *Revista Brasileira de Geografia* 37, 50–84.
Dillinger, W. 1978: A cross-sectional analysis of urban polarization trends. Washington, DC: The World Bank, (mimeo).
Falk, T. 1978: Urban development in Sweden 1960–75: population dispersal progress. In Hansen, N. M. (ed.), *Human settlement systems*. Cambridge, Mass.: Ballinger, 51–83.
Fox, R. W. 1975: *Urban population trends in Latin America*. Washington, DC: Inter-American Development Bank.
Friedmann, J. 1975: The spatial organization of power in the development of urban systems. In Alonso, W. and Friedmann, J. (eds) *Regional policy: Readings in theory and applications*. Cambridge, Mass.: MIT Press, 266–304.

Gaile, G. 1980: The spread–backwash concept. *Regional Studies* 14, 15–25.
Hall, P. and Hay, D. 1980: *Growth centres in the European urban system.* London: Heinemann.
Hansen, N. M. 1980: *Intermediate sized cities as growth centres.* New York: Praeger.
Henderson, J. V. 1980: *A framework for international comparisons of systems of cities.* Urban and Regional Report no. 80–3, Washington, DC: The World Bank.
Henderson, J. V. 1982: The impact of government policies on urban concentration. *Journal of Urban Economics* 12, 280–303.
Keen, D. 1982: *The relationship between urban infrastructure and industrial development among cities of São Paulo State.* National Spatial Policies Working Paper no. 12. Washington, DC: The World Bank.
Linn, J. F. 1979: *Urbanisation trends, polarisation reversal and spatial policy in Colombia.* Urban and Regional Report no. 79-15. Washington, DC: The World Bank.
Mera, K. 1978: Population concentration and regional income disparities: a comparative analysis of Japan and Korea. In Hansen N. M. (ed.), *Human settlement systems.* Cambridge, Mass.: Ballinger, 155–75.
Owen, C. and Witton, R.A. 1973: National division and mobilisation: a reinterpretation of primacy. *Economic Development and Cultural Change* 21, 325–37.
Renaud, B. 1977: *Economic structure, growth and urbanization in Korea.* Multi-disciplinary Conference on South Korean Industrialization, Honolulu (mimeo).
Renaud, B. 1981: *National Urbanisational Policy in Developing Countries.* Oxford: Oxford University Press.
Richardson, H. W. 1977: *City size and national spatial strategies in developing countries.* World Bank Staff Working Paper no. 252. Washington, DC: The World Bank.
Richardson, H. W. 1980: Polarization reversal in developing countries. *Papers of the Regional Science Association.* 45, 67–85 (*see* this volume, Chapter 11).
Ternent, J. A. S. 1976: Urban concentration and dispersal: urban policies in Latin America. In Gilbert A. (ed.), *Development planning and spatial Structure.* London: John Wiley & Sons.
Townroe, P. M. 1979: Employment decentralisation: policy instruments for large cities in less developed countries. *Progress in Planning.* 10,(2) 85–154.
Townroe, P. M. 1983: *Employment decentralization policy for a major metropolis: the case of São Paulo, Brazil.* Urban Development Report no. 83-6. Washington, DC: The World Bank.
Vining, D. R. and Kontuly, T. 1978: Population dispersal from major metropolitan regions: an international comparison. *International Regional Science Review.* 3, 49–73 (*see* this volume, Chapter 6).
Wheaton, W. C. and Shishido, H. 1981: Urban concentration, agglomeration economies, and the level of economic development. *Economic Development and Cultural Change* 30, 17–30.

14 H. Lee,
'Growth Determinants in the Core–Periphery of Korea'

From: *International Regional Science Review* 12 (2), 147–63 (1989)

Introduction

The concept of interdependent development emphasises that the regional economy is an interdependent part of the national economy and that regional growth or development is heavily dependent upon growth in other regions. (Bannister, 1975; Brookfield, 1975; Okabe, 1979; Sheppard, 1977, 1980). Spatial interdependence is a key to explaining the nature of uneven development over space.

Growth brings about a change in the relative importance of concentration and dispersion forces through time. These two forces, pulling in opposite directions, are dynamic, interactive, and in constant flux, changing in relative importance over time. Consequently, evolving spatial patterns can be explained by changes in the relative importance of forces leading to concentration and dispersion within a core–periphery system. In order to understand how a core–periphery system has changed and evolved, factors influencing these forces of concentration and dispersion must be explained.

Most theoretical and empirical work on the spatial aspects of development and the determinants of regional growth has been performed in developed countries. However, prevailing theories may not adequately explain the regional development process in the Third World context. During the past two decades, development of the spatial economy in Third World countries has yielded an extremely dualistic core–periphery structure. In response, some central governments have undertaken forceful policies to ameliorate primacy problems and to control the process of spatial development. The dynamics of regional economic growth in such nations are, then, combinations of autonomous development forces and government intervention (Folmer and Oosterhaven, 1979). In many countries, government policy shaping economic development is viewed as a basic force to which regions respond and adjust (King and Clark, 1978). Without government intervention, the periphery in many developing countries would not be able to compete with the core area, which is fuelled by a process of cumulative causation.

In most developing countries, policies designed to divert new industries away from primate cities have tended to concentrate investment in a few selected larger centres in the periphery, rather than scattering it nation-wide, for efficiency reasons (Alonso, 1968; Mera, 1967; Richardson, 1973b). Often, a limited number of secondary cities, perhaps a subset of the provincial capitals, is promoted on the basis of economic development potential (Richardson, 1981). Since growth within a region tends to be spatially concentrated in this fashion, a concentrated spatial dispersion of economic development emerges. This concentrated spatial dispersion pattern of urban growth is consistent with Richardson's polarisation reversal argument (Richardson, 1973a, 1980).

The emergent pattern of the Korean urban system during the 1970s will be described. An examination of growth determinants influencing the differential growth rate of urban centres located in the core and periphery then follows, with a focus on the interdependencies of development in the urban system. In concluding, some implications of the empirical results for prevailing regional policies and strategies in Korea are discussed.

The concentrated spatial dispersion pattern of urban growth

Since the early 1960s, Korea has experienced rapid economic growth accompanied by an unprecedented rate of urbanisation. The total urban population rose from 7 million, or 29 per cent, in 1960 to 21 million, or 57 per cent, in 1980. This rapid urbanisation was unambiguously associated with an increase in the concentration of population in only a few cities, especially the capital city of Seoul. Accordingly, the primacy index rose from 1.09 in 1960 to 1.51 in 1975.

Seoul has played a dominant role in rapid economic growth of Korea during the past two decades. Cumulative growth forces help explain how and why the Seoul metropolitan area has expanded sharply during the period of rapid industrialisation. Such forces, which have given rise to an urban economy that dominates the rest of the country, illustrate well a spatially polarised development process. Contributing factors include initial advantage (Borchert, 1967; Lampard, 1955, 1968; Pred, 1965; Ullman, 1958), self-perpetuating momentum of growth (Friedmann, 1966; Hirschman, 1958; Lasuen, 1973; Myrdal, 1957; Pred, 1966, 1975), and spatial interaction feedback mechanisms (Bannister, 1976; Bennett, 1975a, 1975b; Cliff and Ord, 1975; Korcelli, 1980; Okabe, 1979; Sheppard, 1977, 1979, 1980).

Seoul city, with its tremendous inertia and historical advantages, has exerted considerable influence on industrial location decisions, inducing agglomeration economies. According to Kwon (1981), the Seoul metropolitan centre accounted for 28 per cent of the nation's value added in manufacturing in 1977, 32 percent of all manufacturing establishments in 1979, and 78 per cent of headquarters of business firms in 1978. Noneconomic factors, including its social and cultural climate, also affect Seoul's pre-eminence.

Faced with an overconcentration in the Seoul metropolitan area, the government has attempted to avoid further heavy concentration of industrial facilities and has set up economically self-sustaining regions throughout the country for the purpose of both national defence and balanced regional development. The government has undertaken strong policies to ameliorate the primacy problem by directing new industry away from Seoul city and towards the periphery.

During the past decade, the government decentralisation strategies have been exerting influence on the spatial distribution of economic growth and population. In order to analyse the pattern of urban growth, data were collected on all urban centres with populations over 20 000 in either 1960 or 1970. The 188 centres so identified were classified into one of four levels in the urban

204 Differential Urbanization

system hierarchy based on population size in 1980. The two largest cities, Seoul and Pusan, are defined as metropolitan centres. The population of Seoul city in 1980 was 8.4 million, and that of Pusan 3.2 million. Regional centres are urban centres with populations greater than 200 000 and include most provincial capital cities and planned industrial cities. Local centres consist of urban centres with a population between 50 000 and 200 000. Rural centres have populations between 20 000 and 50 000 (*see* Figure 14.1).

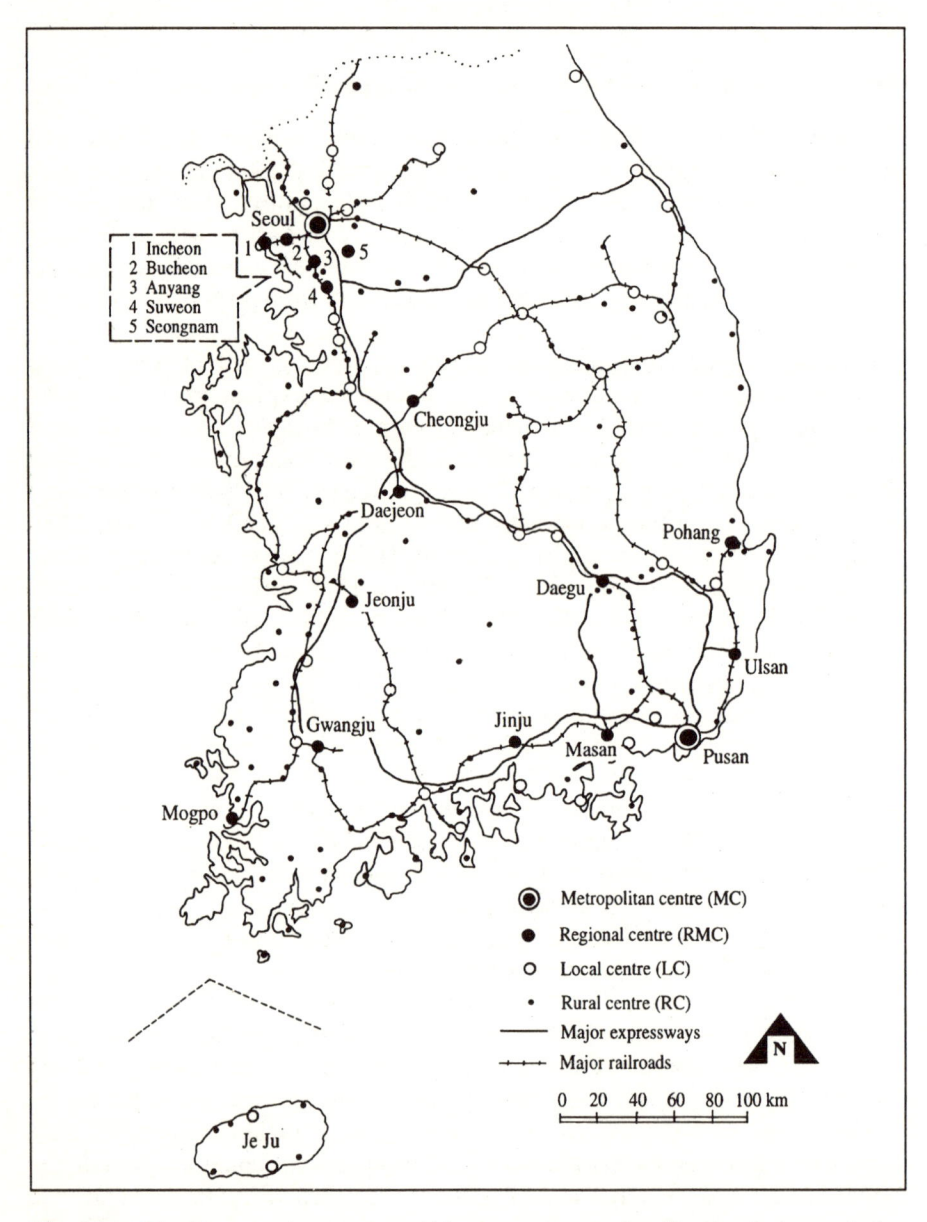

Fig. 14.1 The Korean urban system. Urban centres are classified by their population size in 1980

The patterns of urban change under decentralisation policies are shown in Figure 14.2. The rate of growth of metropolitan centres increased during the 1960s but then started to decline in the 1970s. Meanwhile, the growth rate of regional centres continually rose, with the growth rate for the 1970s exceeding that of the 1960s. The population of all centres increased by nearly 8.1 million persons in the 1970s. Rural centres and local centres expanded by only 0.3 million and 0.9 million, respectively, whereas the corresponding growth in metropolitan centres and regional centres was 4.1 million and 3.2 million, respectively.

The most distinctive phenomenon was the unusually rapid growth of the regional centres, largely due to government policies limiting the further growth of metropolitan centres. Growth impulses diverted from metropolitan centres were transmitted to only a few regional centres. In general, the concentrated spatial dispersion of urban growth can be explained by the location of industrial estates, the most important tool in controlling industrial allocation. The dispersion process was also promoted by diseconomies of the Seoul metropolitan area, including high land prices, physical constraints on development, and a decline in its locational advantages for manufacturing investment when compared with the southern part of the country (Hwang, 1979).

On the regional level, the patterns of urban growth varied considerably. Table 14.1 and Figure 14.3 illustrate the notable differences in growth rates for centres by size, class, and core–periphery status. (A criterion for the latter classification of regions is described in the following section.) In the core area, the growth rate of the metropolitan centre (Seoul) was steadily declining, while the metropolitan centre (Pusan) in the semi-periphery area grew continually through the 1970s. Regional centres, the second tier of urban places, experienced rapid growth in both core and semi-periphery areas. Growth rates in the periphery were typically lowest, and all three size classes experienced decreased growth rates during the 1970s compared with the 1960s. In particular, the rate of growth for rural centres as a whole was negative in the 1970s,

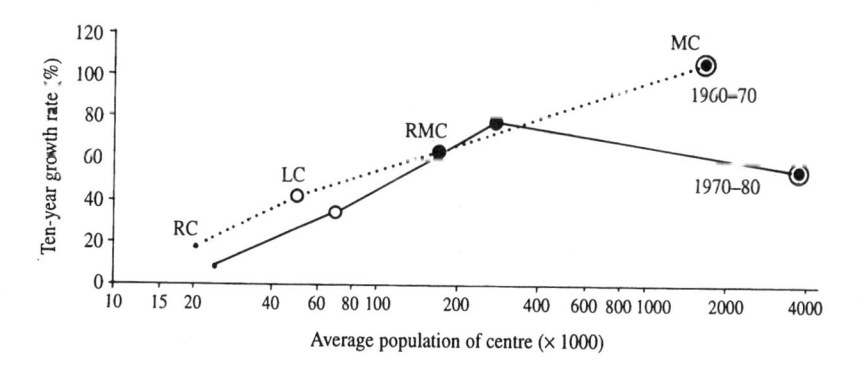

Fig. 14.2 Population growth rates by average population size of each class, 1960–80. (Sources: National Bureau of Statistics, Economic Planning Board, *Population and Housing Census*, 1960, 1970, and 1980)

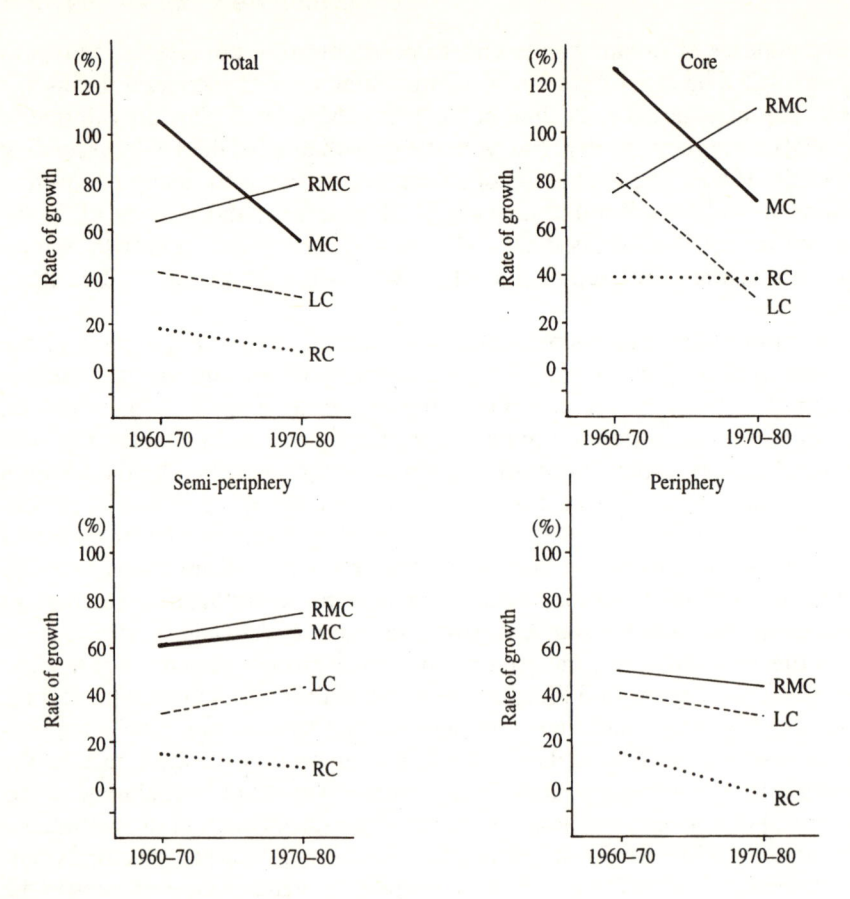

Fig. 14.3 Population growth rates by size and core–periphery status. (Sources: National Bureau of Statistics, Economic Planning Board, *Population and Housing Census*, 1960, 1970, and 1980)

Table 14.1 Population growth rates by size and core–periphery status, 1960–80

	Total	Size class			
		Metropolitan centre	Regional centre	Local centre	Rural centre
Core					
Number of centres	32	1	5	6	20
Change (%)					
1960–70	108	126	76	82	39
1970–80	59	51	115	31	39
Semi-Periphery					
Number of centres	79	1	6	12	60
Change (%)					
1960–70	45	62	65	33	16
1970–80	53	68	75	42	10
Periphery					
Number of centres	77	—	4	18	55
Change (%)					
1960–70	33	—	50	40	16
1970–80	24	—	44	33	−1

Sources: National Bureau of Statistics, Economic Planning Board, *Population and Housing Census*, 1960, 1970, and 1980.

indicating that larger centres were growing, perhaps at the expense of nearby smaller centres. On the other hand, rural centres located in the core area performed relatively well, maintaining their growth rates over the 1960–80 period. Their rapid growth may be viewed as a consequence of spillover effects from nearby larger centres.

Method of analysis

To identify the factors that determine the rate of growth of urban centres, a general function was estimated to relate the growth rate of each urban centre (Y) to changes in the economic structure (S), size of centre (M), inertia of the economy (I), government expenditures (G), and accessibility within the urban system (A). None of these factors alone is a sufficient condition for economic growth. For this reason, multiplicative interaction terms are included in the model. Thus, the functional form is:

$$Y = a + b_1 S + b_2 M + b_3 I + b_4 G \\ + b_1 b_2 S \cdot M + b_1 b_3 S \cdot I + b_1 b_4 S \cdot G + b_2 b_3 M \cdot I \\ + b_2 b_4 M \cdot G + b_3 b_4 I \cdot G + c_1 A(M) + e$$

The rate of growth of a city is hypothesised to depend not only on the variables described above, but also on the economic performance of the region within which it is located. For this reason, the nine provinces are classified into one of three levels of province centrality based on growth performance for the 1960–80 period. Seoul city and its surrounding Gyeonggi province are defined as the core area, which experienced a higher than average growth rate. Semi-periphery areas consist of three provinces that grew moderately during the study period. The periphery consists of the remaining five provinces, which grew slowly or even declined. The province of Jeju Island, in spite of its average growth rate above the national average, is assigned to the periphery, largely because of its remote location. A step-wise multiple regression analysis was undertaken separately for each of these three geographical categories.

The model is tested with variables calculated from published data sources (Economic Planning Board 1960, 1970, 1980; Ministry of Home Affairs 1971 to 1981; and the 1981 Expressway Guide Map). The explanatory variables are lagged in time with respect to the dependent variable in the hope of showing the causes of growth rather than its effects.

The percentage change in population between 1970 and 1980 is used to measure the rate of growth for each centre. During the study period, several cities had their boundaries changed, so it is difficult in these cases to trace the exact change. Data are corrected as far as possible for boundary changes to bring all values into line with the 1980 census-defined areas. It is a useful measure of growth because it reflects growth in employment, retail sales, and so on.

The percentage change in the proportion of manufacturing employment to total employment between 1960 and 1980 is used to measure changes in the economic structure. In developing countries, an increase in the relative importance of employment-generating activities such as manufacturing industries is likely to underpin rapid growth in an urban centre.

The size of centre variable is measured by population in 1970, since by that time most large cities in Korea had achieved the size of self-sustaining growth. The emphasis on size as a positive and strong correlate of growth is contained in the early notion of the 'urban ratchet' (Thompson, 1965).

Inertia is measured by the population growth rate of the 1960–70 decade. Inertia of the economy has been viewed as the sole determinant of the cumulative circular growth process (Hirschman, 1958; Myrdal, 1957; Pred, 1966) in which the immediate past growth rate is crucial for the determination of the current level.

A variety of direct and indirect policy instruments can affect the growth of an urban centre. In particular, direct capital investment and the provision of infrastructure influence growth potential through the multiplier effect. However, data for government expenditures at a disaggregated level are not available. Data could be collected only on cumulative expenditures for public works, such as roads, rivers, water supply, water sewage, bridges, and housing for the 1970–80 period, and on investment for industrial estates. Using the GNP deflator index, the amount of government expenditures was converted to constant 1975 monetary values. Thus, cumulative expenditures (1970–80) of this type per 1000 persons (1970 population figures) is used as a proxy for government expenditures.[1]

Each centre within any urban system has a distinct position and, as a result, has a different accessibility to other centres in the system. The concept of accessibility incorporates the advantages and attractiveness of the location of a centre in terms of its functional proximity. In this study, accessibility is basically a measure of potential for interaction. This potential interaction between two centres is directly proportional to the product of their populations and inversely proportional to the distance between them.

Only centres at higher levels of the hierarchy are assumed to be capable of contributing to any particular centre's growth potential, and the growth of lower-order centres is assumed to be influenced by accessibility to all larger centres rather than to just the nearest larger centre. Taking the example of the accessibility of a local centre, it is measured as:

$$A(M)_J = \frac{\sum_K M_K P_J d_{JK}^{-1} + \sum_L M_L P_J d_{JL}^{-1}}{N}$$

where $A(M)$ is mean accessibility to larger centres for a local centre J, M_K and M_L are sizes of larger centres of level K (regional centres) and L (metropolitan centres) respectively, d_{JK} is distance between two centres J and K, P_J is 1970 population of a local centre J, and N is the total number of regional and metropolitan centres.

Two-way interaction variables are derived by multiplying together pairs of independent variables for structure, size, inertia, and government expenditures. Each variable was first standardised to eliminate scale effects.

The data were examined with respect to typical regression assumptions and problems. Multicollinearity was addressed by calculating multiple correlation coefficients between independent variables. Certain regressors collinear with other ones were subsequently excluded from the regression models. Heteroskedasticity was considered by viewing plots of the residuals against each of the dependent and independent variables in turn, and normality assumptions were tested with Rankit plots and Wilk–Shapira W statistics.

Since regression coefficients may be greatly affected by a few cases, outliners with a standardised residual greater than 3 were tested and eliminated on the basis of the Multreg statistics t-test equation. Finally, a residual map from selected equations was inspected for the existence of spatial autocorrelation.

Growth determinants for centres of different geographical locations

The sharp differences in performance among the rural centres suggest that the growth of rural centres may be highly correlated with that of their nearby larger centres. The variation in growth rates for core, semi-periphery, and periphery areas also implies that factors influencing the rate of growth for each category may vary. Thus, three separate equations were used to estimate determinants of the contrasting patterns of urban development in the core, semi-periphery, and periphery areas.

The core area

The stepwise regression result for the core area was:

$$Y = 734.0 - 161.2M + 1.0I + 96.3 I \cdot G,$$
$$(4.5) \quad (4.3) \quad (4.2) \quad (7.9)$$
$$R^2 = 0.70$$

where Y is the rate of growth from 1970 to 1980, M is the size of a centre in 1970, I is inertia, and $I \cdot G$ is interaction between inertia and government expenditures. The terms in parentheses are t-values. As shown by the equation, urban growth rates in the core area are negatively related to the size of a centre. The effects of inertia and of the interaction term between inertia and government expenditures are positive.

The interaction variable ($I \cdot G$) alone accounts for 44 per cent of the variance in growth rates and possesses the largest t-value. Inertia, as measured by previous growth, combining with government expenditures to promote additional growth, indicates exactly the idea of the circular cumulative growth process. The model for the core area illustrates the contradiction of spatial growth patterns: decentralisation versus inertia of growth. In other words, previously fast-growing centres tend to grow quickly, but that growth is likely to decentralise and hence larger centres tend to have slower growth rates, while smaller centres grow more rapidly.

The map of residuals from the regression shows distinctive clusters of positive and negative residuals (Figure 14.4). Urban centres located north of Seoul city are overestimated, except for two centres. On the other hand, centres

210   Differential Urbanization

located south of Seoul city are consistently underestimated, except for a few less accessible centres whose major economic activities are more agriculture-oriented. This pattern may reflect a combination of political, social, and psychological factors. Growth that might naturally have occurred between Seoul and the northern frontier was deflected to the region immediately south of Seoul for reasons of national security. During the past decade, the most rapidly growing regional centre was Seongnam city, which was planned as one of the satellite cities of Seoul in order to decentralise part of Seoul's population. As might be expected, this city shows the largest positive residual. Among fast-growing regional centres, Anyang city was overpredicted. Considering the extraordinarily fast-growing small rural centres adjacent to Anyang city, the negative residuals for this city may reflect local decentralisation of growth.

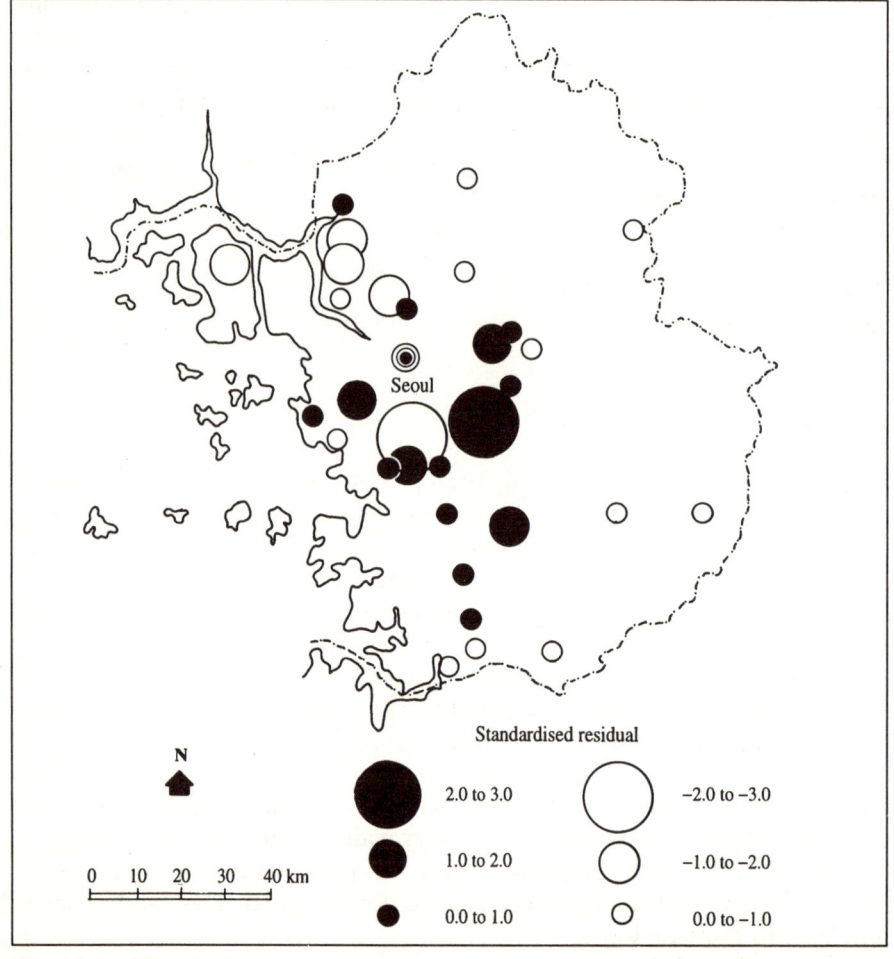

Fig. 14.4 Growth of urban centres in core areas: residuals from regression. Note: places represented by solid circles grew faster than expected. Open circles represent places with growth that fell short of expectations based on the regression

Semi-periphery areas

Eighty per cent of the variance in growth rates for semi-periphery areas is accounted for by four variables statistically significant at the 0.01 level. The equation is:

$$Y = 251.4 + 0.03S + 48.6G + 31.5A + 65.8S \cdot G$$
$$(11.6) \quad (7.4) \quad (8.0) \quad (5.1) \quad (9.7)$$
$$R^2 = 0.80$$

where Y is the rate of growth from 1970 to 1980, S is changes in economic structure, G is government expenditures, A is accessibility to larger centres, and $S \cdot G$ is interaction between economic structure and government expenditures. As shown by the equation, growth was higher if a centre experienced rapid change in its economic structure, received significant government expenditures, and enjoyed superior accessibility. The most crucial factor influencing growth was government expenditure. It alone accounts for 42 per cent of the variance in growth rates. When government expenditures interact with economic structure ($S \cdot G$), they reinforce each other. Money spent on infrastructure and industrial estates seems to have had a positive interaction effect, further stimulating manufacturing activities. From the equation it may be postulated that accessibility is especially important in semi-periphery areas where growth is faster and more polarised, reflecting a growth pole pattern of growth.

The map of residuals from this regression reveals to some degree a spatially autocorrelated pattern (Figure 14.5). Most centres surrounding regional centres, but not adjacent to them, are consistently overestimated, which may indicate that spread effects from regional centres in semi-periphery areas decay very rapidly over distance during the past decade. The benefits from rapid growth of regional centres appear in only a few of the centres closest to regional centres.

Periphery areas

The regression equation for periphery areas is:

$$Y = -184.9 + 24.8M + 0.2I + 18.4G$$
$$(7.2) \quad (3.7) \quad (3.5) \quad (5.0)$$
$$R^2 = 0.64$$

where Y is the rate of growth from 1970 to 1980, M is size of a centre in 1970, I is inertia of the economy, and G is government expenditures. As shown by the equation, growth is positively related to size, inertia, and government expenditures. The growth model for periphery areas seems to conform to the cumulative growth conception. The cumulative growth effect was fuelled by the momentum of previous growth and the advantage of the size effect, with the addition of incentives from government. Thus, previously fast-growing larger centres experienced higher growth rates. There is no apparent spatial diffusion of growth impulses.

The map of residuals has several notable aspects (Figure 14.6). Of the four

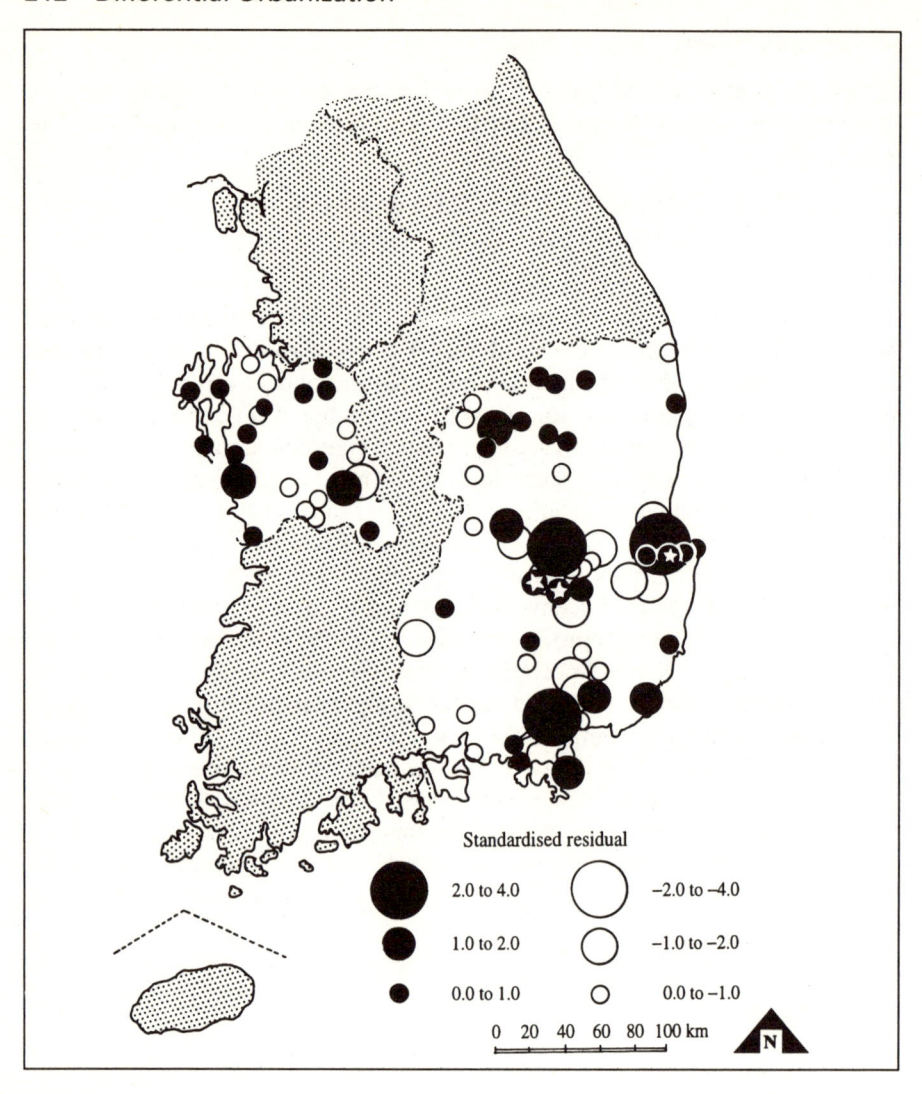

Fig 14.5 Growth of urban centres in semi-periphery areas: residuals from regression. Note: places represented by solid circles grew faster than expected. Open circles represent places with growth that fell short of expectations based on the regression. Urban centres represented by the circled star are deleted extreme outliers

regional centers, three are overestimated. On the other hand, local centres in the same provinces as those three regional centres are underestimated. This result implies the possibility of lateral interaction among regional centers located in core and semi-periphery areas. As Pred (1977) suggests, larger centres may be sufficiently interlinked that lateral interaction as well as hierarchical interaction takes place. This interurban development process within periphery areas casts doubt on the assumption that locating growth poles in less developed regions will stimulate the development of the surrounding hinterland through growth

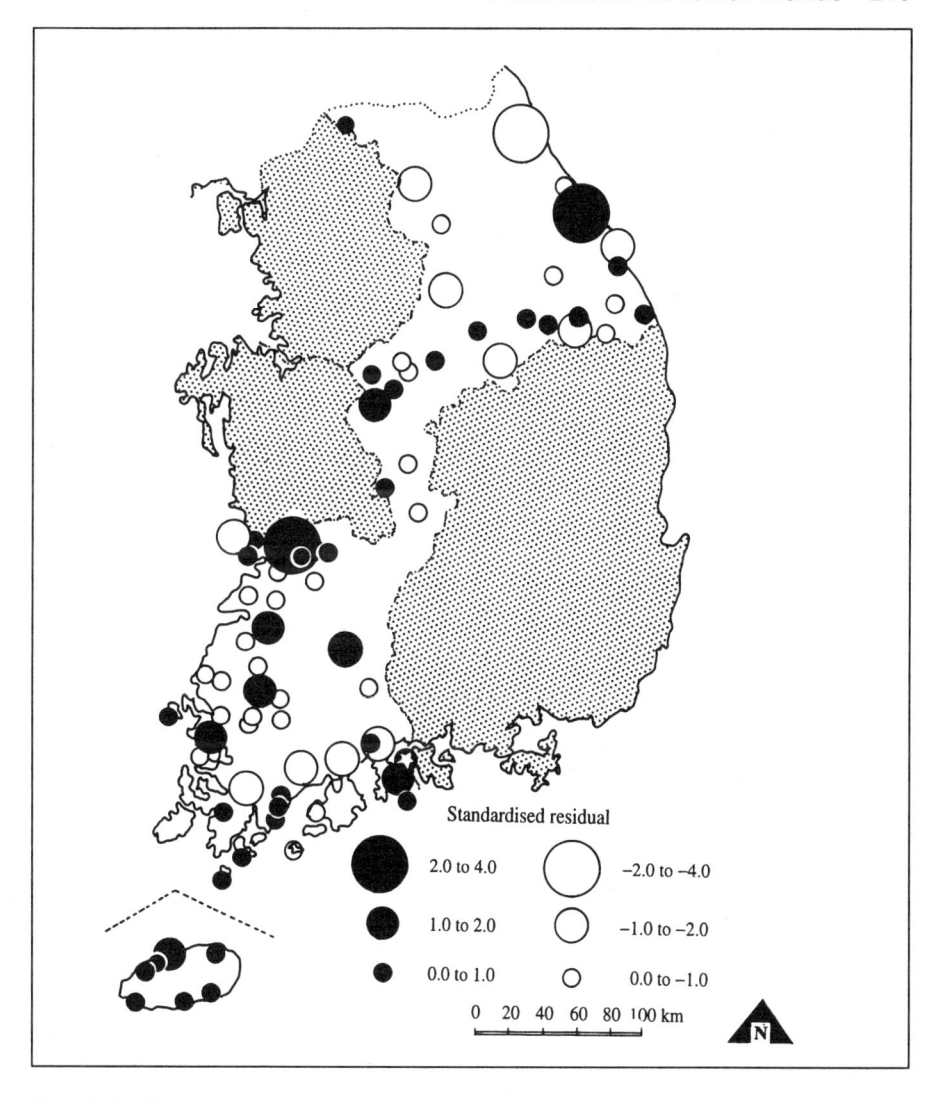

Fig 14.6 Growth of urban centres in periphery areas: residuals from regression. Note: places represented by solid circles grew faster than expected. Open circles represent places with growth that fell short of expectations based on the regression. Urban centres represented by the circled star are deleted extreme outliers

transmission from growth poles. Instead, the effects of growth poles may be transmitted among larger centres in more developed regions. Accordingly, most rural centres are overpredicted in periphery areas because they do not appear to be receiving growth impulses from larger centres.

Conclusions

During the period from 1970 to 1980, rapid urban growth in Korea has been consistent with the policies of government intervention. Quickly growing regional centres resulted in an emergent pattern of concentrated spatial dispersion of urban growth, consistent with Richardson's polarisation reversal hypothesis. The regression results confirm these impressions and reveal some interesting regional contrasts. In the core, the effects of size and inertia reflect a strong process of decentralisation of growth to smaller centers. In semi-periphery areas, growth impulses are transmitted from larger centres in a fashion consistent with the growth pole hypothesis. In periphery areas, however, there is no spillover effect from growth poles but instead a process of cumulative causation and enhanced polarisation.

Such different types of development processes operating in the various regions cast doubt on certain aspects of the growth pole literature, since the impacts of growth poles and the outcomes of government investment vary for different types of regions. Judging from the Korean experience of the interurban development process, the emergence of growth poles in periphery areas did not alleviate stagnation of other small centres in their hinterlands. Therefore, in order to stimulate rural centres in the periphery, government should consider investing directly in rural centres rather than expect autonomous change to occur through the diffusion of growth. Growth pole and regional development theories need to be rethought to consider the different types of interdependent development processes that may occur in various types of regions.

Note that the economic structure variable did not emerge directly as a growth determinant. Only in the regression result for semi-periphery areas did economic structure influence the growth rate. This result suggests that policies that rely on accelerated industrialisation to remedy underdevelopment may not be successful in developing countries.

During the past decade, economic growth, especially the development of manufacturing industries, was a top priority of Korean policymakers. As a result, a substantial structural and developmental imbalance exists across regions of the country and sectors of the economy. Differences in living standards, job opportunities, and social conditions between agricultural and industrial centres are large, owing in part to productivity differences but also to the absence of territorial development strategies. No widespread dispersion of industry has taken place in Korea. Consequently, regional economic diversity and dynamic interdependence have produced highly varied growth rates and employment opportunities across the country.

The main policy implication of this study is the need to examine the anticipated spatial impacts of government investments prior to any decision to implement them. In order to formulate an efficient regional development policy, a government should analyse the prevailing interurban development process and pattern of change and examine the factors that account for observed trends. Such an understanding may provide guidance to policymakers in designing policies to reduce the variability and disparity of regional development.

Acknowledgement

Helpful comments by Eric S. Sheppard are gratefully acknowledged.

Note

1 The government expenditures data are incomplete and imprecise because data for the years 1972, 1977, and 1978 were not available, and data for small rural centres in 1980 could not be obtained for years prior to 1980. For missing data, average government expenditures by province for the corresponding year were retrieved.

References

Alonso, W. 1968: Urban and regional imbalances in economic development. *Economic Development and Cultural Change* 17, 1–14.
Bannister, G. 1975: Population change in southern Ontario. *Annals of the Association of American Geographers* 65, 177–88.
Bannister, G. 1976: Towards a model of impulse transmission for an urban system. *Environment and Planning A* 8, 385–96.
Bennett, R. J. 1975a: Dynamic systems modelling of the north-west region: 1. Spatial-temporal representation and identification. *Environment and Planning A* 7, 525–38.
Bennett, R. J. 1975b: The representation and identification of spatio-temporal systems: an example of population diffusion in north-west England. *Transactions of the Institute of British Geographers* 66, 73–94.
Borchert, J. R. 1967: American metropolitan evolution. *Geographical Review* 57, 301–32.
Brookfield, H. 1975: *Interdependent development*. London: Methuen.
Cliff, A. D. and Ord, J. K. 1975: Space–time modelling with an application to regional forecasting. *Transactions of the Institute of British Geographers* 64, 119–28.
Economic Planning Board, Korea, 1960, 1970 and 1980: *Population and Housing Census*. Seoul Korea: National Bureau of Research and Statistics.
Folmer, H. and Oosterhaven J. (eds) 1979: *Spatial inequality and regional development*. Boston: Martinus Nijhoff.
Friedmann, J. 1966: *Regional development policy: A case study of Venezuela*. Cambridge, Mass.: MIT Press.
Hirschman, A. O. 1958: *The strategy of economic development*. New Haven, Conn.: Yale University Press.
Hwang, M. C. 1979: A search for the capital region of Korea. In Rho, Y. H. and Hwang, M. C. (eds), *Metropolitan planning: Issues and policies*. Seoul: Korea Research Institute for Human Settlements, Working Paper 80–30.
King, L. J. and Clark, G.L. 1978: Government policy and regional development. *Progress in Human Geography* 2, 1–16.
Korcelli, P. 1980: *Urban change: An overview of research and planning issues*. Laxenburg, Austria: International Institute for Applied Systems Analysis.
Kwon, W. Y. 1981: A study of the economic impact of industrial relocation: the case of Seoul. *Urban Studies* 18, 73–90.
Lampard, E. A. 1955: The history of cities in the economically advanced areas. *Economic Development and Cultural Change* 3, 81–137.
Lampard, E. A. 1968: The evolving system of cities in the Unites States: urbanization and economic development. In Perloff, H. S. and Wingo, L. (eds), *Issues in urban economics*. Baltimore: Johns Hopkins University Press, 81–139.
Lasuen, J. R. 1973: Urbanisation and development: the temporal interaction between geographical and sectoral cluster. *Urban Studies* 10, 163–88.
Mera, K. 1967: Trade-off between aggregate efficiency and interregional equity: a static analysis. *Quarterly Journal of Economics* 81, 658–74.

Ministry of Home Affairs. 1971 to 1981: *Municipal yearbook of Korea.*
Myrdal, G. 1957: *Economic theory and underdeveloped regions.* London: Duckworth.
Okabe, A. 1979: Population dynamics of cities in a region: conditions for a state of simultaneous growth. *Environment and Planning A* 11: 609–28.
Pred, A. 1965: Industrialisation, initial advantages, and American metropolitan growth. *Geographical Review* 55, 158–85.
Pred, A. 1966: *The spatial dynamics of US urban industrial growth, 1800–1914.* Cambridge, Mass.: MIT Press.
Pred, A. 1975: On the spatial structure of organisations and complexity of a metropolitan interdependence. *Papers of the Regional Science Association* 35, 115–42.
Pred, A. 1977: *City systems in advanced economies: Past growth, present processes, and future development options.* New York: Wiley.
Richardson, H. W. 1973a: *Regional growth theory.* New York: Wiley.
Richardson, H. W. 1973b: *The economics of urban size.* Farnborough: Saxon House.
Richardson, H. W. 1980: Polarization reversal in developing countries. *Papers of the Regional Science Association* 45, 67–85 (*see* this volume, Chapter 11).
Richardson, H. W. 1981: National urban development strategies in developing countries. *Urban Studies* 18, 267–83.
Sheppard, E. S. 1977: Rethinking the nature of urban and regional interdependencies. Paper presented at the meetings of the Association of American Geographers, Salt Lake City, Utah.
Sheppard, E. S. 1979: Spatial interaction and geographic theory. In Olsson G. and Gale, S. (eds), *Philosophy in geography.* Dordrecht: Reidel, 361–78.
Sheppard, E. S. 1980: *Spatial interaction in dynamic urban systems.* Laxenburg, Austria: International Institute for Applied Systems Analysis, Working Paper 80–103.
Thompson, W. R. 1965: *A preface to urban economics.* Baltimore: Johns Hopkins University Press.
Ullman, E. L. 1958: Regional development and the geography of concentration. *Papers and Proceedings of the Regional Science Association* 4, 179–98.

15   L. A. Brown and V. A. Lawson,
'Polarisation Reversal, Migration-Related Shifts in Human Resource Profiles, and Spatial Growth Policies: A Venezuelan Study'

From: *International Regional Science Review* 12 (2), 165–88 (1989)

Introduction

Since World War II, Third World nations have experienced an increasing concentration of population and economic activity in their core regions or primate cities. Although concentration is initially advantageous, reversal of the trend, or polarisation reversal, is ultimately desirable (Richardson, 1980). Polarisation reversal usually is identified by considering patterns of change in the spatial distribution of aggregate population. As noted by Townroe and Keen (1984), however, the process also involves other elements.

This article examines polarisation reversal in terms of changing human resource profiles brought about by the interaction between migration and national policies affecting the spatial pattern of economic growth. Migrants tend to surpass nonmigrants in human capital attributes such as educational attainment, innovativeness, entrepreneurial skills, and motivation to achieve. This phenomenon, known as migrant selectivity, results in a lessened human resource base in origin areas, enrichment of destination areas, and altered growth prospects for each. Insofar as migration and related movements of human resources reflect broader economic forces, polarisation and its reversal are expected to occur without intervention. Commonly, however, growth policies are implemented to correct economic distortions, to hasten decentralisation, and to satisfy political demands. Whether these policies target economic sectors, geographic locales, or the national economic environment, they generally have spatial impacts, including effects upon the volume and directionality of migration streams and, hence, upon human resource profiles of origin and destination areas.

To address these aspects of polarisation reversal empirically, we initially examine human resource variation among eight Venezuelan urban districts and the rest of Venezuela treated as a single unit. This comparison utilises age, sex, educational attainment, and occupational status variables from individual records of the 1971 Census of Population. Gains or losses from migration for each human capital attribute are related to the economic structure of each place and to government policies promoting growth in the Guayana region and industrial decentralisation overall. A second set of analyses focuses on differences in migrant attributes associated with each sector of employment, thus indicating which elements of urban economic structure prompt shifts in the human resource base.

Previous research of relevance

Background for this study is found in the nexus of three distinct research areas: human resource effects on economic growth; migrant selectivity; and development models which illuminate the reciprocal relationships between migrant selectivity and regional economic growth. Each of these will now be discussed.

Human resources and economic growth

Many regard human resources as a primary determinant of Third World development. Central to the long-standing and widely referenced modernisation paradigm, for example, is the transformation from traditional to modern attitudes and the creation of achievement motivation or entrepreneurial values, which in turn depend upon education, communications, and demonstration effects (Hagen, 1962; Inkeles and Smith, 1974; Lerner, 1958; McClelland, 1961; Rogers, 1969). Indeed, Schultz's (1980) Nobel Prize address states: 'The decisive factors of production in improving the welfare of poor people are not space, energy, and cropland; the decisive factor is improvement in population quality' (p. 640).

Human resource approaches to development see labour productivity as a critical, often catalytic, element of economic growth, and education as an essential mechanism for elevating productivity (Colclough, 1982; Hicks, 1980; King, 1980; Lockeed *et al.*, 1980; Squire, 1979; Todaro, 1985; Yotopolous and Nugent, 1976). Education improves health, nutrition, household living conditions, and other aspects of basic needs which, aside from bettering individual lives, contribute to elevating productivity (Bowman, 1980; Colclough, 1982; Fields, 1980; Todaro, 1985; Wheeler, 1980). Such improvements derive from value shifts as well as from knowledge. The higher incomes that come with educational attainment[1] also play a role by enabling the purchase of better food, housing, medical care, and sanitation services. Furthermore, because education leads to the entrance of females into the labour force and increased social and geographic mobility, it is associated with lower fertility (Berry, 1982, 1985; Cochrane, 1979; Colclough, 1982; Findley, 1977; Findley *et al.*, 1979), which in turn eases non-productive demands on national economies. On the basis of these and similar effects, public expenditures for education, health, and nutrition often are advocated as providing high returns to society (World Bank, 1980).

Others, however, take a less positive view. They argue that education, rather than improving the population at large, serves to screen out individuals who are more motivated, intelligent, and trainable for employment and socialise them accordingly (Blaug, 1973, 1976; Layard and Psacharopoulos, 1974). This perspective further holds that job opportunities and work experience are more critical to productivity, and that scarce public resources could be better allocated to vocational training, job creation, and expansion of job-generating economic activities.

Whether education, work experience, or a combination thereof is seen as more important, it generally is agreed that the skill level of the population, indicated here by occupational status, is a critical resource for economic development (Blaug, 1976; Colclough, 1982; Lockheed *et al.*, 1980; Squire, 1979). Other human resource attributes regarded as important are age (or life cycle stage), since the acquisition of education, work experience, and skills requires the passage of time; and gender because men and women acquire and use different skills. Age and gender also affect individual receptivity to education and skill acquisition. Moreover, younger persons are thought to be more innovative, receptive to change, achievement-oriented, and energetic so that, from a societal point of view, investment in the young permits a long payback period over which gains can be realised. One might note, however, that age and gender effects are usually considered in terms of their interaction with education, rather than singly.

Migrant selectivity

The preceding indicates that certain human resource characteristics are desirable if development is to occur. Interestingly, these align with individual characteristics that differentiate migrants from nonmigrants and form the basis of migrant selectivity.

Educational attainment, for example, varies directly with the propensity to migrate (Connell et al., 1976). Illustrative is Preston et al.'s (1979) study of five areas in highland Ecuador. Among persons with five or six years of schooling, 62 per cent had migrated at some time in their lives, while the figure for those with no schooling was 15 per cent. This differential stems in part from the socialisation experience of education, which broadens individual horizons. The more educated also migrate for employment commensurate with their training, demonstrating a direct relationship between migration propensity and occupational status or skill level. Migration for education also is common (Connell et al., 1976), and many do not return after completing school. Nevertheless, that uneducated and less skilled persons are less likely to migrate is offset by their larger number overall; hence, such individuals still comprise the majority of movers (Connell et al., 1976).

Selectivity by age is also a nearly universal phenomenon. Migration propensity is greatest between the ages of 15 and 30 when career, family, and many other major decisions are made (Connell et al., 1976). Gender, on the other hand, is culture-specific in its relationship to migration, reflecting the differing roles of men and women in various societies (Boserup, 1970; Connell et al., 1976; Khoo et al., 1984; Pittin, 1984; Trager, 1984).

Development, migration and human resource redistribution

Rates and patterns of migration are affected by the locus, range, and mix of job opportunities, by social system characteristics (including degree of 'Westernisation'), by the density of transportation and communications infrastructures, and by the relative importance of origin push factors with respect to destination pulls (Brown and Sanders, 1981). These elements change considerably with development and – more importantly – in predictable spatial patterns that render a distinct order to human resource redistribution resulting from migrant selectivity.

To elaborate, historical experience indicates economic growth is concentrated in a core region over most of a development cycle, reflecting the initial and continuing importance of agglomeration economies. Eventually, rising land values, congestion, needs for market expansion, high labour costs, and other diseconomies of scale prompt decentralisation.[2] This scenario is articulated in the core–periphery conceptualisation wherein a critical occurrence is the shift away from core dominance (Brown, 1981; Friedmann, 1966, 1972; Richardson, 1979), referred to by Richardson (1980) as polarisation reversal.[3]

Migration enters this scenario as follows. The relative strengths of centralisation (or polarisation) at any given time determine where economic conditions, represented in migration models by wage and job opportunity variables, are more attractive (Brown, 1974; Gaile, 1980; Richardson, 1976). Initially, because polarisation effects dominate, origins tend to be in periphery locales, while destinations are in the core. The result is a concomitant shift of quality human resources through migrant selectivity, and an alteration in each place's development base. Accordingly, economic differentials underlying the initial migrations are exacerbated, leading to further migration, increased exacerbation of economic differentials, and so on (Myrdal, 1957).

Change in this centralising tendency, or polarisation reversal, has been sluggish at best in less developed nations. Accordingly, governments frequently employ policy intervention to hasten or stimulate the reversal (Renaud, 1981; Richardson, 1977, 1978, 1980; Rondinelli, 1983; Townroe, 1984). In fact, Townroe and Keen (1984) note that 'Those policies which seek to promote metropolitan decentralisation and/or lagging region development will normally have PR (polarisation reversal), or an acceleration of the reversal once started, as a prime strategic objective' (p. 47). Examples of such intervention include industrial decentralisation incentives or dictates in Korea (Smith *et al.*, 1983) and Venezuela (Blutstein *et al.*, 1977; Ewell, 1984), growth pole and intermediate size city strategies in many countries (Richardson, 1977; Richardson and Richardson, 1975; Rondinelli, 1983; Rondinelli *et al.*, 1984), and the establishment in the periphery of new economic nuclei or national capitals such as Ciudad Guayana in Venezuela, Brasilia in Brazil, and Abuja in Nigeria.

Venezuelan study

The above review indicates a complex set of relationships wherein human resources are catalysts of economic growth; growth prompts migration; migration transports human resources to dynamic locales; and so on. This complexity introduces a risk of oversimplification and misspecification of causal mechanisms, a problem identified elsewhere as simultaneity bias (Gober-Meyers, 1978a; Greenwood, 1975; Willis, 1974). Nevertheless, to further understand the relationship between development, migration, and shifts in the human resource characteristics of subnational political units, a Venezuelan example is considered. The data base and study area are first discussed; attention then turns to empirical analyses.

Data base and study area

The initial data set is 439 815 individual records from Venezuela's 1971 Census of Population, an approximately 5 per cent sample drawn by (and available through) Centro Latinoamericano de Demografia (CELADE). Of these, 116 672 records represent economically active persons; that is, those who indicated employment and were 15 or more years of age at census time, the age criterion used in asking occupation/employment questions. Eliminated from this set were the records of 13 698 economically active persons who had migrated from foreign nations, and 126 records of individuals who did not respond to questions concerning migration.

The remaining 102 848 records were used in analyses reported below. To bring out place differences, they were tabulated according to distrito of present residence into one of eight urban places – Caracas, Maracaibo, Valencia, Barquisimeto, Maracay, San Cristóbal, Ciudad Guayana, Ciudad Bolívar – or into the Rest of Venezuela category, which was treated as a single unit (Figure 15.1). Represented,

Fig. 15.1 The study area

then, are Venezuela's major urban centres plus foci of the Ciudad Guayana Project, a major post-World War II planning effort (Rodwin, 1969; Friedmann, 1966).

The economic structure (employment by sector), population, and average annual growth rate of these geographical units are reported in Table 15.1. Nationally, the largest sectors of employment are (in order) service, agriculture, manufacturing, and commerce. The Rest of Venezuela is dominated by agriculture, which comprises 43.1 per cent of its employment, but also noteworthy are the percentage of the agricultural workforce in San Cristóbal, Ciudad Bolívar, and Barquisimeto, which range from 19.0 to 10.9 per cent.[4] The strongest sector among urban places is service, which includes government employment. It occupies as much as 43.7 and 42.1 per cent of the labour force in San Cristóbal and Caracas, respectively, and no less than 27.6 per cent in Ciudad Guayana. Even in the Rest of Venezuela, service sector employment is 22.6 per cent. Manufacturing is most evident in Maracay, Valencia, Ciudad Guayana, and Caracas, which show percentages ranging from 30.8 to 22.7. Finally, commerce occupies a similar position in the employment structure of all places, ranging from 11.1 per cent in the Rest of Venezuela to 20.0 and 21.7 per cent in Barquisimeto and Maracaibo, respectively.

These patterns reflect Venezuela's decentralisation policies, which emerged in the middle 1950s to offset development-related polarisation. A major component in the decentralisation drive was extensive enlargement of Ciudad Guayana and Ciudad Bolívar, located in the sparsely populated Eastern Plateau region where abundant natural resources provide raw materials for hydro-electric power generation and for steel, aluminium, pulp, paper, and metallurgical goods production (Blutstein et al., 1977; Ewell, 1984; Friedmann, 1966; Morris, 1981; Rodwin, 1969). In addition to employment opportunities, incentives for prospective migrants included public housing, resettlement grants, and a social infrastructure of hospitals, schools, and transportation (MacDonald, 1969; Rodwin, 1969).

222 Differential Urbanization

More spontaneous economic processes underlie growth elsewhere in Venezuela, where only piecemeal policies on industrial decentralisation were implemented. Although this approach might be expected to have minimal impact, the primate city of Caracas, whose rapid growth was a major stimulus of decentralisation policies, has probably grown less than it otherwise might have. Similarly, Valencia and Maracay have probably grown more (Table 15.1), in part because they provided decentralisation opportunities in proximity to Caracas (Figure 15.1).

Table 15.1 Economic structure characteristics of selected Venezuelan places[a]

Economic sector of employment	Caracas	Maracaibo	Valencia	Barquisimeto	Maracay
Agriculture (%)	1.4	6.6	7.0	10.9	3.2
Mining (%)	0.5	2.8	0.5	0.4	0.3
Manufacturing (%)	22.7	15.6	27.2	17.9	30.8
Utilities (%)	2.1	1.7	2.0	1.1	1.8
Construction (%)	6.3	7.9	7.4	8.0	7.5
Commerce (%)	14.9	21.7	16.8	20.0	16.6
Transport & communications (%)	6.4	7.4	4.2	5.2	5.0
Financial (%)	3.6	2.2	1.8	1.7	2.0
Service and government (%)	42.1	34.1	33.1	34.8	32.8
Population[b]	2 144 337	651 574	367 171	330 815	255 134
Average annual growth rate (%)					
1951–71[b]	10.9	8.8	15.7	10.7	14.8
1961–71[b]	5.9	5.4	12.3	6.6	8.9
	San Cristóbal	Ciudad Guayana	Ciudad Bolívar	Rest of Venezuela	All of Venezuela
Agriculture (%)	19.0	9.1	16.5	43.1	23.6
Mining (%)	0.1	6.2	10.0	1.5	1.4
Manufacturing (%)	10.2	24.8	12.7	10.6	16.2
Utilities (%)	0.8	2.5	2.4	1.0	1.4
Construction (%)	6.3	8.2	7.2	5.5	6.2
Commerce (%)	13.8	14.9	12.8	11.1	13.6
Transport and communications (%)	5.1	5.6	3.3	3.9	5.0
Financial (%)	1.0	1.1	1.3	0.7	1.8
Service and government (%)	43.7	27.6	33.8	22.6	30.8
Population[b]	151 717	143 540	103 728	6 573 506	10 721 522
Average annual growth rate (%)					
1951–71[b]	9.1	18.4	11.5	3.8	5.7
1961–71[b]	5.5	8.4	6.4	3.0	4.3

[a] Economic structure characteristics and human resource attributes for all tables are derived from a sample of 102 848 individual records of the 1971 Venezuelan Population Census, with each record weighted to account for sampling variations. This is the Venezuelan-born labour force component af a larger sample of 439 815, drawn by the Centro Latinoamericano de Demografia (CELADE), which represents approximately 5 per cent of the 1971 Venezuelan population
[b] Population and growth rate figures are from 1971 Population Census publications. Average annual growth rate is computed as [(Pop1971/Pop1951)–1]/20 or [(Pop1971/Pop1961)–1]/10

More precisely, traditional polarisation measures indicate virtually no change since 1950. For the 42 urban areas with population greater than 20 000 in 1971, Gini coefficients were 0.739 in 1950, 0.725 in 1961, and 0.723 in 1971. The proportion of this population residing in the Caracas metropolitan area was, during these same years, 0.390, 0.395, and 0.378; while the proportion in the four largest places (Caracas, Maracaibo, Valencia, Maracay) was 0.637, 0.637, and 0.634. Does this mean that changes pertinent to polarisation reversal did not occur over that 20-year period? Analysis of human resource shifts indicate otherwise!

Empirical analyses

Attention now turns to human resource shifts occurring through migration, and in particular, how such shifts are related to economic structure, regional policy, and the overall development process. Human resource characteristics include age; gender, tabulated as 1 if male, 2 if female; educational attainment, tabulated as 0 if without schooling, 1 if primary schooling, 2 if secondary schooling, and 3 if college; and occupational status, tabulated through Treiman's (1977) scale wherein occupations range from 1, the least prestigious, to 100, the most prestigious. Migrants are defined as individuals whose *distrito* of present residence differs from that of birth; this yields 25 921 migrants and 76 927 nonmigrants.[5]

Two sets of analyses are discussed. The first set pertains to human resource gains and losses from migration and their relationship to the economic structure of each place. The second set pertains to differences in migrant attributes associated with each sector of employment. The principal methodology employed in this study is cross tabulation and tests for significant differences between means. Logit analysis was considered, but limitations of available computer programs, both on the number of variables that can be considered and on the range of values for each, were deemed too restrictive.

Human resource gains and losses from migration

A typical member of the Venezuelan labour force in 1971 was 33.7 years old, educated slightly less than the primary level (mean = 0.9), employed in a job with an occupational status of 34.6, and more likely to be male than female (mean = 1.3) (Table 15.2). However, human resource profiles of urban *distritos* and the Rest of Venezuela differ markedly, with the Rest of Venezuela showing a greater proportion of males, considerably less educational attainment, and lower occupational status. At the opposite extreme is Caracas, which has more females and greater educational attainment than other urban *distritos*.

Concerning migration effects, the Rest of Venezuela has nearly 12 times more outmigrants than inmigrants. The reverse is true for Caracas, and high ratios favouring inmigration also are found for Maracaibo, Valencia, Maracay, and Ciudad Guayana. Overall, outmigrants tend to surpass stayers in educational attainment and occupational status, which is expected from previous

Table 15.2 Means and inmigrant versus outmigrant differences for human resource attributes by place

Human resource attributes		Caracas	Maracaibo	Valencia	Maracay	Ciudad Guayana	Ciudad Bolívar	Barquisimeto	San Cristóbal	Rest of Venezuela
Gender[a] (population mean = 1.25)	S^b	1.36	1.24	1.30	1.31	1.31	1.25	1.28	1.32	1.18
	I	1.38	1.27	1.32	1.29	1.22	1.26	1.33	1.28	1.23
	O^c	1.28	1.29	1.35	1.31	1.36	1.37	1.29	1.37	1.35
	t	13.89*	−1.72	−2.78*	−1.83	−6.10*	−5.96*	4.01*	−3.63*	−20.58*
Age (population mean = 33.69)	S	29.99	32.64	31.74	30.56	31.61	34.20	33.57	33.12	34.86
	I	33.66	36.98	32.37	33.08	31.86	34.99	33.88	34.46	32.76
	O	30.33	31.38	36.15	35.30	34.15	36.27	35.21	32.37	33.75
	t	19.45*	20.45*	−12.64*	−6.91*	−3.70*	−2.35*	−4.28*	3.67*	−6.58*
Educational attainment (population mean = 0.93)	S	1.38	1.16	1.15	1.07	1.19	0.91	0.92	1.03	0.67
	I	1.20	1.03	1.15	1.15	1.13	1.09	1.25	1.29	1.15
	O	1.45	1.50	1.23	1.21	1.33	1.46	1.09	1.21	1.15
	t	−23.17*	−26.08*	−4.22*	−3.20*	−5.53*	−10.48*	8.66*	2.15*	0.37
Occupational status (population mean = 34.55)	S	38.97	38.56	36.25	36.97	37.19	35.33	35.15	35.74	31.81
	I	36.54	36.79	36.25	37.91	36.88	38.30	39.22	42.81	38.25
	O	41.23	41.11	38.35	38.17	38.60	42.59	36.89	36.07	36.35
	t	−24.02*	−14.18*	−6.37*	−0.74	−2.67*	−6.87*	7.05*	9.83*	11.42*
Stayers		11 949	4 975	2 363	1 072	343	669	3 000	1 133	51 423
Inmigrants		16 495	1 500	2 163	1 908	863	274	841	112	1 765
Outmigrants		1 363	716	478	433	93	286	818	987	20 747

* Indicates (inmigrant mean − outmigrant mean) is significantly different at 0.05 level
[a] See text for scales associated with each human resource attribute in Tables 15.2–15.6
[b] S = stayer mean, I = inmigrant mean, O = outmigrant mean, t = t-value for inmigrants versus outmigrants
[c] The gender, age, educational attainment, and occupational status of outmigrants are as of 1971, not as of the time of outmigration

research, but also tend to be older and female, an unexpected finding. These differences between outmigrants and stayers are especially marked in the Rest of Venezuela; yet its inmigrants tend to have as much or more education and higher occupational status than those leaving. One emergent pattern, then, is human resource depletion in the Rest of Venezuela due to outmigration volume, but partial replacement by inmigrants of better or similar quality. This indicates that *incipient polarisation* reversal may be under way.

Three patterns emerge among urban *distritos*, and the order of columns in Table 15.2 reflects these. First, Barquisimeto and San Cristóbal are similar to the Rest of Venezuela in that outmigrants are noticeably better off than stayers in educational attainment and occupational status. Both groups are surpassed by inmigrants in these qualities; and, at least in San Cristóbal, outmigration considerably exceeds inmigration. Given that agriculture is important in both Barquisimeto and San Cristóbal (Table 15.1), these places can be viewed as part of Venezuela's periphery. Like the Rest of Venezuela, a gain in human resource ingredients pertinent to polarisation reversal is in evidence.

Caracas and Maracaibo provide a second pattern. Outmigrants tend to be younger and more educated and possess higher occupational status than either inmigrants or stayers, but the number of outmigrants is relatively small. A similar pattern is in evidence for Valencia, Maracay, Ciudad Guayana, and Ciudad Bolívar, except that outmigrants tend to be older than either inmigrants or stayers. Gender shows no consistent pattern among these groups; migration adds females in Caracas and Barquisimeto, but males in Valencia, Ciudad Guayana, Ciudad Bolívar, and San Cristóbal.

In terms of these data, then, the urban *distritos* of Barquisimeto and San Cristóbal are similar to Venezuela's periphery in both human resource and economic structure aspects. Emerging as core areas, on the other hand, are Caracas, Maracaibo, Valencia, Maracay, Ciudad Guayana, and Ciudad Bolívar, which share a different set of human resource dynamics and have economic structures (Table 15.1) that better conform with advanced economy norms. The qualifying phrase 'In terms of these data,' which initiates the paragraph, is important, however. Under broader criteria, researchers have tended to see Barquisimeto as core, and Ciudad Guayana and Ciudad Bolívar as periphery.

That core places gain numbers through migration, but lose in educational attainment and occupational status qualities, indicates they serve development by supplying human resources elsewhere, thus contributing to polarisation reversal. Also noteworthy are age and gender differences between the longer established core *distritos* of Caracas and Maracaibo and the recently prominent cities of Valencia, Maracay, Ciudad Guayana, and Ciudad Bolívar, where migration provides an impetus toward a younger and more male population. This may be typical of newly emerging urban places, which are characterised as having dynamic economic structures and, hence, greater opportunity. That newly emerging places have relatively greater strength in manufacturing, rather than finance or service activity (Table 15.1), also may be typical.

Given the prominence of Venezuela's decentralisation policies, it is important to examine their contribution to the relationships just noted.

Accordingly, inmigrants to Ciudad Guayana were compared to those elsewhere for each human resource attribute (Table 15.3). An unambiguous pattern is found for age and gender, with the Ciudad Guayana sample tending to be younger and more male than any other group of inmigrants. In educational attainment and occupational status, however, Guayana migrants show no consistent pattern of differentiation from inmigrants to other core *distritos* and are less endowed than inmigrants to periphery places such as Barquisimeto, San Cristóbal, and the Rest of Venezuela. In spite of directly focused policy efforts, then, Ciudad Guayana does not differ markedly from other recently emerging urban *distritos*; that is, although its dynamic growth and industrial orientation attract more venturesome migrants, the more skilled continue to gravitate towards spontaneously growing periphery locales.

These findings, together with the knowledge (from earlier, exploratory analyses) that inmigrants to Ciudad Guayana tend to come from surrounding locales, suggest two scenarios of planned growth impact. First, Ciudad Guayana might perform a staging function by training quality persons who originate locally and then migrate elsewhere; this would explain our finding that periphery locales benefit from migration-related shifts in human resources. Alternatively, the majority of quality human resources may be absorbed rather than exported, thus furthering the build-up of Guayana's integrated growth nucleus. This hypothesis is suggested by its high volume of inmigration relative to outmigration. These scenarios, which will be examined

Table 15.3 Inmigrant human resource attributes of Ciudad Guayana compared to other places

Human resource attributes		Ciudad Guayana	Caracas	Maracaibo	Valencia	Maracay
Gender	I^a	1.22	1.38	1.27	1.32	1.29
	t	—	−18.81*	−5.77*	−10.76*	−8.02*
Age	I	31.86	33.65	36.98	32.37	33.08
	t	—	−8.34*	−19.72*	−2.11*	−4.93*
Educational attainment	I	1.13	1.20	1.03	1.15	1.15
	t	—	−5.39*	6.21*	−1.41	−1.25
Occupational status	I	36.88	36.54	36.79	36.25	37.91
	t	—	1.40	0.30	2.45*	−3.89*
		Ciudad Bolívar	Barquisimeto	San Cristóbal	Rest of Venezuela	
Gender	I	1.26	1.33	1.28	1.23	
	t	−2.86*	−10.33*	−3.05*	−1.47	
Age	I	34.99	33.88	34.46	32.76	
	t	−7.74*	−6.92*	−4.58*	−3.55*	
Educational attainment	I	1.09	1.25	1.29	1.15	
	t	1.50	−7.10*	−4.63*	−1.34	
Occupational status	I	38.30	39.22	42.81	38.25	
	t	−3.31*	−7.43*	−9.71*	−4.89*	

* Indicates (Ciudad Guayana mean − other place mean) is significantly different at the 0.05 level
[a] I = Inmigrant mean, and t = t-value for Ciudad Guayana versus place designated by column

in the following section, are associated with, respectively, exploitative and developmental tendencies in urban places (Hoselitz, 1960; Lawson, 1984, 1986; Rondinelli, 1983). As used in this literature, exploitative refers to urban areas that drain human and natural resources from their sphere of influence, while returning few positive local impacts. Developmental centres use these resources in a manner that establishes a base for future economic growth.

Economic sector differences in human resource attributes

Tables 15.4, 15.5, and 15.6 report the average age, educational attainment, and occupational status, by economic sector, for inmigrants to each urban *distrito*, the Rest of Venezuela, and all of Venezuela. Analysis of variance statistics (Table 15.7) indicates that the human resource composition of migration streams is better accounted for by economic sector than by geographic place effects, but that both are highly significant. Interaction effects between economic sector and geographic place also are significant.

To further explore economic sector influences, first compare the percent of migrants in each sector across all of Venezuela (Tables 15.4, 15.5, and 15.6) with similar figures for the total population (Table 15.1). Disparity is most evident in agriculture, which employs 23.6 per cent of the Venezuelan labour force but only about 3 per cent of its migrants. Alternatively, service employment is approximately 13 percentage points more prevalent among migrants than among the overall population, and manufacturing approximately 5 points more prevalent. Commerce employs nearly the same proportion in both populations. Clearly, then, migrants differ from the general population in employment preferences.

Migrants also differ from the general population in human resource attributes, as established in the preceding section, but there is additional variation, within the migrant group itself, by economic sector. Migrants employed in manufacturing, for example, average 30.7 years of age, compared to 37.4 for agriculture and 34.0 years for commerce, finance, and service (Table 15.4). In educational attainment (Table 15.5), agriculture is lowest with a 0.6 average; manufacturing and commerce show 1.1, while migrants in finance and service average 1.5 and 1.3, respectively. In occupational status (Table 15.6), agriculture is again the least endowed, with an average score of 27; manufacturing and service carry prestige averaging approximately 36; and finance and commerce score 40.

To place these findings in a broader context, consider the developmental–exploitative distinction introduced above. If an area's economic activity is concentrated in sectors which attract quality migrants, such as manufacturing and service, its human resource base and related growth potential will be greater. These activities, then, have developmental impacts on urban and regional dynamics. Activities which attract less endowed migrants, such as agriculture in the present example, engender exploitative impacts. Note, however, that an economic sector exploitative in human resource terms may be developmental by other criteria. Commercial export agriculture, for example, generates needed foreign capital. Further, an economic sector's ability to attract

Table 15.4 Mean inmigrant age by economic sector and destination place

Economic sector of employment	Caracas	Maracaibo	Valencia	Maracay	Ciudad Guayana	Ciudad Bolívar	Barquisimeto	San Cristóbal	Rest of Venezuela	All of Venezuela	
Agriculture	40.02	35.81	38.38	34.16	36.05	40.74	34.07	35.18	36.97	37.38	3.20%
	(139)[a]	(65)	(97)	(42)	(83)	(24)	(32)	(10)	(269)	(761)	
Mining	40.22	44.70	37.26	35.31	38.48	37.34	39.43	—	38.02	39.89	1.05%
	(90)	(43)	(9)	(6)	(44)	(21)	(7)		(31)	(251)	
Manufacturing	30.57	34.86	30.43	30.89	30.14	32.95	31.17	30.22	30.06	30.70	21.21%
	(3 115)	(161)	(554)	(534)	(204)	(35)	(121)	(14)	(310)	(5 048)	
Utilities	38.35	38.50	35.17	36.82	33.12	36.21	38.91	32.82	39.68	37.91	2.21%
	(350)	(36)	(34)	(33)	(21)	(7)	(15)	(3)	(27)	(526)	
Construction	34.31	39.73	34.54	34.23	34.88	33.56	33.73	40.47	36.47	34.82	6.57%
	(970)	(111)	(139)	(129)	(65)	(20)	(42)	(4)	(84)	(1 564)	
Commerce	33.47	37.88	33.92	34.91	32.54	35.35	36.54	34.51	33.86	34.12	13.85%
	(1 961)	(273)	(295)	(254)	(113)	(24)	(165)	(13)	(197)	(3 295)	
Transport and communication	38.04	41.12	37.39	36.74	34.26	31.88	37.24	40.21	36.80	37.94	6.12%
	(963)	(118)	(86)	(90)	(40)	(8)	(52)	(4)	(95)	(1 456)	
Financial	34.86	37.53	32.24	31.19	32.76	35.85	33.20	—	28.28	34.30	2.62%
	(480)	(22)	(29)	(36)	(8)	(4)	(19)		(26)	(624)	
Services, including government	34.14	35.12	31.45	33.20	29.53	33.66	33.19	34.32	31.08	33.68	43.17%
	(7 122)	(522)	(704)	(629)	(229)	(111)	(320)	(51)	(554)	(10 272)	
Total observations	15 190	1 381	1 947	1 753	807	254	773	99	1 593	23 797	100.00%

[a] Numbers in parentheses are the number of observations in each sample

Table 15.5 Mean inmigrant educational attainment by economic sector and destination place

Economic sector of employment	Caracas	Maracaibo	Valencia	Maracay	Ciudad Guayana	Ciudad Bolívar	Barquisimeto	San Cristóbal	Rest of Venezuela	All of Venezuela	
Agriculture	0.97	0.54	0.67	0.93	0.51	0.25	0.37	0.93	0.52	0.64	3.11%
	(132)[a]	(54)	(90)	(40)	(82)	(19)	(28)	(10)	(251)	(706)	
Mining	1.51	1.16	1.11	1.04	1.13	0.99	1.60	—	1.48	1.31	1.03%
	(85)	(41)	(9)	(6)	(43)	(18)	(7)	—	(26)	(235)	
Manufacturing	1.12	1.07	1.16	1.11	1.28	1.17	1.08	1.00	1.21	1.13	21.28%
	(2 992)	(147)	(527)	(513)	(203)	(33)	(119)	(13)	(289)	(4 836)	
Utilities	1.13	0.98	1.27	1.13	1.17	0.79	1.30	1.36	1.28	1.14	2.22%
	(336)	(32)	(33)	(33)	(21)	(7)	(14)	(3)	(26)	(505)	
Construction	0.89	0.80	0.76	0.84	0.92	0.82	0.97	0.41	0.95	0.87	6.55%
	(922)	(104)	(134)	(127)	(64)	(18)	(38)	(4)	(77)	(1 488)	
Commerce	1.17	0.87	1.11	1.12	1.09	1.18	1.19	1.89	1.10	1.13	13.85%
	(1 883)	(263)	(276)	(240)	(112)	(22)	(161)	(13)	(178)	(3 148)	
Transport and communication	1.13	1.01	1.08	1.11	1.12	1.42	1.21	1.32	1.16	1.13	6.12%
	(923)	(107)	(80)	(87)	(40)	(8)	(51)	(4)	(90)	(1 390)	
Financial	1.47	1.49	1.63	1.36	1.12	1.14	1.18	—	1.72	1.47	2.64%
	(465)	(21)	(27)	(35)	(8)	(4)	(18)	—	(23)	(601)	
Services, including government	1.26	1.16	1.28	1.25	1.24	1.33	1.45	1.63	1.44	1.27	43.20%
	(6 828)	(515)	(660)	(611)	(226)	(105)	(306)	(51)	(516)	(9 818)	
Total observations	14 566	1 284	1 836	1 692	799	234	742	98	1 476	22 727	100.00%

[a] Numbers in parentheses are the number of observations in each sample

Table 15.6 Mean inmigrant occupational status by economic sector and destination place

Economic sector of employment	Caracas	Maracaibo	Valencia	Maracay	Ciudad Guayana	Ciudad Bolívar	Barquisimeto	San Cristóbal	Rest of Venezuela	All of Venezuela	
Agriculture	30.29	26.95	28.44	32.09	24.37	28.27	27.34	29.98	25.66	27.43	3.19%
	(139)[a]	(65)	(97)	(42)	(83)	(24)	(32)	(10)	(269)	(761)	
Mining	42.64	41.04	36.82	34.96	36.12	35.55	41.93	—	42.20	40.18	1.05%
	(90)	(43)	(9)	(6)	(44)	(21)	(7)		(31)	(251)	
Manufacturing	35.40	36.41	37.39	36.88	39.27	38.74	35.92	37.97	38.38	36.14	21.21%
	(3 115)	(161)	(554)	(534)	(204)	(35)	(121)	(14)	(310)	(5 048)	
Utilities	36.64	33.90	36.13	38.57	36.94	37.11	35.91	33.09	42.60	36.83	2.21%
	(350)	(36)	(34)	(33)	(21)	(7)	(15)	(3)	(27)	(526)	
Construction	36.32	36.49	35.17	35.65	36.73	34.10	36.30	35.53	37.68	36.24	6.57%
	(970)	(111)	(139)	(129)	(65)	(20)	(42)	(4)	(84)	(1 564)	
Commerce	38.79	40.24	39.63	39.00	40.27	40.04	39.00	48.17	40.76	39.20	13.86%
	(1 962)	(273)	(295)	(254)	(113)	(24)	(165)	(13)	(197)	(3 296)	
Transport and communication	34.54	33.76	33.68	37.73	33.44	35.45	35.55	31.16	34.71	34.63	6.12%
	(964)	(118)	(86)	(90)	(40)	(8)	(52)	(4)	(95)	(1 457)	
Financial	39.80	45.66	40.32	41.57	42.61	39.90	45.79	—	47.36	40.65	2.62%
	(480)	(22)	(29)	(36)	(8)	(4)	(19)		(26)	(624)	
Services, including government	36.22	35.96	35.30	38.57	37.63	42.06	42.46	48.34	43.06	36.96	43.17%
	(7 125)	(552)	(704)	(629)	(229)	(111)	(320)	(51)	(554)	(10 275)	
Total observations	15 195	1 381	1 947	1 753	807	254	773	99	1 593	23 802	100.00%

[a] Numbers in parentheses are the number of observations in each sample

Table 15.7 Economic sector versus geographic place effects for age, educational attainment, and occupational status of inmigrants: F-statistics from analysis of variance[a]

	Age	Educational attainment	Occupational status
Economic sector effects	359.80	492.36	338.87
Geographic place effects	61.36	37.26	84.35
Interaction effects	8.25	10.63	13.11

[a] The basic data for these analyses are summarised in Tables 15.4, 15.5 and 15.6. All statistics are significant at the 0.001 level with (8,8) degrees of freedom

and retain quality human resources is linked to its macro-economic environment; for example, the returns to peasant or small-scale agriculture and its attractiveness as an economic enterprise have historically been lessened by foreign trade policies (Richardson, 1980; Swift, 1978).

Because development is a spatial process that differentiates areas from one another, geographic place effects also enter into the relationship between economic sector and human resource attributes. To address this, a distinction noted in the preceding section is utilised by jointly treating Caracas and Maracaibo as established core; Valencia, Maracay, Ciudad Guayana, and Ciudad Bolívar as recent core; and Barquisimeto, San Cristóbal, and the Rest of Venezuela as periphery. In terms of age (Table 15.4), financial and service sector migrants to the Rest of Venezuela tend to be considerably younger, and those to the established core older than elsewhere; other patterns are not evident. In education (Table 15.5), attainment among agricultural migrants tends to be higher in the established and recent cores, and lower in the periphery. The reverse pattern holds for service sector migrants. Also noteworthy is high educational attainment among manufacturing migrants to Ciudad Guayana, and among financial migrants to Valencia and the Rest of Venezuela. Occupational status (Table 15.6) of commerce migrants to the periphery somewhat exceeds that to core areas, whereas service migrants to the periphery are much higher in status than their core counterparts. Also noteworthy are low prestige scores for agricultural migrants to the Rest of Venezuela and Ciudad Guayana; high scores for manufacturing migrants to Ciudad Guayana, Ciudad Bolívar, San Cristóbal, and the Rest of Venezuela; and high status for commerce migrants to San Cristóbal and financial migrants to Maracaibo, Barquisimeto, and the Rest of Venezuela.

If these findings on geographic place effects are collected together, two additional points emerge. First, incipient polarisation reversal is again evident in that migrants to periphery locales, in several economic sectors, are significantly better endowed than their core counterparts. Second, Ciudad Guayana, as a planned growth centre, does not appear to behave differently than other places. Hence, policies which promote spontaneous or diffuse growth seem no less effective than focused regional development efforts in influencing human resource shifts.

Concluding observations

Three broad points may be drawn from this research regarding the interrelationships between migrant selectivity, human resource distribution, urban economic structure, regional policy, and polarisation reversal in the development process of Venezuela. This study, more generally concerned with migration effects upon development, complements studies of development influences on migration (Brown and Jones, 1985; Brown and Lawson, 1985; Brown and Sanders, 1981; Morrison, 1983; Peek and Standing, 1982). As such, it contributes to understanding reciprocal migration–development–urbanisation relationships, a topic of increasing concern (Brown and Stetzer, 1984; Gober-Meyers, 1978a, 1978b; Greenwood, 1978, 1981; Salvatore, 1981).

First, Venezuela appears to have been in the early stages of polarisation reversal in 1971. Core urban places continue to receive high migration volumes and thus maintain their dominance in absolute numbers, but migration-induced shifts of quality human resources clearly favour periphery locales. It seems, then, that a base for economic growth is being formed in the periphery, but that its magnitude in 1971 was not yet sufficient to alter the overall directionality of migration streams. This finding is important because polarisation reversal has been examined largely in terms of aggregates such as population distribution or net migration balances (Richardson, 1980; Townroe and Keen, 1984) and not in terms of the human resource shifts involved. An important consideration for future research is the time lapse between changes in human resource profiles and changes in the aggregate population measures more commonly employed to gauge polarisation reversal.

Second, the finding that both economic structure and geographic place effects influenced human resource distribution in Venezuela carries implications for the design of public policy, which often has polarisation reversal as an objective. Specifically, policy oriented towards a particular economic sector, such as steel or automobile production, may fail to generate expected trickle-down effects because of inadequacies in regional spatial structures and a lack of complementarity with local skills, factor endowments, or market demands (Belsky *et al.*, 1983; Evans, 1982; Rondinelli and Evans, 1983). Alternatively, incorporating a place dimension in development policy raises the issue of whether to choose a spatially concentrated focus, as in Venezuela's Guayana Project, or a dispersed focus, as in Venezuela's industrial decentralisation efforts. The former, essentially a growth centre strategy, has been given ample support (Friedmann, 1966; Hansen, 1972; Richardson and Richardson, 1975; Rodwin, 1969), but findings here are ambiguous as to its merit.[6] Ciudad Guayana gained in terms of net migration flows, but not in the human resource characteristics of its migrants. On the other hand, Ciudad Guayana's impact upon national consciousness and in fomenting economic shifts favouring the geographic periphery should not be underestimated. With regard to policies aimed at spatially dispersing economic activity, the findings here suggest these are equally, if not more, effective in distributing human resources throughout the national territory.

Also relevant to this policy debate is the finding that service or manufacturing activity attracts quality migrants, thus operating to build up human resources and enhance regional growth potentials. On the other hand, agriculture attracts less endowed migrants and is thus more likely to have exploitative, rather than developmental, impacts on urban and regional dynamics. On its face, this conclusion supports an economic sector focus in regional development policy. However, recent intermediate-size city research argues that sector success or failure is not an absolute. Instead, it must be set against local conditions. Rondinelli and Evans (1983), for example, explain differential urban growth in terms of local services, urban functions, transportation linkages, articulations of urban places into a hierarchical urban system, and so on.

Finally, macro policies which are not spatial in articulation frequently have distinct spatial implications (Glickman, 1980; Lentnek, 1980; Renaud, 1981; Richardson, 1977; Ruane, 1983). Because such policies condition both economic sector and regional attribute effects, they are an integral element of polarisation reversal and must be addressed by future research. To illustrate, consider currency exchange regulations facilitating export of abundant resources. Third World nations such as Venezuela competing in world petroleum markets tended to maintain fairly constant exchange rates from the 1960s through 1980, even though their economies experienced inflation. As a result, locally produced goods became increasingly expensive relative to imported ones and real wages deflated. Hence, domestic markets for local artisans, small-scale manufacturers, and agriculturalists were dampened. Devaluation in the early 1980s, forced by shrinking petroleum markets and recessionary economies, brought the reverse effect. During these policy eras, growth or stagnation in specific regions or urban areas depended on the way their economic structure interfaced with, and thus was affected by, existing currency policy. Hence, policies that are apparently aspatial may have distinct spatial implications. This fact must be considered when one evaluates the relationships among economic sectors, regional attributes, polarisation, and its reversal.

Notes

1. A multi-country survey by Psacharopoulos and Hinchliffe (1973, updated in Psacharopoulos, 1980), for example, indicates that primary education is associated with an average 24 per cent increase in lifetime earnings, although the range is from 6.6 per cent for Singapore to 82.0 per cent for Venezuela.
2. Recent empirical accounts of such decentralisation include Townroe (1984), Townroe and Keen (1984), Vining (1982), and Vining and Kontuly (1978).
3. Core–periphery can be seen as a spatial counterpart to the dual economy or two-sector growth formulation of economics (Brown and Jones, 1985; Brown and Lawson, 1985). This is an interesting interpretation in that the dual economy's driving force is the presence of human resources in the form of a virtually unlimited supply of labour at a low wage; for elaboration, *see* Todaro (1976, 1985).
4. Urban *distrito* boundaries frequently encompass rural areas, and the degree to which this occurs influences the percentage employed in agriculture. Although not ideal, *distrito* is the smallest areal unit for which adequate data are available. Further, that an urban place is overbounded, and thus includes rural land, is itself indicative of areal character.

5 Place of birth is the only variable in the 1971 Venezuelan Census of Population that allows identification of *distrito* of origin. This variable is often criticised as a basis for migration research, but its use is less problematic in the present study. First, our sample is restricted to employed persons aged 15 or more; thus, the young and the elderly are excluded. Second, the concern here is only that certain individuals possess a human resource potential and shift location rather than remain in their place of birth; the timing of migration is not addressed, nor is the geography of human resource acquisition. Third, this study is concerned with long-term trends, which are better indicated by place of birth than by short-term migration data.

6 In fact, growth centre policy has been subject to extensive critical review in recent years, often by the same people who supported it earlier (e.g. Conroy, 1973; Richardson, 1978; Rodwin, 1978).

References

Belsky, E., Hackenberg, R., Karaska, G. and Rondinelli, D. 1983: The role of secondary cities in regional development. Worcester, Mass: Clark University/IDA Cooperative Agreement on Settlement and Resource Systems Analysis and Management, Concepts Paper.

Berry, E. H. 1982: Migration, fertility, and development: A conceptual note and proposal for research on Ecuador. Columbus: Ohio State University Department of Geography, Studies on the Interrelationships Between Migration and Development in Third World Settings, Discussion Paper 3.

Berry, E. H. 1985. Regional structure effects on migration and fertility in Ecuador. Columbus: Ohio State University Department of Geography, Studies on the Interrelationships between Migration and Development in Third World Settings, Discussion Paper 15.

Blaug, M. 1973: *Education and the employment problem in developing countries*. Geneva: International Labour Organisation.

Blaug, M. 1976: Human capital theory: a slightly jaundiced survey. *Journal of Economic Literature* 14 (3), 827–56.

Blutstein, H. I., Edwards, J. D., Johnston, K. T., McMorris, D. S. and Rudolph, J. D. 1977: *Area handbook for Venezuela*. Washington, DC: United States Government Printing Office.

Boserup, E. 1970: *Woman's role in economic development*. New York: St Martin's Press.

Bowman, M. J. 1980: Education and economic growth: an overview. In King, T. (ed.), *Education and income*. Washington, DC: World Bank, Staff Working Paper 402, pp. 1–71.

Brown, L. A. 1974: Diffusion in a growth pole context. In Helleiner, F. and Stöhr, W. (eds), *Proceedings of the Commission on Regional Aspects of Economic Development of the International Geographical Union*. Vol. II: *Spatial aspects of the development process*. Toronto: Allister 243–57.

Brown, L. A. 1981: *Innovation diffusion: A new perspective*. London and New York: Methuen.

Brown, L. A. and Jones, J.P. III. 1985: Spatial variation in migration processes and development: a Costa Rican example of conventional modelling augmented by the expansion method. *Demography* 22 (3), 327–52.

Brown, L. A. and Lawson, V. A. 1985: Migration in Third World settings, uneven development, and conventional modelling: a case study of Costa Rica. *Annals of the Association of American Geographers* 75 (1), 29–49.

Brown, L. A. and Sanders, R. L. 1981: Towards a development paradigm of migration, with particular reference to Third World settings. In DeJong, G. F. and Gardner, R.W. (eds), *Migration decision making: Multidisciplinary approaches to microlevel studies in developed and developing countries.* New York: Pergamon Press, 149–85.

Brown, L. A., and Stetzer, F. C. 1984: Development aspects of migration in Third World settings: a simulation, with implications for urbanisation. *Environment and Planning A* 16 (12), 1583–603. (*see* this volume, Chapter 18)

Cochrane, S. H. 1979: *Fertility and education: What do we really know?* Baltimore: Johns Hopkins University Press.

Colclough, C. 1982: The impact of primary schooling on economic development: a review of the evidence. *World Development* 10 (3), 167–85.

Connell, J., Dasgupta, B., Laishley, R. and Lipton, M. 1976: *Migration from rural areas: Evidence from village studies.* Oxford: Oxford University Press.

Conroy, M. E. 1973. On the rejection of 'growth centre' strategies in Latin American regional development planning. *Land Economics* 49 (4), 371–80.

Evans, H. 1982. *Urban functions in rural development: The case of the Potosi region in Bolivia.* Washington, DC: United States Agency for International Development, Bureau for Science and Technology, Office of Multisectoral Development, Regional and Rural Development Division, Project Report.

Ewell, J. 1984: *Venezuela.* Stanford, Calif.: Stanford University Press.

Fields, G. S. 1980: Education and income distribution in developing countries: a review of the literature. In King, T. (ed.), *Education and income.* Washington, DC: World Bank, Staff Working Paper 402, pp. 231–315.

Findley, S. E. 1977: *Planning for internal migration: A review of issues and policies in developing countries.* Washington, DC: United States Government Printing Office.

Findley, S. E., Gundlach, J., Kent, D. P. and Rhoda, R. 1979: *Rural development, migration, and fertility: What do we know?* Washington, DC: United States Agency for International Development, Office of Rural Development and Development Administration, Rural Development and Fertility Project, Final Report.

Friedmann, J. 1966: *Regional development policy: A case study of Venezuela.* Cambridge, Mass.: MIT Press.

Friedmann, J. 1972: A general theory of polarised development. In Hansen, N. M. (ed.), *Growth centres in regional economic development.* New York: Free Press, 82–107

Gaile, G. I. 1980: The spread–backwash concept. *Regional Studies* 14 (1), 15–25.

Glickman, N. J. (ed.) 1980: *The urban impacts of federal policies.* Baltimore: Johns Hopkins University Press.

Gober-Meyers, P. 1978a: Employment motivated migration and growth in post-industrial market economies. *Progress in Human Geography* 2 (2), 207–29.

Gober-Meyers, P. 1978b: Interstate migration and economic growth: a simultaneous equations approach. *Environment and Planning A* 10 (11), 1241–52.

Greenwood, M. J. 1975: Simultaneity bias in migration models: an empirical examination. *Demography* 12 (3), 519–36.

Greenwood, M. J. 1978: An econometric model of internal migration and regional economic growth in Mexico. *Journal of Regional Science* 18 (1), 17–31.

Greenwood, M. J. 1981: *Migration and economic growth in the United States: National, regional, and metropolitan perspectives.* New York: Academic Press.

Hagen, E. E. 1962: *On the theory of social change: How economic growth begins.* Homewood, Ill.: Dorsey Press.

Hansen, N. M. (ed.) 1972: *Growth centers in regional economic development*. New York: Free Press.
Hicks, N. 1980: *Economic growth and human resources*. Washington, DC: World Bank, Staff Working Paper 408.
Hoselitz, B. F. 1960: *Sociological aspects of economic growth*. Glencoe, Ill.: Free Press.
Inkeles, A. and Smith, D.H. 1974: *Becoming modern: Individual change in six developing countries*. Cambridge, Mass.: Harvard University Press.
Khoo, S., Smith, P. C. and Fawcett, J. J. 1984: Migration of women to cities: the Asian situation in comparative perspective. *International Migration Review* 18 (4), 1247–63.
King, T. (ed.) 1980: *Education and income*. Washington, DC: World Bank, Staff Working Paper 402.
Lawson, V. A. 1984: *Intermediate cities and developmental/exploitative impacts on regional development*. Columbus: Ohio State University Department of Geography. Studies on the Interrelationships between Migration and Development in Third World Settings, Discussion Paper 19.
Lawson, V. A. 1986: National economic policies, local variation in structure of production, and uneven regional development: The case of Ecuador. Ph.D. dissertation, Ohio State University.
Layard, R. and Psacharopoulos, G. 1974: The screening hypothesis and returns to education. *Journal of Political Economy* 82 (5), 985–98.
Lentnek, B. 1980: Regional development and urbanisation in Latin America: the relationship of national policy to spatial strategies. In Thomas, R. N. and Hunter, J. M. (eds), *Internal migration systems in the developing world, with special reference to Latin America*. Cambridge: Schenkman, 82–113.
Lerner, D. 1958: *The passing of traditional society: Modernising the Middle East*. Glencoe, Ill.: Free Press.
Lockheed, M. E., Jamison, D. T. and Lau, L. J. 1980: Farmer education and farm efficiency: a survey. *Economic Development and Cultural Change* 29 (1), 37–76.
MacDonald, J. S. 1969: Migration and the population of Ciudad Guayana. In Rodwin, L. (ed.), *Planning urban growth and regional development: The experience of the Guayana program of Venezuela*. Cambridge, Mass.: MIT Press, 109–25.
McClelland, D. C. 1961: *The achieving society*. Princeton, NJ: Van Nostrand. Reprinted 1976, New York: Irvington Press.
Morris, A. 1981: *South America*, 2nd edn. Ottawa: Barnes & Noble.
Morrison, P. A. (ed.) 1983: *Population movements: Their forms and functions in urbanisation and development*. Liège: Ordina Editions.
Myrdal, G. 1957: *Economic theory and underdeveloped regions*. London: Duckworth.
Peek, P., and Standing, G. (eds). 1982: *State policies and migration: Studies in Latin America and the Caribbean*. London: Croom Helm.
Pittin, R. 1984: Migration of women in Nigeria: the Hausa case. *International Migration Review* 18 (4), 1293–314.
Preston, D. A., Taveras, G. A. and Preston, R. A. 1979: *Rural emigration and agricultural development in Highland Ecuador: Final report*. Leeds: University of Leeds School of Geography, Working Paper 238.
Psacharopoulos, G. 1980: Returns to education: an updated international comparison. In King, T. (ed.), *Education and income*. Washington, DC: World Bank, Staff Working Paper 402, pp. 73–109.
Psacharopoulos, G. and Hinchliffe, K. 1973: *Returns to education: An international comparison*. Amsterdam: Elsevier; San Francisco: Jossey-Bass.

Renaud, B. 1981: *National urbanisation policies in developing countries*. Oxford: Oxford University Press.
Richardson, H. W. 1976: Growth pole spillovers: the dynamics of backwash and spread. *Regional Studies* 10 (1), 1–9.
Richardson, H. W. 1977: *City size and national spatial strategies in developing countries*. Washington, DC: World Bank, Staff Working Paper 252.
Richardson, H. W. 1978: Growth centres, rural development and national urban policy: a defence. *International Regional Science Review* 3 (2), 133–52.
Richardson, H. W. 1979: *Regional economics*. Urbana: University of Illinois Press.
Richardson, H. W. 1980: Polarization reversal in developing countries. *Papers of the Regional Science Association* 45, 67–85 (*see* this volume, Chapter 11).
Richardson, H. W. and Richardson, M. 1975: The relevance of growth centre strategies to Latin America. *Economic Geography* 51 (2), 163–78.
Rodwin, L. (ed.) 1969: *Planning urban growth and regional development: The experience of the Guayana Program of Venezuela*. Cambridge, Mass.: MIT Press.
Rodwin, L. 1978: Regional planning in less developed countries: a retrospective view of the literature and experience. *International Regional Science Review* 3 (2), 113–31.
Rogers, E. M. 1969: *Modernization among peasants: The impact of communications*. New York: Holt, Rinehart, & Winston.
Rondinelli, D. A. 1983: *Secondary cities in developing countries: Policies for diffusing urbanization*. Beverly Hills: Sage.
Rondinelli, D. A. and Evans, H. 1983: Integrated regional development planning: linking urban centres and rural areas in Bolivia. *World Development* 11 (2), 31–53.
Rondinelli, D. A., Nellis, J. R and Cheema, G. S. 1984: *Decentralization in developing countries: A review of recent experience*. Washington, DC: World Bank, Staff Working Paper 581.
Ruane, F. P. 1983: Trade policies and the spatial distribution of development: a two sector analysis. *International Regional Science Review* 8 (1), 47–58.
Salvatore, D. 1981: *Internal migration and economic development: A theoretical and empirical study*. Washington, DC: University Press of America.
Schultz, T. W. 1980: The economics of being poor. *Journal of Political Economy* 88 (4), 639–51.
Smith, W. R., Huh, W. and Demko, G. J. 1983: Population concentration in an urban system: Korea 1949–1980. *Urban Geography* 4 (1), 63–79.
Squire, L. 1979: *Labor force, employment, and labor markets in the course of economic development*. Washington, DC: World Bank, Staff Working Paper 336.
Swift, J. 1978: *Economic development in Latin America*. New York: St Martin's Press.
Todaro, M. P. 1976: *Internal migration in developing countries: A review of theory, evidence, methodology, and research priorities*. Geneva. International Labour Office.
Todaro, M. P. 1985: *Economic development in the Third World*. 3rd edn. New York and London: Longman.
Townroe, P. M. 1984. Spatial policy and metropolitan economic growth in São Paulo, Brazil. *Geoforum* 15 (2), 143–65.
Townroe, P. M., and Keen, D. 1984: Polarisation reversal in the State of São Paulo, Brazil. *Regional Studies* 18 (1), 45–54 (*see* this volume, Chapter 13).
Trager, L. 1984: Family strategies and the migration of women: migrants to Dagupan City, Philippines. *International Migration Review* 18 (4), 1264–77.
Treiman, D. J. 1977: *Occupational prestige in comparative perspective*. New York: Academic Press.

Vining, D. R. Jr. 1982: Migration between the core and the periphery. *Scientific American* 247 (6), 45–53.
Vining, D. R. Jr and Kontuly, T. 1978: Population dispersal from major metropolitan regions: an international comparison. *International Regional Science Review* 3 (1), 49–73 (*see* this volume, Chapter 6).
Wheeler, D. 1980: *Human resource development and economic growth in developing countries: A simultaneous model.* Washington, DC: World Bank, Staff Working Paper 407.
Willis, K. G. 1974: *Problems in migration analysis.* Lexington, Mass.: Lexington Books.
World Bank 1980: *World development report, 1980.* New York: Oxford University Press.
Yotopolous, P. A. and Nugent, J. B. 1976: *Economics of development: Empirical investigations.* New York: Harper & Row.

16 H. S. Geyer,
'Implications of Differential Urbanisation on Deconcentration in the Pretoria–Witwatersrand–Vaal Triangle Metropolitan Area, South Africa'

From: *Geoforum* 21 (4), 385–96 (1990)

Concentration vs deconcentration in developed and less developed countries

Since the middle 1970s when the beginning of a new era in population migration patterns was announced in the USA (Berry, 1976), a considerable body of empirical evidence has evolved debating the relevancy of the concept of counterurbanisation in a number of developed countries (DCs) (Vining and Strauss, 1977; Vining and Kontuly, 1978; Vining and Pallone, 1982; Ogden, 1985; Dean, 1986; Kontuly *et al.*, 1986; Champion, 1989). Gordon (1979), however, argued that what has been presented as the beginning of a new epoch in migration dynamics in the USA was nothing more than a continuation of the urbanisation process as formulated by Clark (1967). Koch (1980), too, described claims of early signs of counterurbanisation in Western Europe as either the result of the exodus of a particular sector of the population (mostly the elderly) from metropolitan to nonmetropolitan regions, or the result of a negative economic conjuncture. Both Gordon (1979) and Koch (1980) regard the claim of the existence of counterurbanising forces in certain DCs as a *non sequitur*, claims based on the general overbounding of the functional spaces (Yeats, 1980) of metropolitan areas. Despite this controversy and the fact that census figures of the mid 1980s have indicated a decrease generally in the rates of counterurbanisation in the USA and north-west Europe (Cochrane and Vining, 1988), the concept of counterurbanisation in certain DCs seems to be well established (Hall, 1987).

Almost simultaneously to Berry's (1976) announcement of counterurbanisation, Richardson (1977) introduced the concept of polarisation reversal in a less developed spatial economic milieu. This study also aroused considerable interest (Linn, 1978; Richardson, 1980; Lo and Salih, 1981; Townroe and Keen, 1984; Lee, 1985). In parallel to the controversial criticism surrounding the relevancy of the concept of counterurbanisation in a developed economic milieu, Vining (1986) also questions certain claims of the beginning of the process of urban deconcentration in certain less developed countries (LDCs). According to him, these claims of deconcentration in specific LDCs may also be ascribed to an underbounding of the functional areas of metropolitan areas.

Differential urbanisation in DCs and LDCs

The indications are that while mainstream population migration patterns point to the beginning of the process of counterurbanisation in a number of DCs, substream migration patterns prevail in these countries which differ fundamentally from the former. While the mainstream migration tendency in the USA, for example, was counterurbanisation during the 1970s, Blacks and other minority groups were continuing to concentrate in the larger metropolitan areas (Berry, 1976, p. 21). In the [former] FRG and France, a large proportion of the younger, more economically active sector of the population was concentrating in core regions during the 1970s, while more of the older economically less active persons tended to migrate to rural areas (Koch, 1980, p. 63).

Similar patterns of differential urbanisation are manifested in LDCs. Since 1960 the relative concentration of different sectors of the population of South Africa has changed in a differentiated manner (Table 16.1).[1] It is clear from the results presented in Table 16.1(A)–(D) that two distinct, but opposing, population migration patterns are evolving in South Africa. On the one hand, Blacks are concentrating in the larger metropolitan areas. On the other hand, the shares of Whites and 'Coloureds' are increasing relatively in the core fringe zones and intermediate city regions. While the former mainstream migration pattern suggests a continuation of the urbanisation process, the latter migration pattern suggests early signs of the beginning of Richardson's (1977) process of polarisation reversal in South Africa.

According to Table 16.1(E), the share of Blacks has also tended to increase proportionately in the outer peripheral areas over the years. A large proportion of these Blacks have, however, accumulated on the borders inside Bantustans adjacent to the Preteria–Witwatersrand–Vaal traingle (the PWV) and Durban–Pinetown–Pietermaritzburg – the two largest metropolitan areas of the country (Smit, 1985; Dewar *et al.*, 1986; Geyer, 1987, 1989a, 1989b). Typical of the dualistic economic situation found in many LDCs, where the 'airplane and the mule' are used side by side (Hirschman, 1972, p. 124), these Bantustans form part of the outer periphery of the South African economic space despite their location adjacent to the metropolitan areas. Although these Bantustans are regarded as outer peripheral development areas (Geyer *et al.*, 1988), the geographical location of these large Black semi-urbanised

Table 16.1 Population migration in South Africa, 1960–85 (%)

Year	Blacks	Whites	'Coloureds'	Indians	Total
(A) *Inner core*					
1960	39.89	35.99	10.86	13.26	100.00
1970	40.90	36.93	11.23	10.94	100.00
1980	42.94	36.26	10.72	10.08	100.00
1985	44.71	35.13	10.82	9.34	100.00
(B) *Outer core*					
1960	47.51	37.32	12.53	2.64	100.00
1970	42.71	36.12	15.21	5.96	100.00
1980	38.73	37.76	16.56	6.95	100.00
1985	40.14	36.19	17.42	6.25	100.00
(C) *Core fringe*					
1960	60.80	23.21	15.92	0.07	100.00
1970	58.23	24.35	17.36	0.06	100.00
1980	52.45	27.14	20.34	0.07	100.00
1985	48.56	29.45	21.93	0.06	100.00
(D) *Intermediate city*					
1960	59.06	27.96	12.15	0.83	100.00
1970	58.35	27.13	13.76	0.76	100.00
1980	52.96	29.58	16.48	0.98	100.00
1985	52.96	29.06	17.38	1.00	100.00
(E) *Outer periphery*					
1960	78.40	10.72	9.36	1.52	100.00
1970	81.30	9.26	8.24	1.20	100.00
1980	82.78	7.67	8.46	1.09	100.00
1985	83.18	7.94	7.88	1.00	100.00

agglomerations adjacent to the core regions represent *de facto* urban concentration – a settlement pattern which indirectly supports mainstream urbanisation in the country. The implementation over many years of a comprehensive web of racially discriminatory legislation in South Africa, which prevented Blacks from migrating freely to the core regions (Geyer, 1989a, 1989b), has played a major role in the development of First and Third World economic regions next to each other in the country.

Not all the cities which were regarded as 'intermediate' in the national study (Table 16.1) are proportionally of the size which Richardson (1980, pp. 68, 80) seems to have had in mind with his concept of polarisation reversal. This matter will be investigated in greater detail in the following section.

Differential urbanisation in and around the PWV

For the purpose of this part of the study the PWV and adjacent areas were geographically subdivided into five zones: an inner core zone, an intermediate core zone, an outer core zone, a core fringe zone and an intermediary city zone (Figure 16.1). The proportional distribution of the different population groups and of economic production volumes in the formal urban economic sectors in and around the PWV is shown in Figures 16.2 and 16.3, respectively.

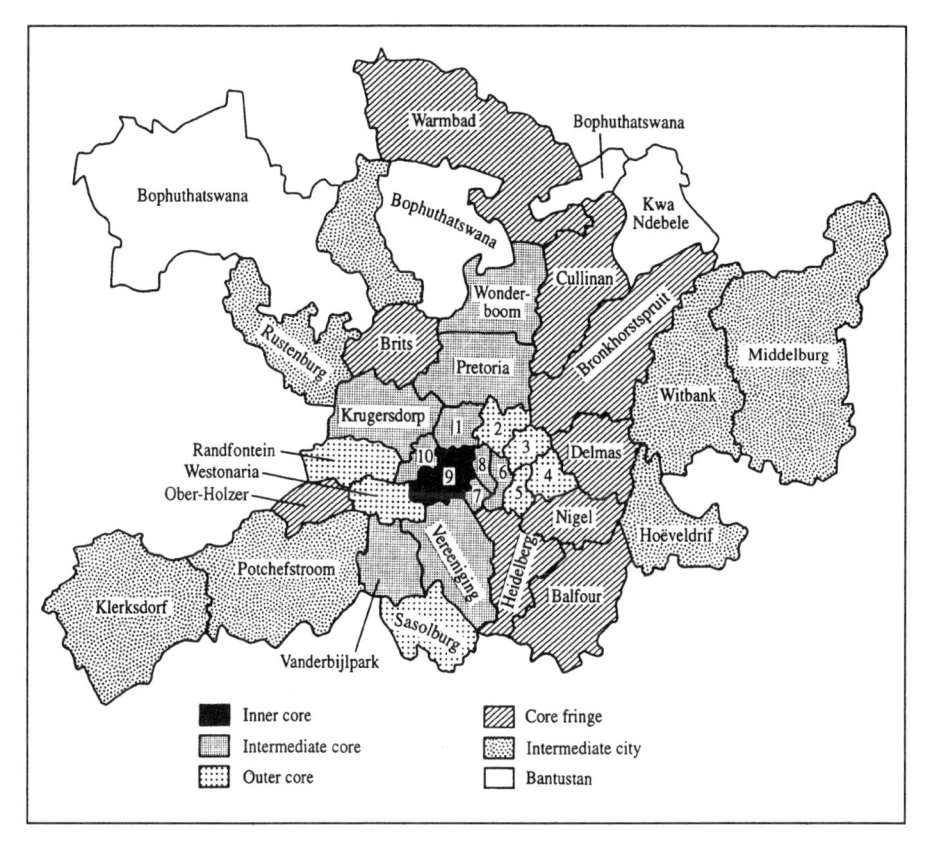

Fig. 16.1 Study area. 1, Randburg; 2, Kempton Park; 3, Benoni; 4, Springs; 5, Brakpan; 6, Boksburg; 7, Alberton; 8, Germiston; 9, Johannesburg; 10, Roodepoort

It is clear from the information contained in Figure 16.2 that there has been a proportional increase in the number of Blacks, 'Coloureds' and Indians in the inner core zone (Figure 16.2A) since 1970, while the share of Whites has decreased relatively in this zone over the same period. The proportions of 'Coloureds' and Indians relative to the Blacks and Whites are insignificant in this zone, however. At the same time, a relative decrease in the share of Blacks and a relative increase in the share of Whites have been observed in the intermediate core zone, while the shares of 'Coloureds' and Indians have remained relatively stable (Figure 16.2D). In the outer core zone (Figure 16.2C), the share of Blacks increased relatively up to the 1970s, then decreased, while that of Whites showed exactly the opposite trend. The shares (and changes in the proportions) of 'Coloureds' and Indians in this zone are insignificant. In both the core fringe zone and the intermediate city regions (Figure 16.2D and E, respectively) the proportion of Blacks has decreased consistently over the years recorded, while that of Whites has increased consistently during the same period. Once again the proportions and changes in the shares of 'Coloureds' and Indians were insignificant in these zones, although a slight proportional increase has been observed in the share of 'Coloureds'.

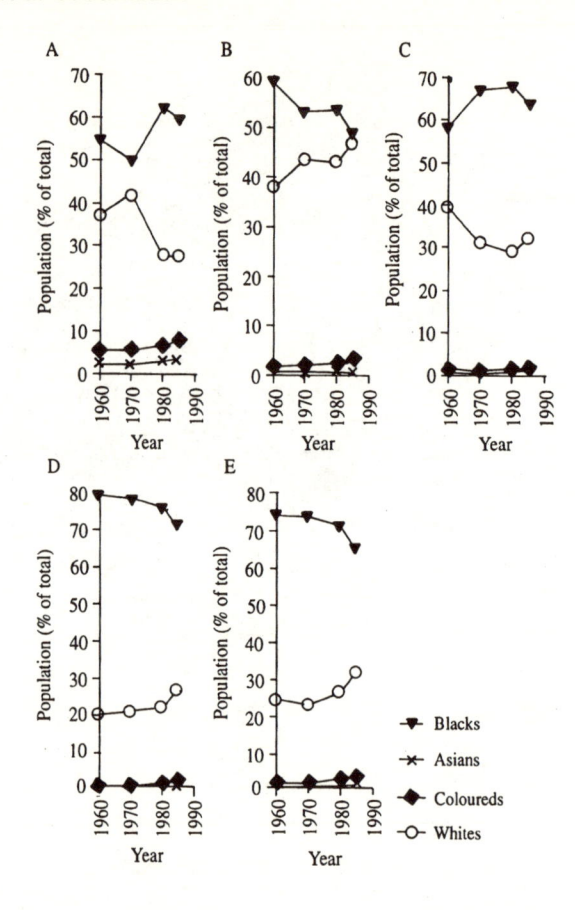

Fig. 16.2 Distribution of population in the Pretoria–Witwatersrand–Vaal triangle (the PWV) and immediate surrounding regions, 1960–85. A, Inner core; B, intermediate core; C, outer core; D, core fringe; E, intermediate city region. (Source: South Africa, 1986a)

When the proportional distributions of production values in the formal urban economic sectors in the different zones are compared (Figure 16.3), it is clear that the relative importance of the various economic sectors differs significantly between the zones. In the inner and intermediate core zones (Figure 16.3A and B, respectively) the primary (mining) sector plays an insignificant role in the gross geographical product (GGP) values, but it is a greater contributor in the outer core zone (Figure 16.3C). Mining is by far the dominant economic sector in the core fringe zone (Figure 16.3D), while it is also the dominant sector in the intermediate city region (Figure 16.3E). The proportional contribution of the service sector to the total GGP remains relatively low in comparison with that of the other sectors in all the zones. The proportional contribution of commerce decreases relatively from the inner core zone outwards, while the proportional contribution of secondary industry is the highest in the intermediate and outer core zones.

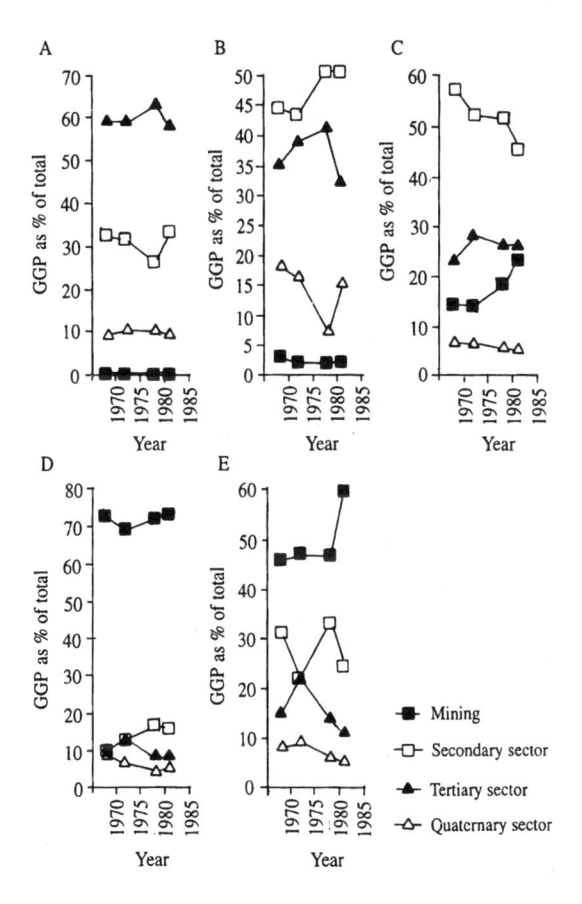

Fig. 16.3 Distribution of gross geographical product in the PWV and immediate surrounding regions, 1968–81. A, Inner core; B, intermediate core; C, outer core; D, core fringe; E, intermediate city regions. (Source: South Africa, 1974a, 1985a, 1988)

As might be expected, the performances of the various economic sectors vary somewhat over the years. This can be attributed to a variety of factors such as cyclical changes in the economy of the country, changes in local and international markets as well as implicit and explicit local, provincial and central government action. It is not only the relative growth or decline of an economic sector's contribution over time which is of importance, however, but also the proportional significance of that economic sector's contribution in the various zones in comparison with the others.

Before the significance of these differential urbanisation results for the PWV can be assessed, existing views on the development of the area must be outlined first.

Governmental views on the development of the PWV

Owing to the effect of the policy of apartheid on the national industrial development policy in South Africa (Geyer, 1989b) and the effect of the latter, in turn, on the government's present spatial development policy for the PWV area (Geyer, 1989a), the government's policy of apartheid cannot be separated from its spatial policy for the PWV.

In 1956 a 'Preliminary Guide Plan' for the PWV was announced by the Natural Resources Development Council, a body established in 1947 by the central government (South Africa, 1956). The main objective with the guide plan was to control development in areas where natural resources are being exploited. In terms of this guide plan the areas to be utilised for the different urban functions, especially for industrial purposes, were restricted. Large areas between and around the Pretoria, Witwatersrand and Vaal Triangle areas were indicated as green 'belts'. All industrial expansion was restricted to the inner metropolitan areas.

The 1956 Preliminary Guide Plan was never finalised, but after a silence of almost two decades was followed up in 1974 by *Proposals for a Guide Plan for the PWV* (South Africa, 1974b). Emphasis in industrial development was placed on the far eastern, the far western, and the southern parts of the PWV. Only very rudimentary indications of future urban development in relation to open spaces were given (Figure 16.4). For the first time the proposals suggested an obvious assessment by the government of the role of urban development in the PWV in a wider regional context. The emergence of development axes between Pretoria, Rustenburg and Witbank–Middelburg was acknowledged with Rustenburg, Brits, Witbank–Middelburg and Potchefstroom–Klerksdorp mentioned as possible 'deconcentration points'. The importance of the development axes in a national development context was confirmed in the *National Physical Development Plan of South Africa* (South Africa, 1975), but now intermediate centres such as Potchefstroom–Klerksdorp, Witbank–Middelburg and Rustenburg were designated as 'growth poles' (Figure 16.5). These 'growth poles' were described as 'towns or complexes of towns which will, without much stimulus, command sufficient growth potential to develop and support a large population'.

Once again these proposals were not transformed into official policy, but instead were followed up by another 'Draft Spatial Development Strategy for the PWV Complex' (Figure 16.6) (South Africa, 1981a). Apart from the tentative segregationist residential development policy indicators of the 1974 proposals which were repeated in the 1981 Draft Strategy, industrial development in and around the PWV was for the first time explicitly linked with Black settlement patterns (Geyer, 1989a). At a meeting in 1981 where this Draft Strategy was introduced to a selected group of people, it was severely criticised. Instead of altering the Draft Strategy in accordance with suggestions made by the interested parties at the meeting, another meeting was held during November 1981 where the so-called '*Good Hope Plan*' was announced (South Africa, 1981b, 1985a).

As a national industrial development strategy for the country, the *Good Hope Plan* contained specific policy suggestions for the PWV. These included six 'deconcentration points', all within the border areas of two Bantustans, one north-east and one north-west of the PWV (Figure 16.7). Although these 'deconcentration points' were described as 'physically well situated and

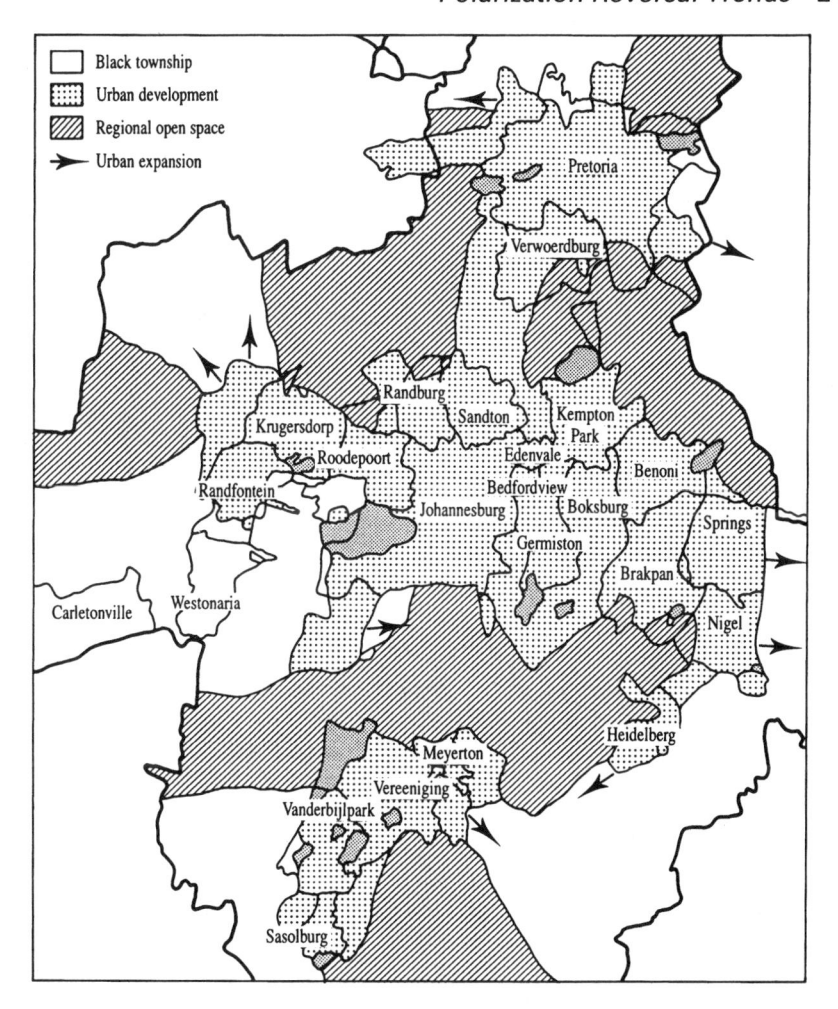

Fig. 16.4 Proposed regional open spaces and medium-short-term urban development. (Source: South Africa, 1974b)

economically supported by natural growth tendencies', they have an obvious spatial political function in terms of the 'Homeland policy' of the government (Geyer, 1989a, 1989b). They clearly play a central role in the promotion of development in these outer peripheral Bantustan areas and in the diversion of Black urbanization away from the PWV.

None of the 'deconcentration points' suggested for the PWV in the *Good Hope Plan* is supported by significant natural endowments or by spontaneous economic growth tendencies other than the accumulation of large numbers of Blacks at some points within the Bantustans, directly or indirectly as a result of the policy of apartheid. In fact, economic incentive measures have to be applied to attract industrial development to these 'deconcentration points', which otherwise would have been located in the PWV. It has been shown that

246   Differential Urbanization

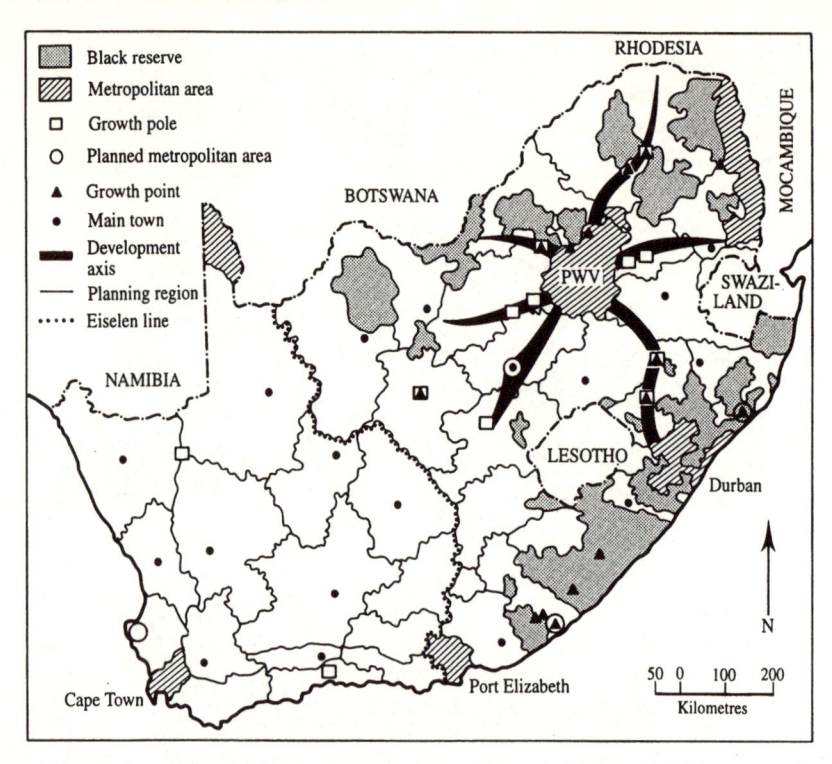

Fig. 16.5 National Physical Development Plan of South Africa, 1975. (Source: South Africa, 1975)

industrial incentives are commonly abused by industrialists in South Africa. Some industrialists use incentives to compete with international and local industries inside the larger metropolitan areas in the sphere of international divisions of labour, while other larger companies utilise the financial incentives available at such centres through branch activity to bolster company liquidity during economic slumps (Dewar, 1987; Wellings and Black, 1986; Bell, 1973, 1987).

The suitability of these deconcentration points as employment centres for the largely underdeveloped Black communities residing in the Bantustans must also be questioned (Geyer, 1989a, 1989b). Large numbers of Blacks travel from the deconcentration points within the Bantustans to the PWV and back each day in search of employment (Dewar *et al.*, 1986). Many of these Blacks are unskilled, however, and simply cannot be accommodated within the technically sophisticated formal industrial sector envisaged at these deconcentration points and in the metropolitan area (Geyer, 1987). This is for instance borne out by differences in the attitudes of workers in the Bantustans generally to the disciplined formal work environment compared to those in the major metropolitan areas, which are reflected in productivity differences (Addleson and Tomlinson, 1987). Alternative methods of employment within the informal urban sector should therefore be devised in order to provide

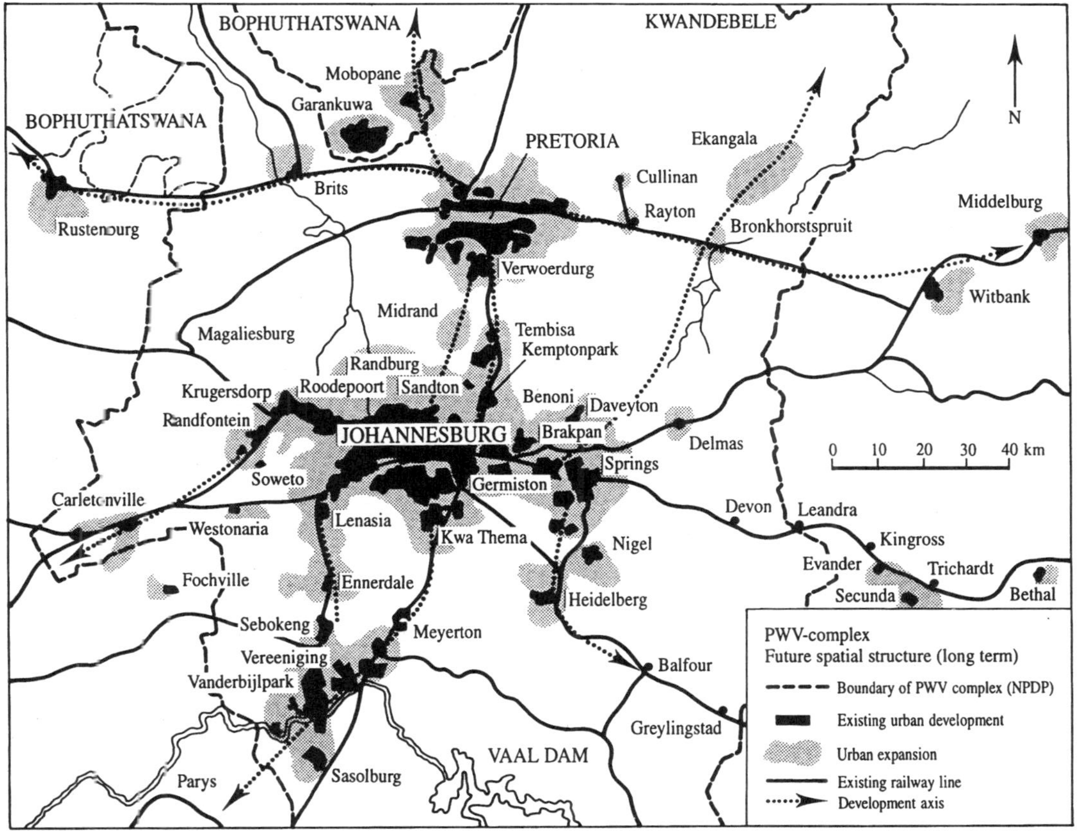

Fig. 16.6 Spatial development strategy for the PWV complex, 1981

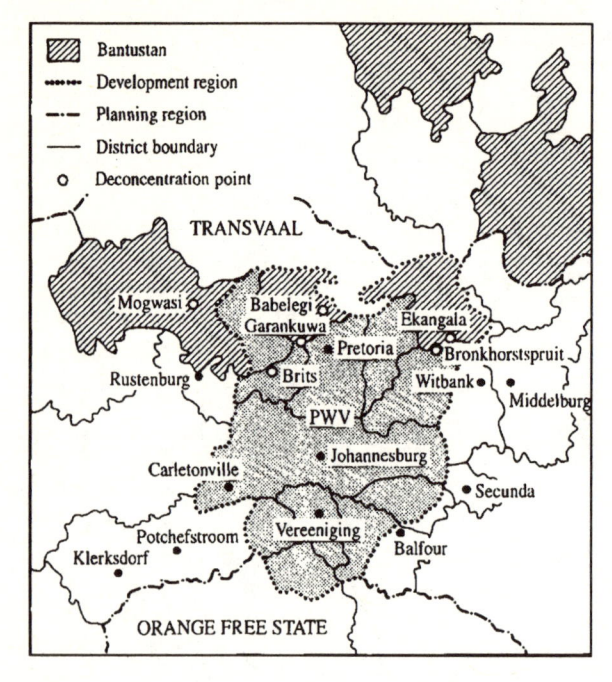

Fig. 16.7 Good Hope Plan proposal for the PWV, 1981. (Source: South Africa, 1981b)

employment for these people in the PWV (Geyer, 1988). This includes methods to integrate the formal and informal sectors within the present urban areas (Geyer, 1989c).

Subsequently, 'guide plans' have been compiled by the central government for the various urban subcentres in the PWV. These plans, which are in different phases of completion, include the *Vaal River Guide Plan* (South Africa, 1982), the *Greater Pretoria Guide Plan* (South Africa, 1984b), the *Draft Guide Plan for the East Rand/Far East Rand* (South Africa, 1984a), and the *Draft Guide Plan for the Central Witwatersrand* (South Africa, 1986b). Contrary to what the term 'guide plan' normally implies in planning, the extent to which land may be used for the various land uses in the PWV is strictly prescribed in terms of cadastral boundaries by these 'guide plans' and they can, therefore, be regarded as supra-urban town planning schemes, rather than guide plans in the true sense of the word. As paternalistic bureaucratic instruments these spatially inflexible 'guide plans' inhibit town planning decision-making at local authority level, and do not give a total perspective of developmental guide lines for the PWV as a whole.

It is clear, therefore, that changes are necessary in the state's present approach to development planning for the PWV – changes in terms of objectives, format, time frame and flexibility.

Extra-governmental views on the development of the PWV

Various development approaches have been suggested for the PWV in extra-governmental circles. They vary from an approach of even dispersal through various forms of centralised-cum-concentrated decentralised development to a centralised approach where development in the PWV is allowed to continue largely unhindered.

In an evaluation study by the Development Bank of Southern Africa (DBSA) of the present national spatial policies of the state the DBSA suggests an approach in which a differential scale of industrial incentives applies nationwide; an approach in which only the PWV does not qualify for incentives. The other major metropolitan areas should, according to the DBSA, qualify for some incentive measures, while smaller centres in the inner, intermediate and outer periphery (Geyer *et al.*, 1988) should qualify for more support. The onus rests on the local authority, however, to compete for support. It is an approach where market forces are allowed to play a dominant role in the development process.

In the formulation of this approach, it is quite obvious that the DBSA was heavily influenced by Richardson's (1987, p. 208) view that, of the three factors which historically influenced population distribution most – i.e. market forces, implicit spatial policy, and explicit spatial policy – the latter has proven to be potentially the weakest. The question arises, however, whether the South African economy can support such a 'blanket approach' to industrial development. The criticism that the many industrial development points which are presently being recognised by the State exceed the available resource capacity of the country (Bell, 1987; Dewar, 1987; MacCarthy, 1987; Geyer, 1987, 1989b) applies even more to the suggested approach, even if allowance is made for designated fund limitations. Meanwhile, an embargo has been placed on DBSA's study results while they are still being 'considered' by the government. The State has subsequently announced that it will later decide 'if and how' the findings of the DBSA study will be made public. Based on the track record of the State on similar matters in the past, one must expect certain differences between the DBSA report and the government's official policy to be concealed in the final reports to be made public.

Another school of thought holds that economic development should be encouraged away from the PWV with the grafting of incentives onto intermediate centres around it. In the creation of incentives at these intermediate centres, care should be taken, though, not to subsidise firms in order to survive, because the value added in production would then, according to Addleson and Tomlinso (1987), be negative and the success of the industrial decentralisation be based on the continual transfer of resources from more prosperous areas. It is suggested that growth be encouraged at specific growth points in the periphery which have under-utilised potential, and that the emphasis be put on grass roots industrial development, implemented from the bottom upwards at these peripheral urban centres (Dewar, 1987; Geyer, 1987, 1989b).

Maasdorp (1985) suggests 'suburbanisation' of industry as a policy alternative to industrial incentives, with the spontaneous leapfrogging of industrial development from the PWV to deconcentration points as an expected result. He links the development of centres such as Brits, Bronkhorstspruit, Rustenburg, Witbank–Middelburg and Secunda with the PWV–Durban development axis. It is expected that spontaneous industrial leapfrogging from the inner PWV towards deconcentration points at its periphery may not be realised without direct or indirect external inducement, however. In their assessment of the present officially recognised 'deconcentration points', Dewar *et al.* (1986) have detected the presence of diseconomies which mitigate against such 'spontaneous' reasoning.

The development axis may fulfil a useful role as an indirect instrument to stimulate industrial leapfrogging from the PWV towards deconcentration points, but only if the PWV, its surrounding intermediate cities, and its deconcentration points form a truly viable 'system of cities' (Geyer, 1987, 1989a, 1989b; Bos, 1989). In such an approach, careful consideration should be given to the correct location of potential deconcentration points in relation to the PWV and its surrounding intermediate cities. Development should only be stimulated at those intermediate centres which, through their functional ties with the PWV, can lead to spontaneous development overflow towards deconcentration points which are located on such development axes. Only then can a considerable degree of spontaneous leapfrogging of industries be expected to occur from the PWV to surrounding deconcentration points.

Another school of thought supports the continued growth of the PWV (Tomlinson, 1988). This view is based on the principle that the metropolis is still relatively small in terms of world standards and that economic development in South Africa is hampered unnecessarily by the diversion of growth away from the PWV. The latter view is based on the fact that the decentralisation policy pursued by the government over the past two decades, especially the forced diversion of certain types of industries away from the PWV by means of the Physical Planning Act, No. 88 of 1967, as well as the unnecessary infringements placed on development by the comprehensive apartheid legislation in the country, has hampered economic development in general.

In the following section, the effect of the differential urbanisation study results on the development of the PWV will be investigated.

Implications of differential urbanisation for the PWV

In view of the overall proportional increase of the Black (low-income) population relative to the other population groups in the inner core of the PWV over the past 15 years (Figure 16.2A), special attention needs to be directed towards housing and the creation of job opportunities for this component of the population in this zone. As indicated earlier, it can be expected that a large part of the Black population, which presently resides in Bantustans on the 'Black' side of the boundaries, will also migrate to the PWV as soon as discriminatory legislation and settlement policies, which are at present applicable in the 'White' areas, are finally removed. This would further increase the considerable (and constantly increasing) levels of unemployment presently prevailing amongst the Blacks in the country's major metropolitan areas (Levin and Du Plessis, 1986; Geyer, 1987, 1988).

Owing to the implementation of the apartheid policy of the government of South Africa, wide-ranging ostracising action has been taken by the international economic community against the country over the past 28 years.

The effect of increasing punitive economic action taken against South Africa over the past decade[2] can clearly be seen in the relative decrease in commercial activities in all the zones of the PWV (cf. Figure 16.3). The relative decrease in the white population – i.e. the high-income group[3] – in the inner core zone since 1980 has also contributed to the proportional decrease in commercial activities in the inner core zone.

The decrease in commercial activity in the inner core zone implies not only less buying power in this zone, but also a decrease in employment opportunities for Blacks. All possible channels should be explored, therefore, through which employment opportunities can be created for Blacks in this area. Programmes which are currently in progress, involving Blacks in self-help housing schemes in the inner core zone, should continue and expand. Similarly, work which is currently being done to initiate the development of informal commercial and industrial activities amongst the lower-income groups in this zone must continue. Much can still be done, however, to systematise the hierarchical structure of this sector and to integrate it with the formal sector in this zone (Geyer, 1988, 1989c).

The proportional increase in the shares of Whites in the intermediate core, outer core, core fringe zones, and intermediate city regions confirms the early signs of the beginning of the process of deconcentration amongst the higher-income group which is detected in the national figures. In view of the tendencies revealed in Figures 16.2B–E, and 16.3B–E, respectively, emphasis should be placed on the deconcentration of manufacturing industries to the intermediate core zone, while maximum expansion in both the manufacturing and mining sectors should be considered in the outer core, core fringe and intermediate city regions to accommodate the increasing proportion of the skilled and highly skilled population component in these zones.

The location of the (politically motivated) designated 'deconcentration points' within the Bantustans north-east and north-west of the PWV does not correspond with the differential population migration tendencies in and around the metropolitan area. Bantustans need to be depoliticised. They should become ordinary development regions without any apartheid political connotations attached to them, while the development of the informal sector should be maximised in the semi-urbanised settlements in these areas instead of the formal industrial sector. If intermediate centres such as Potchefstroom–Klerksdorp, Witbank–Middelburg and Rustenburg are regarded as decentralisation points in a system of cities with the PWV, a geographically more balanced distribution of deconcentration points will have to be created to support the system of cities. More deconcentration points, some within the intermediate/outer core zone and some within the core fringe zone, need to be created to reinforce the system of cities. Centres such as Midrand and Meyerton within intermediate/outer core zone, and Brits, Bronkhorstspruit, Secunda and Carletonville within the core fringe zone, all of which are centres with considerable growth potential, and all of which are located on primary communication axes inside the PWV and between the PWV and the intermediate cities, should be considered.

None of these policy matters will, however, make much sense if they are not accompanied by the complete political and economic liberation of all races of colour in the South African society, and that can happen only when persistent governmental inertia is finally overcome.

Conclusions

Indications are that patterns of differential urbanisation occur in certain DCs and LDCs. While mainstream population migration patterns indicate the beginning of the process of counterurbanisation in DCs such as the USA, the FRG and France, substream migration typical of a totally different phase of development is taking place simultaneously. In an LDC such as South Africa, Blacks are increasing proportionally in the larger metropolitan inner areas, while the number of Whites is increasing proportionally in the metropolitan fringe regions and intermediate city regions.

Study results show similar migration tendencies in regions in and around the PWV. The share of Blacks is increasing proportionally in the inner core zone of the PWV, while the share of Whites is increasing relatively in the outer metropolitan areas, metropolitan fringe areas and surrounding intermediate city regions. This differential urbanisation pattern calls for significant changes in the South African government's present politically biased deconcentration strategy for the PWV. These changes include: (1) the inclusion of the intermediate cities around the PWV as part of a system of cities with the metropolis; (2) the stimulation of development in these intermediate cities; (3) the elimination of segregation-oriented 'deconcentration points' in the Bantustans; and (4) the identification of alternative deconcentration points inside and on the periphery of the PWV to support the system of cities. Special attention should also be given to the stimulation of the informal urban sector in the PWV, in order to meet the consumer and job requirements of the increasing numbers of semi- and unskilled Blacks who are migrating to this area. Ways should be devised to increase the integration of the informal and formal urban economic sectors within the PWV. These developmental problems cannot be addressed effectively, however, as long as the apartheid policy, or elements of it, remains intact in South Africa.

Acknowledgement

The results presented in this paper were made possible by a grant from the University of Potchefstroom.

Notes

1 The national space was subdivided into five zones: inner core zones, outer core zones, core fringe zones, intermediate city zones, and an outer periphery zone. The inner core zones refer to the mother cities of the major metropolitan areas of the country, i.e. Johannesburg in the PWV, Durban in the Durban–Pinetown–Pietermaritzburg area, and Cape Town in the Cape Town–Bellville area. The outer

core zones refer to the urbanized and semi-urbanized areas immediately surrounding the mother cities. The core fringe zones refer to the towns at the outskirts of these major metropolitan areas. The intermediate city regions refer to regions containing cities which, on a national scale, act as regional centres (Geyer, 1987, pp. 283–4, 293–5; Geyer *et al.*, 1988, pp. 329–30; Bos, 1989), and the outer periphery to all other areas.

2 The rand–dollar exchange rate has dropped from $1.28/R1.00 in 1980 to $0.35/R1.00 (commercial) and $0.23/R1.00 (financial) in May 1989.

3 In 1980 the median income of the Blacks was R1843, of the Indians, R2253, of the 'Coloureds', R2317, and of the Whites, R10 534.

References

Addleson, M. and Tomlinson, R. 1987: Responses of manufacturers to decentralisation incentives. In Tomlinson, R. and Addleson, M. (eds), *Regional Restructuring under Apartheid*. Johannesburg: Raven Press.

Bell, R. T. 1973: Some aspects of industrial decentralisation in South Africa. *South African Journal of Economics* 41, 401–31.

Bell, T. 1987: Is industrial decentralisation a thing of the past? In Tomlinson, R. and Addleson, M. (eds), *Regional Restructuring under Apartheid*. Johannesburg: Raven Press.

Berry, B. J. L. 1976: The counterurbanisation process: urban America since 1970. *Urban Affairs Annual Review* 11, 17–30 (*see* this volume, Chapter 1).

Bos, D. J. 1989: Prospects for the development of intermediate size cities as part of a decentralisation programme for Southern Africa. *Development Southern Africa* 6, 58–81.

Champion, D. H. 1989: Counterurbanisation in Europe. *Geographical Journal* 155, 52–80.

Clark, C. 1967: *Population growth and land use*. New York: St Martin's Press.

Cochrane, S. G. and Vining, D. R. 1988: Recent trends in migration between core and peripheral regions in developed and advanced developing countries. *International Regional Science Review* 11, 215–43 (*see* this volume, Chapter 7).

Dean, K. G. 1986: Counterurbanisation continues in Brittany. *Geography* 71, 151–4.

Dewar, D. 1987: An assessment of industrial decentralisation as a regional development tool, with special reference to South Africa. In Tomlinson, R. and Addleson, M. (eds), *Regional Restructuring under Apartheid*. Johannesburg: Raven Press.

Dewar, D., Todes, A. and Watson, V. 1986: Industrial decentralisation policy in South Africa: rhetoric and practice. *Urban Studies* 5, 363–76.

Geyer, H. S. 1987: The development axis as a development instrument in the Southern African Development Area. *Development Southern Africa* 4, 271–301.

Geyer, H. S. 1988: On urbanisation in South Africa. *South African Journal of Economics* 56, 154–72.

Geyer, H. S. 1989a: Apartheid in South Africa and industrial deconcentration in the PWV area. *Planning Perspectives* 4, 251–69.

Geyer, H. S. 1989b: Industrial development policy in South Africa: the past, present, and future. *World Development* 17, 379–96.

Geyer, H. S. 1989c: The integration of the formal and informal urban sectors in South Africa. *Development Southern Africa* 6, 29–42.

Geyer, H. S., Steyn, H. S., Uys, S. and Van der Walt, F. 1988: Development management regions in South Africa: an empirical evaluation. *Development Southern Africa* 5, 307–35.

Gordon, P. 1979: Deconcentration without a 'clean break'. *Environment and Planning A*, 11, 281–90 (*see* this volume, Chapter 4).

Hall, P. 1987: Metropolitan settlement strategies. In Rodwin, L. (ed.), *Shelter, settlement and development*. Boston: Allen & Unwin.

Hirschman, A. O. 1972: *The strategy of economic development*. New Haven, Conn.: Yale University Press.

Koch, R. 1980: 'Counterurbanisation' auch in Westeuropa? *Informationen zur Raumentwicklung* 2, 59–69 (*see* this volume, Chapter 8).

Kontuly, T., Wiard, S. and Vogelsang, R. 1986: Counterurbanisation in the Federal Republic of Germany. *Professional Geographer* 38, 170–81.

Lee, K. S. 1985: Decentralisation trends of employment location and spatial policies in LDC cities. *Urban Studies* 22, 151–62.

Levin, M. and Du Plessis, A. P. 1986: *The Unemployment rate of Blacks in the Port Elizabeth metropolitan area*. Vista University of Port Elizabeth, Research Report No. 4.

Linn, J. F. 1978: Urbanisation trends, polarisation reversal, and spatial policy in Colombia. Institute für Siedlungs- und Wohnungswesen der Westfälischen Wilhelms-Universität Münster, Paper no. 12, Münster.

Lo, F. and Salih, K. 1981: Growth poles, agropolitan development, and polarisation reversal: the debate and search for alternatives. In Stohr, W. B. and Taylor, D. R. F. (eds) *Development from above or below? The dialectics of regional planning in developing countries*. New York: John Wiley.

Maasdorp, G. 1985: Co-ordinated regional development: hope for the Good Hope proposals? In Giliomee, H. and Schlemmer, L. (eds), *Up against the fences. Poverty, passes and privilege in South Africa*. Cape Town: David Philip.

MacCarthy, C. L. 1987: Industrial decentralisation and employment creation. In Tomlinson, R. and Addleson, M. (eds), *Regional restructuring under apartheid*. Johannesburg: Raven Press.

Ogden, P. E. 1985: Counterurbanisation in France: the results of the 1982 population census. *Geography* 70, 24–35.

Richardson, H. W. 1977: *City size and national spatial strategies in developing countries*. World Bank Staff Working Paper no. 252, Washington, DC.

Richardson, H. W. 1980: Polarization reversal in developing countries. *Papers of the Regional Science Association* 45, 67–85 (*see* this volume, Chapter 11).

Richardson, H. W. 1987: Spatial strategies, the settlement pattern, and shelter and services policies. In Rodwin, L. (ed.), *Shelter, settlement and development*. Boston: Allen & Unwin.

Smit, P. 1985: The process of black urbanisation. In Giliomee, H. and Schlemmer, L. (eds), *Up against the fences: poverty, passes and privilege in South Africa*. Cape Town: David Philip.

South Africa (Republic of), Department of Statistics. 1974a: *Gross Geographical Product by Magisterial District, 1968*. Report no. 09-14-01. Pretoria: Government Printer.

South Africa (Republic of), Department of Planning and the Environment. 1974b: *Proposals for a Guide Plan for the PWV*. Pretoria: Government Printer.

South Africa (Republic of), Department of Planning and the Environment 1975: *National Physical Development Plan*. Pretoria: Government Printer.

South Africa (Republic of), Office of the Prime Minister. 1981a: *A Spatial Development Strategy for the PWV Complex*. Pretoria: Government Printer.

South Africa (Republic of), Department of Foreign Affairs 1981b: *The Good Hope Plan for Southern Africa*. Pretoria: Government Printer.

South Africa (Republic of), Office of the Prime Minister 1982: *Vaal River Complex Guide Plan*. Pretoria: Government Printer.

South Africa (Republic of), Department of Constitutional Development and Planning. 1984a: *Draft Guide Plan for the East Rand/Far East Rand*. Pretoria: Government Printer.

South Africa (Republic of), Department of Constitutional Development and Planning. 1984b: *Greater Pretoria Guide Plan*. Pretoria: Government Printer.

South Africa (Republic of), Central Statistical Services. 1985a: *Gross Geographic Product at Factor Incomes by Magisterial District, 1978*. Report No. 09-14-05, Pretoria: Government Printer.

South Africa (Republic of), Secretariat for Multilateral Co-operation in Southern Africa. 1985b: *Manual on the Implementation of the Regional Industrial Development Incentives Introduced on 1 April 1982*. Johannesburg: Perskor.

South Africa (Republic of), Central Statistical Services. 1986a: Population Census: *Geographical Distribution of the Population with a Review for 1960–1985*. Report no. 02-85-01. Pretoria: Government Printer.

South Africa (Republic of), Department of Constitutional Development and Planning. 1986b: *Draft Guide Plan for the Central Witwatersrand*. Pretoria: Government Printer.

South Africa (Union of) Natural Resources Development Council. 1956: *Pretoria–Witwatersrand–Vereeniging Area Guide Plan*. Pretoria: Government Printer.

Tomlinson, R. T. 1988: Pretoria–Witwatersrand–Vereniging Region: can/should regional policy divert growth from South Africa's core? Paper presented at the Development Society of Southern Africa's biennial conference, 6–8 July 1988. University of Durban–Westville, Durban.

Townroe, P. M. and Keen, D. 1984: Polarisation reversal in the state of São Paulo, Brazil. *Regional Studies*, 18, 45–54 (*see* this volume, Chapter 13).

Vining, D. R. 1986: Population redistribution towards core areas of less developed countries, 1950–1980. *International Regional Science Review* 10, 1–45 (*see* this volume, Chapter 12).

Vining, D. R. and Kontuly, T. 1978: Population dispersal from major metropolitan regions: an international comparison. *International Regional Science Review* 3, 49–73 (*see* this volume Chapter 6).

Vining, D. R. and Pallone, R. 1982: Migration between core and peripheral regions: a description and tentative explanation of the patterns in 22 countries. *Geoforum*, 13, 339–410.

Vining, D. R. and Strauss, A. 1977: A demonstration that the current deconcentration of population in the United States is a clean break with the past. *Environment and Planning A*, 9, 751–8 (*see* this volume, Chapter 3).

Wellings, P. A. and Black, A. 1986: Industrial decentralisation under apartheid: the relocation of industry to the South African periphery. *World Development* 14, 1–38.

Yeats, M. 1980: *North American urban patterns*. London: Edward Arnold.

PART THREE
Long-term Migration Trends in Developed and Less Developed Countries

SECTION ONE
MIGRATION CYCLES

17 B. J. L. Berry,
'Migration Reversals in Perspective: The Long-Wave Evidence'

From: *International Regional Science Review* 11 (3), 245–51 (1988)

In 1982, Vining and Pallone announced that in the developed world, the century-long migration towards high-density core regions was over. Echoing my earlier formulation of the counterurbanisation concept (Berry, 1970, 1976), they argued that we were entering a new epoch of polarisation reversal. Now, Cochrane and Vining (1988) retract their epochal view: core–periphery dispersal appears to be ending, they argue, and core regions are once again attracting migrants. And just as authors scrambled for an explanation of counterurbanisation and polarisation reversal a decade ago, one sees a similar scramble to rationalise the backswing. What I fear is that the mavens who watch the census for the latest trends may have been too close to their statistics. In this respect, I, too, must plead *mea culpa*. A longer-term view is needed to set these shorter-term statistical swings and the puppetlike reactions of the mavens into perspective.

Such a view is provided by the much-abused but newly rehabilitated long-wave concept. A few simple graphics will illustrate, drawn from my forthcoming book in which I review the massive literature pro and con on long waves, provide new evidence about their existence, and present a new conceptual synthesis.[1] Echoing Samuelson's remark that long waves are science fiction, many economists have rushed to reject the idea. They may, I fear, have thrown out a vigorous child with the earlier, admittedly murky, bath water.

Burns and Mitchell (1946) argued that if long waves exist, they should be discernible in accelerations and decelerations of economic growth. In keeping with their suggestion I calculated, among many other things, the average annual rates of growth of real gross national product and real gross national product per capita in the US for each decade from 1790 to the present. The results are shown in Figure 17.1. There are clear waves of approximately 50 years' duration. From each low-growth episode there is an era of accelerating growth. An upper turning point is reached, and growth decelerates to another low. At the lowest points, growth of real per capita gross national product slows to zero.

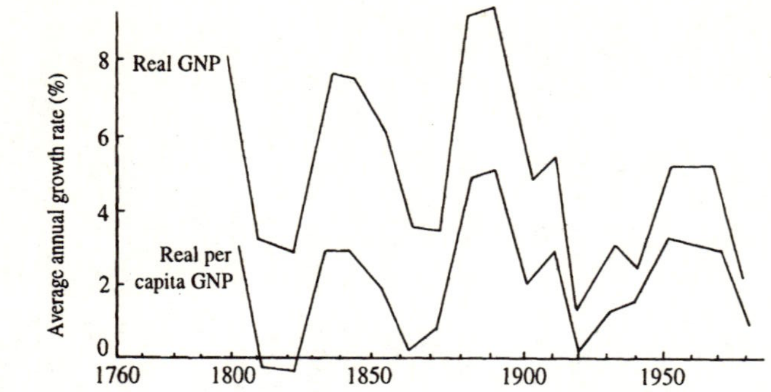

Fig 17.1 Accelerations and decelerations of US economic growth, 1790–1980

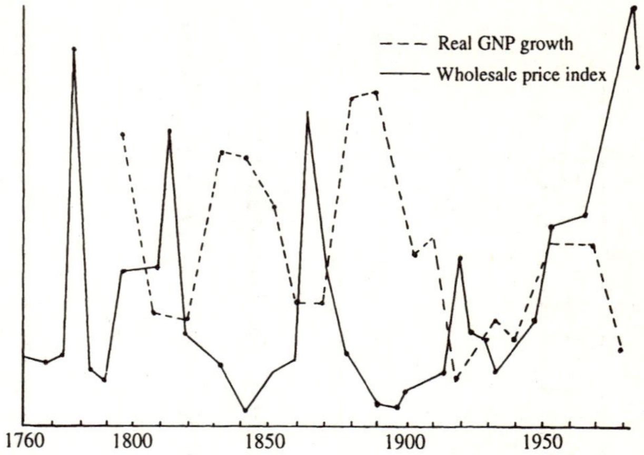

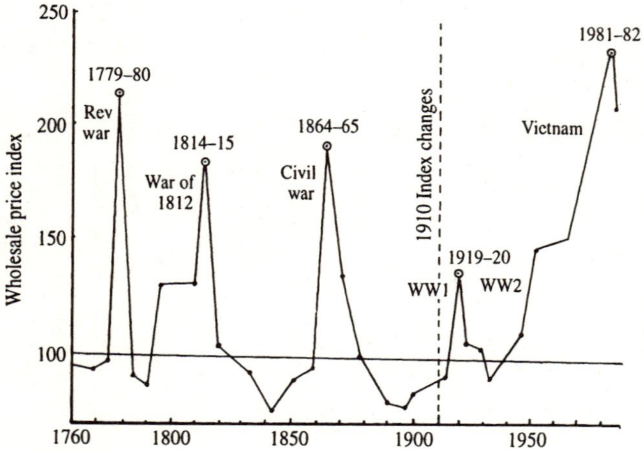

Fig 17.2 US wholesale prices, 1760–1982

Fig 17.3 Superimposition of Figures 17.1 and 17.2 reveals countervailing rhythms of prices and growth

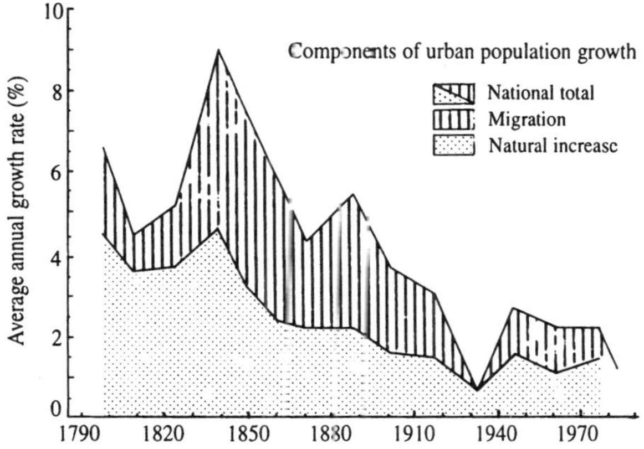

Fig 17.4 Accelerations and decelerations of US urban growth, 1790–1980

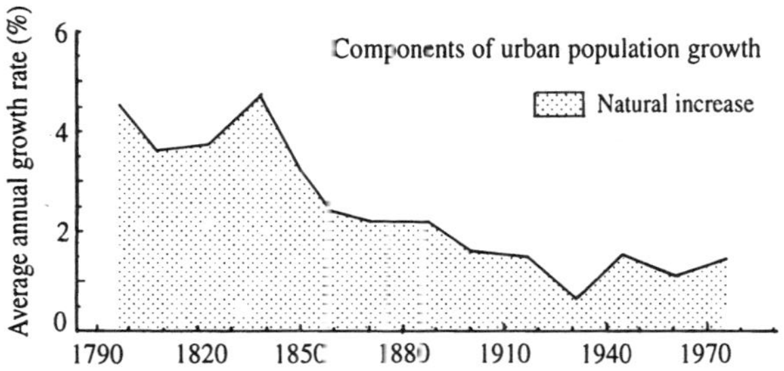

Fig 17.5 Contribution of natural increase to the US urban growth rate

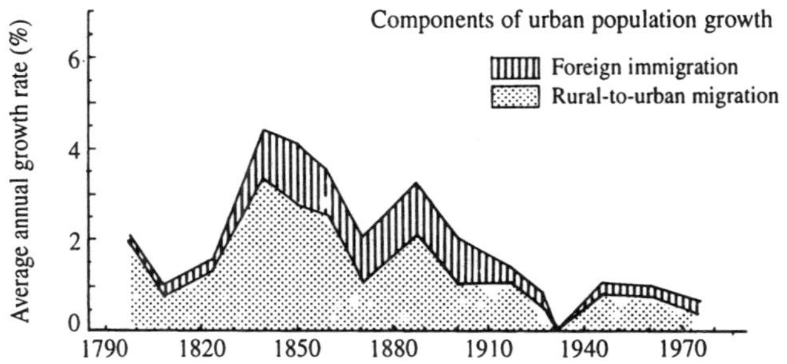

Fig 17.6 Migration's contribution to the US urban growth rate has been long wave in character, 1790–1980

262 Differential Urbanization

The lowest growth rates coincide with peaks of inflation. In Figure 17.2 the familiar 55-year Kondratiev long waves of prices are seen and attention is drawn to the inflation crises marked by the Kondratiev peaks of 1814–15, 1864–65, 1919–20, and 1981–82. Kondratiev troughs occurred in the 1840s, 1890s, and 1930s.

Figure 17.3 reveals that growth and prices have countervailing rhythms. Punctuated by stagflation crises, periods of Kondratiev price downswings coincide with periods of accelerated growth; Kondratiev price upswings coincide with growth deceleration (for analysis of this relationship *see* Hartman and Wheeler, 1979).

What have these contrapuntal rhythms to do with counterurbanisation and polarisation reversal? The US urban growth rate has accelerated and decelerated in the past two centuries (Figure 17.4). Extracting the secularly declining contribution of natural increase (Figure 17.5), migration's contribution remains (Figure 17.6). It is this pattern that is strongly cyclical, which is as it should be if migration proceeds in lockstep with the rhythms of economic growth (Figure 17.7). Each wave of economic growth has produced a wave of urbanward migration (on the question of causality, *see* Tabuchi, 1988). Waves of migration have produced accelerations of urban population growth that have been strong enough to produce a sequence of secondary pulses along the nation's overall urbanisation logistic (Figure 17.8). Such waves of urban growth are but one manifestation of the increasing regional inequality that accompanies surges in economic growth (for further discussion refer to Burns, 1987). Symmetrically, migration slowdown in the direction of polarisation reversal should be a feature of economic slowdown until, in the Kondratiev troughs, both big cities and core regions are brought to a standstill, growth-ward migrants stay home, and the net migration statistic reverses.

There is evidence that the 1970s and 1980s have not been unique. Thirty years ago C. Warren Thornthwaite lectured at the University of Chicago on his research on the reverse migration to rural America in the 1930s. During the nineteenth century growthward migration to the United States from overseas

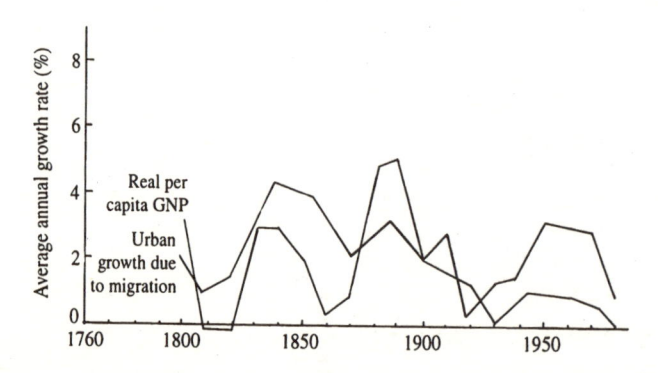

Fig. 17.7 The cyclical relationship between economic growth and urbanward migration, 1790–1980

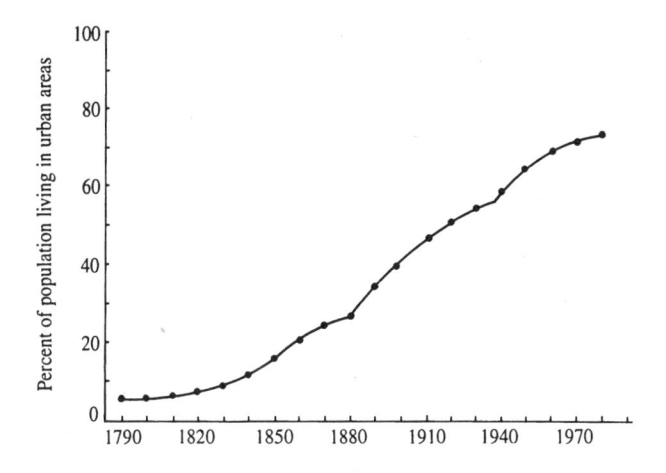

Fig. 17.8 Long-wave fluctuations along the US urbanisation curve, 1790–1980

resulted in the countercyclicality of many economic indicators of the participants in the Atlantic trading economy (Thomas, 1972).[2]

The moral is clear: The mavens have tracked the transition that accompanies a Kondratiev stagflation crisis. Stepping back and exploiting the full richness of emergent long-wave theory produces no surprises, simply the documentation of one small, albeit important, segment of a longer-term rhythmic sequence. If explanation is to be found, it is in the dynamics of this long-wave sequence, not in reactive speculation about the proximate causes of changes in trend. I hope that my forthcoming books will not only place our understanding of these longer-term dynamics on a firm statistical base, but will provide a theoretical synthesis that embodies the core of explanation.

Notes

1. The forthcoming book is entitled *The clocks that time development*. In it, I develop a theory that meets all of Rosenberg and Frischtak's (1983) requirements for an acceptable long-wave theory. One reviewer noted: 'In the first two chapters you build a persuasive empirical case for the existence of cyclicality – build it to the point where it stands as an immovable barrier in the path of anyone who would like to brush the possibility aside' as many are too ready to do.
2. A second book in preparation, *Interdependent urbanization: Migration streams and the countercyclicality of growth in open urban systems*, explores these patterns worldwide for the past two hundred years.

References

Berry, B. J. L. 1970: The geography of the United States in the year 2000. *Transactions* 51, 21–53.

Berry, B. J. L. 1976: *Urbanization and counterurbanization*. Beverly Hills, Calif.: Sage Publications.

Berry, B. J. L. Forthcoming: *The clocks that time development.*
Burns, A. F. and Mitchell, W. C. 1946: *Measuring business cycles.* New York: National Bureau of Economic Research.
Burns, L. S. 1987: Regional economic integration and national economic growth. *Regional Studies* 21 (4), 327–42.
Cochrane, S. G. and Vining, D. R. Jr. 1988: Recent trends in migration between core and peripheral regions in developed and advanced developing countries. *International Regional Science Review* 11 (3), 215–45 (*see* this volume, Chapter 7).
Hartman, R. S. and Wheeler, D. R. 1979: Schumpeterian waves of innovation and infrastructure development in Great Britain and the United States: the Kondratieff cycle revised. *Research in Economic History* 4, 37–85.
Rosenberg, N. and Frischtak, C. R. 1983: Long waves and economic growth: a critical appraisal. *American Economic Review* 73, 146–51.
Tabuchi, T. 1988: Interregional income differences and migration: their interrelationships. *Regional Studies* 22 (1), 1–10.
Thomas, B. 1972: *Migration and urban development.* London: Methuen.
Vining, D. R. Jr. and Pallone, R. 1982: Migration between core and peripheral regions: a description and tentative explanation of the patterns in 22 countries. *Geoforum* 13 (4), 339–410.

18 L. A. Brown and F. C. Stetzer,
'Development Aspects of Migration in Third World Settings: A Simulation, with Implications for Urbanisation'

From: *Environment and Planning A* 16, 1583–603 (1984)

Introduction
In this paper we report a simulation of development–migration interrelationships in Third World settings and their implications for urbanisation.[1] Its design is guided by the 'development paradigm of migration' (Brown and Sanders, 1981), an extension of earlier work by Mabogunje (1970) and Zelinsky (1971), and the most fully elaborated framework linking these phenomena. In the paradigm, place-to-place movements are viewed in the context of areal evolution, the argument being that factors influencing relocation shift in role according to prevailing 'development milieu'. For example, growth processes affect social and economic conditions, government policies, infrastructure, technological achievement, and other macro-aspects of regional systems, which in turn affect conditions directly pertinent to migration. As a result, modern sector wage rates and job opportunities might play a dominant role in advanced settings whereas migration-chain or resource-push effects would be more important under less advanced conditions.[2] Furthermore, this comparison may apply to a nation at different times in its evolution; or, alternatively, to subnational regions at the same time, with the implication that

some are less advanced than others and therefore experience different 'processes' of migration.

Articulation of the paradigm has been primarily in terms of temporal rather than spatial dimensions of development; and in a nonmathematical form. These imbalances in elaboration are addressed in the simulation reported here, which mathematically portrays a hypothetical nation of 21 urban places (or regions) evolving through a complete development cycle. This enables identification of patterns and trends implicit in paradigm assumptions, including shifts in the relative importance of pertinent variables, both locationally and over time, which then are compared with expectations from previous research. Consequences for population distribution also are examined, in part through altering the initial urban system to which the model is applied.

Such tasks provide a basis for extending and further specifying the development paradigm of migration, and for exploring its policy implications. One area of immediate gain is the urbanisation link, which was not identified in the earlier formulations of the paradigm. With advantage of hindsight, however, this omission seems myopic in that urbanisation stems both from the absolute transfer of people from rural to urban areas and from the higher rate of migrant fertility compared with that of other urban residents (Berry, 1982, 1984, 1985; Conway, 1983; Goldstein and Goldstein, 1981; Todaro and Stillkind, 1981; United Nations, 1973, Chapters 4 and 6).

We proceed as follows. In the next section we discuss previous research and related issues pertinent to model formulation. An overview of the general characteristics of the simulation follows; and we then detail the way each model component, and therefore the totality, was defined and made operational. We next focus attention on application to an initially primate urban system, then to a rank-order arrangement and a primate city system with enhanced diseconomies of urban size, thus demonstrating how migration patterns, population distribution, and urban system characteristics are affected by aspects of initial landscapes. Last, we discuss our findings in terms of broader social science concerns.

Previous research, model formulation, and related issues

Of concern here is previous research reflected in the simulation, as well as our rationale for choosing between alternative assumptions (or formulations). We first turn attention to migration, then to its interrelationships with development, and last, to implementation considerations.

Although seen to flow from less to more advantageous places, migration is usually studied as independent of links with development or urbanisation. Guiding factors include wage and job opportunity differentials in the modern sector; employment opportunities in the informal and rural nonfarm or small-scale enterprise sectors; migration chains based upon family, extended family, and acquaintance relationships; circular and seasonal migration strategies; individual or household characteristics such as age and family size; and resource-push factors related to the origin town or village: its economic well-being, pattern of resource distribution among social classes, local social norms, and its integration into the urban network (Brown and Sanders, 1981). Statistically estimated models, consisting primarily of 'labour-force adjustment' and 'human capital' formulations, employ a selection of these factors – typically place-specific wage rates, job opportunities, and amenity levels, together with a surrogate for place-to-place flows, or place-specific levels, of information – and are implemented in a neo-classical framework (Todaro, 1976, 1981).

Our migration component consists of factors typical of statistical formulations, but with two important additions. First, it includes an explicit 'migration-chain' effect, that is, migrants tend to follow in the paths of relatives, friends, and acquaintances who have migrated earlier. Underlying the widespread importance of this in Third World settings (Connell *et al.*, 1976; Levy and Wadycki, 1973; Rempel, 1980, 1981) are the paucity of formal communication mechanisms, the perceived reliability of informal communications sent through migration chains, and the role of friends, relatives, and acquaintances in destinations who make adjustment easier should migration occur.

The second factor given more than usual attention is 'resource push'. Previous research emphasises pull factors such as urban amenities, job opportunities, and wages, in part reflecting the 'urban bias' noted by Lipton (1976) and Todaro and Stillkind (1981). However, considerable evidence indicates that push factors also play an important and often dominant role (Brown, 1981, Chapter 8; Connell *et al.*, 1976; Conway and Shrestha, 1981; Findley, 1981; Havens and Flinn, 1975; Lipton, 1976, 1980; Roberts, 1978; Saint and Goldsmith, 1980; Thapa and Conway, 1983). Underlying this may be excessive population pressure on rural resources, partially aggravated by high rates of population increase; modernisation or innovation diffusion, often resulting in land loss and curtailment of local employment opportunities; or catastrophes such as earthquakes, flooding, drought, and revolution. In more theoretical contexts, political economy conceptualisations relate resource push, and factors fomenting it, to actions of metropolitan powers (Gregory and Piche, 1981; Riddell, 1981; Roberts, 1978; Swindell, 1979; Taylor, 1980).

The above examples indicate that resource push often derives from societal structure or development processes; other factors relevant to individual migration behaviour are similarly affected. This interface is central to the development paradigm of migration, but a structural perspective also is implicit in labour-force adjustment and human capital models through market-condition effects on migration decisions. Further, explicit elaboration of the interface occurs in 'dual economy' conceptualisations of development (Todaro, 1976, 1981; Yotopoulos and Nugent, 1976) wherein movements from traditional to modern locales are motivated by wage and job-opportunity differentials between them; such differentials persist until the modern sector develops sufficiently to transform the traditional sector and/or to absorb its excess labour; and meanwhile, economic growth proceeds on the basis of extensive human resources provided by traditional areas. Similar arguments relating development and migration are in core–periphery or growth-pole conceptualisations (Richardson, 1979, Chapters 6 and 7), where balance between 'trickle-down' and 'polarisation' effects, both presently and historically, determines whether economic conditions, represented by wage and job-opportunity differentials, are more attractive in core or periphery locales.

Despite such long-standing recognition, and seemingly obvious importance, development effects upon migration have rarely been an object of investigation. Furthermore, when they are investigated, as in models with simultaneous equations (Gober-Meyers, 1978a, 1978b; Greenwood, 1975a, 1975b, 1978; Salvatore, 1981) or in other modelling (Casetti, 1981b; Ledent, 1980, 1982a), development generally is represented in simple terms (for example, as regional income levels); and migration is represented as being governed by identical relationships in all locales. The latter is contrary to the development paradigm of migration and parallel observations (for example, Urzua, 1981), which postulate differences in the role of a variable according to the milieu (or set of local conditions) within which it operates.

Creating a more complex scenario involves balancing operational ability (a *sine qua non*) with substantive content, both in component design and selection and in the model overall.[3] As a result, in our simulation we embody widely articulated relationships such as demographic transition representations of population dynamics, the view that development is channelled through urban systems in hierarchical and neighbourhood-effect fashion, and gravity model portrayal of place-to-place interactions.[4]

Use of these components is based primarily on their generating patterns often found in Third World settings, but this bias is partly a response to the myriad of alternative, and often conflicting, interpretations given to operational elements. For example, hierarchical and neighbourhood-effect tendencies in development diffusion through urban systems are interpreted as representing both system efficiency (Berry and Kasarda, 1977; Mabogunje, 1980; Renaud, 1981; Richardson, 1977, 1979) – a neo-classical view – and colonial expansion or control (Fair, 1982; Mabogunje, 1980; Riddell, 1981; Santos, 1979; Slater, 1978) – a political economy view. Similar observations pertain to interpretation of demographic transition trends (Beaver, 1975; Caldwell, 1981, 1982; Teitelbaum, 1975; van de Walle and Knodel, 1980; Wrigley and Schofield, 1981, 1983), but more relevant to our effort is whether the 'convergence' of fertility and mortality rates evidenced historically in Europe and the United States of America will occur in Third World settings. Although sentiment is divided (Teitelbaum, 1975), recent evidence indicates that a downward trend in fertility levels, leading to convergence, has begun (Hackenberg and Magalit, 1983; Tsui and Bogue, 1978).

Another issue is the philosophical orientation of our model. Although prima facie neo-classical, by embracing the development paradigm of migration, which is structural in approach and thus compatible with political economic-historical considerations, the simulation blends both perspectives. For example, a central concern of political economy research is societal structure effects on individual decisions, paralleled by our focus on development, and particular elements such as resource push, found explicitly in the simulation. Equally important, operationalisation of political economy concepts would probably yield a form similar to that used here, rendering interpretation the more basic issue. Alternatively, however, neo-classical approaches represent the dominant tradition of Third World studies, today as well as years ago, and thus provide a common base from which to argue the importance of development (and societal structure) to migration.[5]

In summary, our simulation conforms to traditional formulations by embodying standard assumptions concerning development diffusion through urban or central place systems, as in work by Hudson (1969) or Berry (1972); demographic transition trends for population dynamics in the context of development; essentially neo-classical migration components; and an assumption that development processes ultimately lead to convergence, rather than to divergence, in place characteristics. Although scepticism persists, there is yet no proof of incorrectness in this framework, nor an alternative that could be embraced in an operational mode. Further, by focusing analytical attention on development and related structural effects, the end product bridges neo-classical and political economy perspectives.

Model overview

As noted, the simulation portrays spatial and temporal patterns of migration and population distribution associated with passage through a 'typical' development cycle by 21 urban places (or regions) constituting a hypothetical nation. Its migration component embodies a push–pull process intermediated by information flows. Specifically, push for out-migration from region i stems from population pressure upon a limited resource base, whereas pull for in-migration to region j depends upon its 'development-generated' employment opportunities.[6] Potential migrants learn about the latter both through interpersonal or informal communication channels and through formal ones such as the mass media.

Further, push, pull, and information flows change over time, in part because of an exogenous 'development' process altering population growth rates, employment opportunities, and formal communication channels. Development spread is guided by population-potential mechanisms embedded within logistic diffusion equations; hence, it generally begins at the primate city or core area, moves outwards in accordance with place population size and distance from the core, and, once initiated in a given locale, follows an S-shaped trajectory over time. Within this broad frame of reference, however, each exogenous factor is treated differently, rendering distinct diffusion patterns.

Population growth rates, one exogenous factor, reflect the assumption that each place passes through a demographic transition, the initiation of which, as well as the period of high population growth, follow the core-to-remote-locale pattern described immediately above. Accordingly, place-specific birth and death rates begin high and relatively in balance; decline in death rates accompanies the initial development wave with its improvements in medical technology and health-related conditions; birth rates follow a similar but lagged pattern, awaiting changes in socio-economic structure; and eventually, both rates come again into balance at an equally low level.

Development-generated employment opportunities, another exogenous factor, derive in part from place-specific and time-specific growth multipliers, but also interact with endogenous factors affecting population distribution. Specifically, an overabundance of persons in core or other locales creates diseconomies of urban size and hastens decentralisation of employment opportunities; until this happens, however, a greater than proportional share of employment opportunities goes to more populated places.

Formal communication channels, such as media, trade, and business interactions, the third exogenous development factor, provide information about employment opportunities in one place to persons in another. Although initially reflecting population size of places and intervening distances, the connections grow differentially over time in a manner favouring higher levels of development. Thus, links to core areas strengthen first, whereas those to remote locales lag in proliferation.

The model also contains three endogenous factors: place-specific resource bases, informal communication channels, and migration itself. These are discussed in turn.

Each place-specific resource base (R_{it}), comprising both traditional (R_{it}^T) and development-generated (R_{it}^D) sectors, is measured in terms of numbers of people supported by the local economy of region i in time t. Over time, three changes may occur. First, traditional sector use may intensify and thus support more persons. Second, that effect may also occur through resource transfer from traditional to development-generated modes of production; for example, by shifting to appropriate technology in agriculture. Third, resource transfer may also diminish local support; for example, by shifting to plantation or capital-intensive agriculture. The last change depends upon rates of economic growth in neighbouring regions and their distance from region i. To the extent that the resource base is insufficient, 'push' for out-migration occurs. Thus,

$$F_{it}^{PUSH} = f(R_{it}^T, R_{it}^D, P_{it}) \qquad (1)$$

where P_{it} is the population size of region i.

Given push from region i, information on job opportunities in potential destinations j is provided by formal communications, an artefact of development discussed above, and by informal communications representing chain effects from previous migrations. Specifically, informal communications from region j to region i in time t (C_{jit}^I) reflect the number of migrants residing in region j who previously resided in region i ($M_{ijt'}$) (where t' indicates all time periods prior to t) relative to the present population of region i (P_{it}). That is,

$$C_{jit}^I = f(M_{ijt'}, P_{it}) \qquad (2)$$

Migration, the third endogenous element, derives both from push factors, as portrayed in equation (1), and from new development-generated employment opportunities in potential destinations (E_{jit}^{NEW}) as communicated to origins through formal (C_{jit}^F) and informal channels. That is,

$$F_{jit}^{PULL} = f(E_{jt}^{NEW}, C_{jit}^F, C_{jit}^I) \qquad (3)$$

Collecting functional relationships, then, we find that migration between region i and region j in time t is

$$M_{ijt} = f(F_{it}^{PUSH}, F_{jt}^{PULL})$$

$$= f(R_{it}^T, R_{it}^D, P_{it}, E_{jt}^{NEW}, C_{jit}^F, C_{jit}^I)$$

$$= f(R_{it}^T, R_{it}^D, P_{it}, E_{jt}^{NEW}, C_{jit}^F, M_{ijt'}) \qquad (4)$$

In summary, a highly interrelated system has been specified. An initial distribution of population, job opportunities, and formal communication channels provides the nexus for a round of new job creation and population growth, which in turn alters local resource bases. This gives rise to migration flows, thus creating new channels of informal communications, altering population distribution, and more generally affecting conditions pertaining to development and migration in future rounds. Having overviewed the system, we now turn attention to a more detailed elaboration.

Model specification

To implement the model (Stetzer, 1983) we employed a hypothetical spatial system of 21 urban places (or regions) arranged on a slightly displaced triangular (central place) lattice. Three simulations were carried out. The first, reported in the next section, began with one place designated as primate and an initial population of 50 000; the remaining 20 were assigned populations of 1000 to 20 000 in increments of 1000. The second simulation specified an initial population distribution in terms of rank-size assumptions; the third employed the primate city distribution with considerably larger initial populations to enhance diseconomies of urban size. Simulations 2 and 3 are reported in the section 'Application under alternative systems'.

Now that we have established a base urban system, exogenous development is the next concern. Using potential measures embedded within logistic diffusion equations, this takes account of birth, death, employment growth, and formal communications.

Specifically, the birth rate for i at time t (B_{it}) is determined by the initial rate assigned all places (B^{MAX}); the maximum possible decline (B^{DECL}) with development and the time (T_i^B) at which the birth rate in region i begins to fall from B^{MAX}, estimated by its interaction potential with the primate city. That is,

$$B_{it} = B^{MAX} - B^{DECL} \{1 + a^B \exp[-b^B (t - T_i^B)]\}^{-1} \qquad (5)$$

where a^B and b^B are parameters. Similarly, the death rate for i at time t (D_{it}) is

$$D_{it} = D^{MAX} - D^{DECL} \{1 + a^D \exp[-b^D (t - T_i^D)]\}^{-1} \qquad (6)$$

The rate of development-generated job creation (E_{it}^{DG}), which varies between 0 (no growth in employment) and 1 (maximum growth), is a logistic function of the time such employment begins in region i (T_i^{EDG}) estimated by its interaction potential with the primate city, and the effect of urban size economies or diseconomies (b_{it}^{EDG}), allowing employment growth at increasing rates in places of 1 million or less and attenuating growth rates of larger places. Hence,

$$E_{it}^{DG} = \{1 + a^{EDG} \exp[-b_{it}^{EDG} (t - T_i^{EDG})]\}^{-1} \qquad (7)$$

Formal communications from region j to region i, the third exogenous element, is an amalgam of two others. One, their potential connectivity, is identical for both places and is determined by the time that improvement in formal communications begins between regions j and i (T_{ji}^{CF}) and the rate of improvement (a_{ji}^{CF}), both of which relate to i–j–primate city interaction potentials. The second component creates asymmetry in communications based upon the population size of region j relative to that of region i, thus recognising the greater incidence of information flows from larger to smaller places. Hence,

$$C_{jit}^F = \{1 + [\exp(-a_{ji}^{CF}(t-T_{ji}^{CF}))]\}^{-1}[1 + \exp(-b^{CF}(P_{jt}/P_{it}))]^{-1} \tag{8}$$

Exogenous factors represent the diffusion of development effects from core areas through the urban system. Interaction of these with endogenous factors – resource base, amount of employment, and level of informal communications – produces migration. We now turn our attention to this facet of the model.

The resource base of region i at time t (R_{it}), measured as the number of persons supported, consists of a traditional (R_{it}^T) and development-generated (R_{it}^D) sector. The latter increases in accordance with jobs created in time t (E_{it}^{NEW}) and the number of persons supported by each job (a^{RD}); that is,

$$R_{it}^D = R_{i,t-1}^D + a^{RD} E_{it}^{NEW} \tag{9}$$

Changes in traditional employment reflect three processes. First, its initial support level is incremented by a proportion, a^{RT}, representing more efficient resource use in response to population pressure. Second, each new development-generated job (E_{it}^{NEW}) reallocates resources from traditional ones, reducing them by b^{RT}. Third, development-generated job-creation rates in other places (E_{jt}^{DG}) relative to region i (E_{it}^{DG}), but attenuated by distance (d_{ij}), also reduce the traditional resource base of region i, by c^{RT}. Thus,

$$R_{it}^T = R_{i,t-1}^T + a^{RT} R_{i,t-1}^T - b^{RT} E_{it}^{NEW} - c^{RT} \sum_j \max[(E_{jt}^{DG} - E_{it}^{DG})/d_{ij}; 0] \tag{10}$$

The total resource base (R_{it}), then, is the sum of development-generated and traditional levels of employment, that is,

$$R_{it} = R_{it}^D + R_{it}^T \tag{11}$$

New jobs in time t (E_{it}^{NEW}) are determined by the rate of development-generated job creation in region i (E_{it}^{DG}), an exogenous element, and the magnitude of its economy, measured as the number of people supported at the beginning of time t ($R_{i,t-1}$). Thus,

$$E_{it}^{NEW} = a^{ENEW} E_{it}^{DG} R_{i,t-1} \tag{12}$$

where a^{ENEW} is a proportionality factor.

Informal communications from region j to region i in time t (C_{jit}^I), representing migration chain effects, are determined by i-to-j flows of previous times ($M_{ijt'}$) relative to the population size (P_{it}) of region i. Hence, with t' representing all times t prior to the present,

$$C_{jit}^I = \sum_{t'} a_{t'}^{CI} (M_{ijt'}/P_{it}) \tag{13}$$

where $a_{t'}^{CI}$ is a proportionality factor, the value of which inversely relates to $t-t'$, reflecting the progressively diminishing effect of earlier migrants. Total communications (C_{jit}), then, is a weighted average of formal and informal,

$$C_{jit} = a^c C_{jit}^F + (1-a^c) C_{jit}^I \qquad (14)$$

where, to capture the greater importance of migration chain effects, a^c is relatively small.

With the above components in place, migration flows from region i to region j in time (M_{ijt}) are computed in several steps. First, the migration pool (M_{it}^P) of region i is determined as a proportion a^{MP} of its population (P_{it}) in excess of total resource base (R_{it}); that is,

$$M_{it}^P = a^{MP}(P_{it} - R_{it}) \qquad (15)$$

Second, the strength of push factors at region i (F_{it}^{PUSH}), which impel potential migrants to become actual ones, is derived from resource base size (R_{it}) relative to total population (P_{it}):

$$F_{it}^{PUSH} = 1 - (R_{it}/P_{it}) \qquad (16)$$

Third, pull effects (F_{it}^{PULL}) reflect job opportunities in other places at previous times ($E_{j't'}^{NEW}$) and the likelihood their existence is communicated to region i ($C_{j'it'}$), where j' is all places except region i and t' all time periods prior to the present. Thus,

$$F_{it}^{PULL} = \sum_{j'} \sum_{t'} a_{t'}^M C_{j'it'} E_{j't'}^{NEW} \qquad (17)$$

where $a_{t'}^M$ is a proportionality factor, with values inversely related to $t-t'$, reflecting the progressively diminishing credibility, and hence pull, of older information.

Last, out-migration ($M_{i.t}$)' and its allocation to other places (M_{ijt}) is determined. The first represents combined effects of pushes and pulls on the migration pool of region i, that is,

$$M_{i.t} = M_{it}^P F_{it}^{PUSH} F_{it}^{PULL} \qquad (18)$$

where F_{it}^{PUSH} and F_{it}^{PULL} have been constrained between 0 and 1 to render them combinable. Allocation of migrants to particular destinations j is determined by the pull of each relative to aggregate pulls impacting region i at time t, or

$$M_{ijt} = M_{i.t} [(\sum_{t'} c_{t'}^M C_{jit'} E_{jt'}^{NEW})/F_{it}^{PULL}] \qquad (19)$$

where the term in innermost parentheses is the contribution of region j to equation (17), and the equation overall is an expression of Luce's (1959) choice axiom. Note also that the model takes account of the decision to leave a residence, the decision of where to relocate, and the way these interact to determine final migration patterns, thus rendering it congruent with current conceptualisations (Brown and Moore, 1970; Brown and Sanders, 1981).

Application to an initially primate city system

To explore further the world delineated by our model, it was applied to an initially primate city landscape, which many think typical of Third World, lesser-developed nations (Berry and Kasarda, 1977) As noted, this consisted of one place with a population of 50 000 and 20 places with populations of 1000 to 20 000 in increments of 1000 (Figure 18.l(a)). Implementation was for 120 iterations or time periods, representing a complete development cycle of 300 years.[8] For generalisation purposes, places were aggregated and averaged according to geographic characteristics: the primate city (PC, one place); intermediate urban places close to the primate city (IC, three places); intermediate urban places far from the primate city (IF, two places); small urban places close to the primate city (SC, four places); and small urban places far from the primate city (SF, four places) (Figure 18.l(a)).[9]

In aggregate, the country experienced a 36-fold increase in population, from 260 000 to 9.4 million. Within this, the primate city increased in population from 50 000 to 3 375 000, almost doubling population share (from 19 per cent to 36 per cent), while intermediate places (IC, IF) began with 16 000–20 000 and ended with 571 000–756 000, thus maintaining a population share of approximately 35 per cent as a group and 7.0 per cent individually. Also noteworthy is a shifting in rank among ICs and IFs, with places closer to the PC advancing at the expense of those more distant. Small urban places (SC, SF) generally retained initial rankings, but ended with a wide range of sizes, from 975 to 378 000, the smallest actually losing 25 people over the development cycle. Thus, PC growth was counterbalanced by population loss from small places; yet some SCs actually grew to be similar in size to intermediate places (Figures 18.1 and 18.2).

These changes in population distribution arise partly from migration. Specifically, net migration rates by type of place (Figure 18.3) indicate that the PC gains far more persons from migration than do other places; intermediates gain somewhat; and small places are net losers. Further, PC proximity provides benefits in that IC places gain more than IF, and SC lose less than SF. However, timing of these impacts varies. Migration has little effect on population distribution in the first 30 time periods. In approximately time periods 30 to 67, the PC experiences increasingly higher rates of gain, while others lose at increasing rates, but losses are much greater for small places than for intermediate places and, within each, for places far from the PC than for places close to the PC. In approximately time periods 68 to 90, the PC gains from migration decrease rapidly, with the PC eventually becoming a net loser; intermediate places improve position and emerge as gainers; while small places continue losing until time period 80, whereupon fortunes reverse. Last, there is convergence to low net migration for all places.[10]

In general, then, larger places and places closer to the PC experience greater overall migration effects, and experience them earlier. Since this conforms with the pattern of development diffusion, the role of development is evident. To elaborate the theme further, we turn attention to migration directionality.

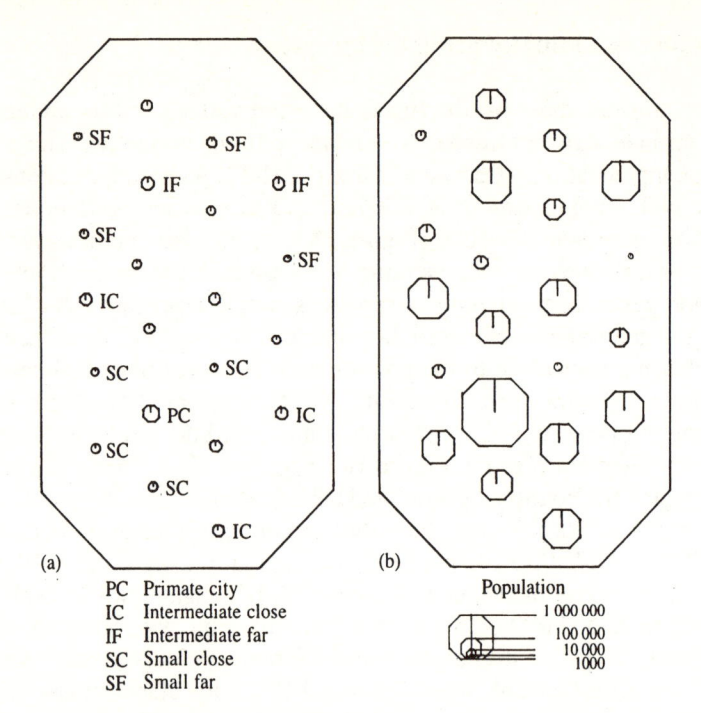

PC Primate city
IC Intermediate close
IF Intermediate far
SC Small close
SF Small far

Fig. 18.1 Population distribution for the primate city case; (a) initial population distribution, (b) terminal population distribution

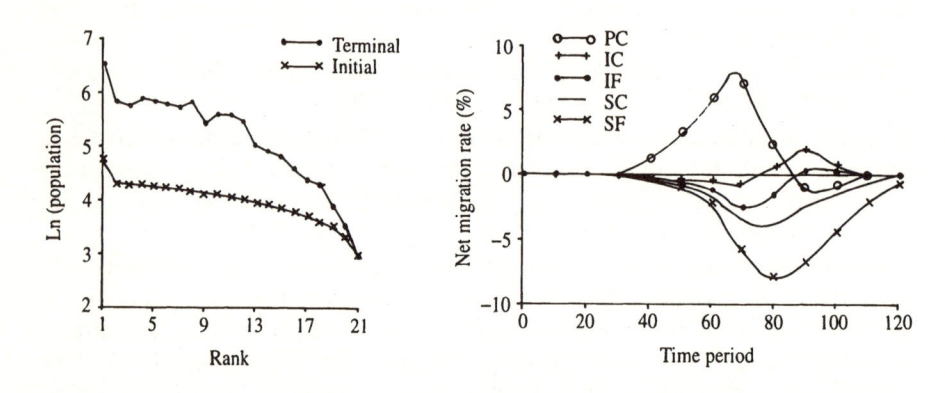

Fig. 18.2 Graphs of city size distribution for the primate city case

Fig. 18.3 Net migration rates for the primate city case (*see text* for explanation of abbreviations)

Rates of out-migration to the PC (Figure 18.4(a)) are near 0 for the first 20 time periods, rise slowly over the next 35, then sharply to a peak between time periods 75 and 80, and decline linearly to 0 by the end of the development cycle. Differentiation among places occurs only after time period 65. Specifically, rates of out-migration for ICs and IFs peak sooner and at lower levels than for SCs and SFs; rates for ICs and IFs decline to 0 by time period 110, rates for SCs reach 0 at time period 120, and SFs are left with positive although low rates of PC-directed migration. Overall, migration to the PC is greatest for SFs and least for ICs.

IC, IF, SC, and SF rates of migration to places other than PC (Figure 18.4(b)) are patterned similarly to their PC-directed counterparts, but somewhat lower. More noteworthy is PC out-migration to other places, which starts earlier, at time period 10; is somewhat higher until time period 50; peaks later, in time period 90; and drops to 0 by time period 110. Further, PC out-migration rates peak nearly as high as rates for SFs and considerably higher than others.

These patterns indicate that out-migration rates are primarily affected by place size and location, particularly in the second half of the development cycle. The tendency is offset, however, by urban-size economies or diseconomies; most visibly in the temporal pattern of PC out-migration, but also in lower rates for close places (IC, SC) compared with far places (IF, SF). Further, a consideration of out-migration and net migration rates together suggests the PC initially receives population both from small and from intermediate places, and subsequently redistributes it to the latter. Particularly intriguing is the congruence of this observation with the polarisation reversal phenomenon predicted

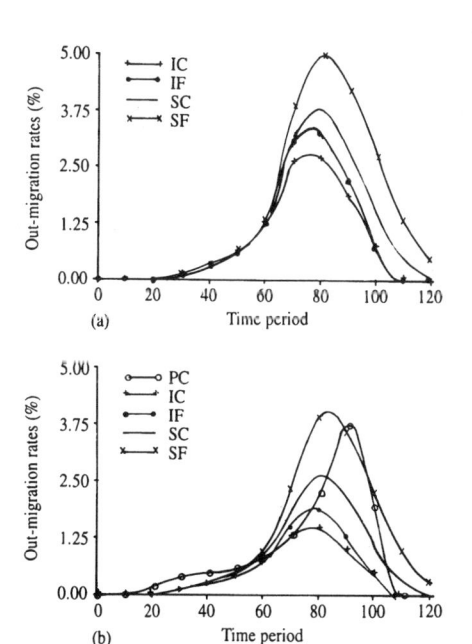

Fig. 18.4 Out-migration rates for the primate city case; (a) out-migration rates to primate city, (b) out-migration rates to other cities (*see text* for an explanation of the abbreviations)

by core–periphery and growth-pole paradigms of development (Richardson, 1979, Chapters 6 and 7; 1980).

Another concern is factors underlying migration patterns; in particular, resource push, development-generated employment pull, and informal and formal communications.

Resource push (F^{PUSH}) (Figure 18.5), exhibiting a bell-shaped curve form, begins early in response to population growth, peaks about midway through the development cycle when regional disparities are marked, and falls towards 0 as economic convergence approaches. However, several divergencies should be noted. Resource push at the PC, for example, levels off at about time period 35, but continues rising elsewhere, reflecting the emergent economic growth of the PC, and the related drain of resources from other areas while their population is rapidly increasing. About time period 75, however, resource push at the PC reaches a higher peak, but fails elsewhere, which is caused by employment decentralisation to intermediate places while the PC population growth continues, in large part from in-migration. Last, PC, IC, and IF resource push are 0 at the end of the development cycle, but the resource push for SCs is slightly positive and that for SFs reasonably high, given the existing economic equilibrium.

Development-generated employment, in terms of total and new jobs per capita (Figure 18.6), generally follows an exponential trend, with low levels through the first half of the development cycle, and steady but gradual increases after that. Distributionally, trends of ICs, IFs, SCs, and SFs are similar, but intermediate places terminate with identical and high employment levels, whereas smaller places are less endowed and further differentiated by location, SFs ending least favourably. By contrast, PC employment breaks to a noticeably higher level between time periods 20 and 60; then reverses to become, from time periods 80 through 90, lower than every place except SF; and finally ends with levels that are high but somewhat less than intermediate-size places.

Loss of PC employment after the midpoint of the development cycle coincides with its highest rate of in-migration, leading to diseconomies of urban

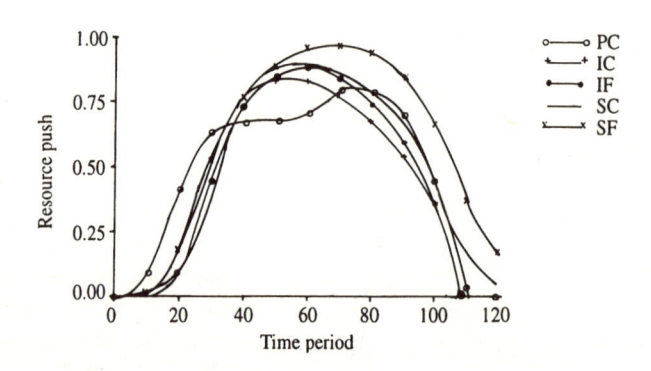

Fig. 18.5 Resource push for the primate city case (*see text* for an explanation of the abbreviations)

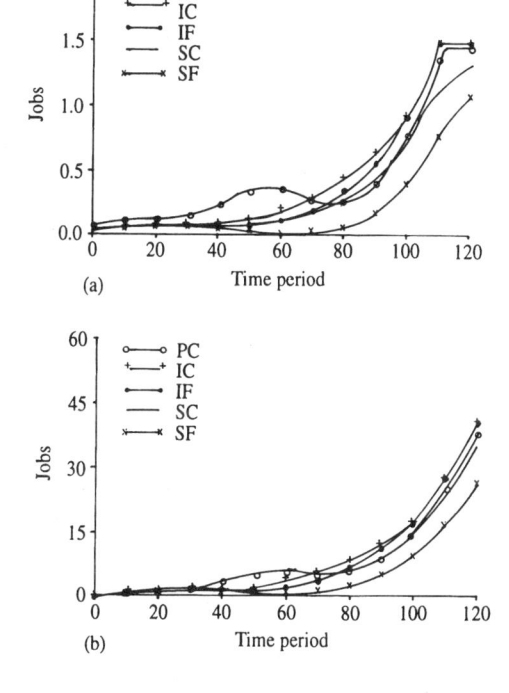

Fig. 18.6 Employment opportunities per hundred population for the primate city case; (a) new jobs, (b) total jobs (*see text* for an explanation of the abbreviations)

size, as population accumulates and multiplies. More generally, a consideration of findings thus far discussed (and Figures 18.2–18.6) indicates that development prospects depend upon a threshold level of economic agglomeration. To elaborate, smaller places, especially remote ones, end the development cycle with a considerably worse position in the urban hierarchy, lingering resource push, and continuing out-migration, the last needed to maintain balance in the local economy. This occurs because SF resource bases are drained before development-generated employment is initiated, thereby destroying conditions for a 'natural' transition to a modern economy. Development also drains ICs and IFs, but less severely. Hence, smaller places are trapped in a poverty relieved only by out-migration, whereas larger (or more advantageously located) ones conserve sufficient resources to provide a base for growth, once the opportunity is present. This is reminiscent of 'low-level equilibrium trap' phenomena (Casetti, 1981a, 1982; Leibenstein, 1957; Nelson, 1956), but extended to include a spatial frame of reference.

Last, we consider the role of informal and formal communications (Figure 18.7), which link resource push and development-generated employment pull. Initially, until approximately time period 30, virtually all communications are formal, reflecting both a dearth overall and low levels of migration through which informal links are established. Between time periods 30 and 85, informal channels proliferate more rapidly than formal ones, and come to play an equal or greater role. The fulcrum then shifts back towards formal but, although not evident from Figure 18.7, at high levels of interaction overall.

278 Differential Urbanization

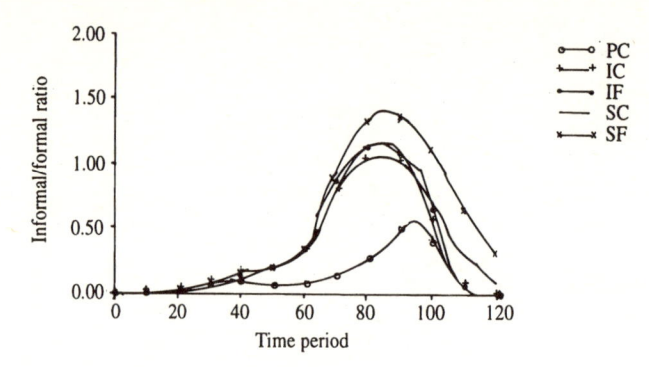

Fig. 18.7 Ratio of informal to formal communication for the primate city case (*see text* for an explanation of the abbreviations)

Distributionally, this pattern is most characteristic of SF places and only slightly less characteristic for SCs, ICs, and IFs; but formal communications dominate the PC throughout. By the end of the development cycle, they also dominate intermediate places, whereas in smaller ones, informal communications noticeably persist. Lingering of informal interactions is especially strong in SF-to-PC links which generate migration even when not warranted by PC employment opportunities, with obvious consequences for its poverty level.

Application under alternative urban systems

In this section we report the application of the simulation to an initially rank-ordered urban system and to an initially primate city system with enhanced diseconomies of urban size. Of particular concern is the effect of those landscapes upon migration patterns, population distribution, and urban system characteristics.

Effects of an initially rank-ordered system of cities

This simulation begins with a landscape of 260 000 persons distributed in rank-order fashion; hence, population (P_k) of the kth ranked place is given by (P_1)/k, allocating 71 324 to the first ranked, 35 662 to the second, 23 775 to the third, and 3396 to the smallest. Compared with the primate city landscape discussed earlier, this has the same total population, more in the rank-one place, more variation among intermediate-size places, and less variation among small places (compare Figure 18.8(a) with Figure 18.1(a)).

The temporal trends of net migration under rank-order assumptions are similar to the primate city simulation (Figure 18.9 compared with Figure 18.3). However, movement to the rank-one city is visibly higher, and small places in its proximity experience considerably more population drain, whereas remote ones are affected little by this change in initial landscape. Further, ICs benefit less from development than IFs, the opposite to primate city findings. This is mainly because the IFs are larger, an artefact of central-place assumption.[11]

PC Primate city
IC Intermediate close
IF Intermediate far
SC Small close
SF Small far

Population
1 000 000
100 000
10 000
1000

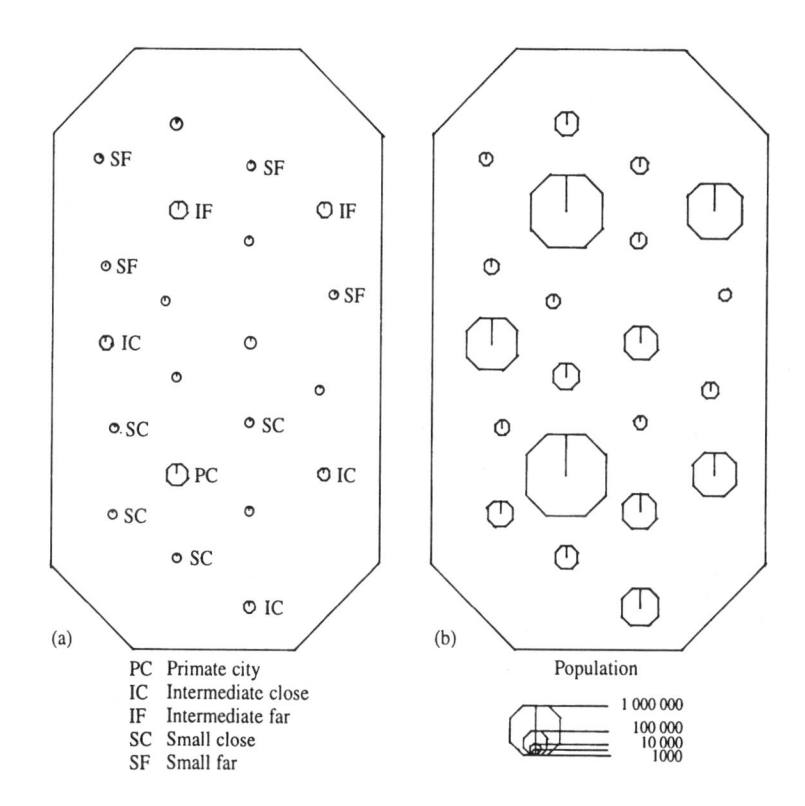

Fig. 18.8 Population distribution for the rank-order case; (a) initial population distribution, (b) terminal population distribution

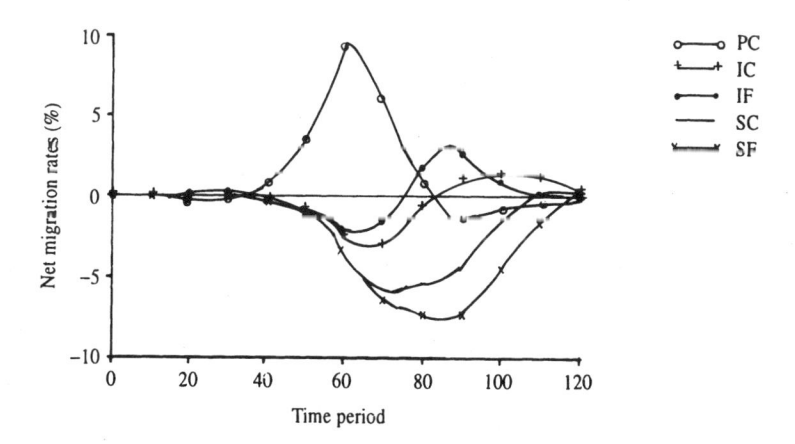

Fig. 18.9 Net migration rates for the rank-order case (*see text* for an explanation of the abbreviations)

It follows that initial landscape characteristics affect final population distribution. Specifically, primate-city assumptions produce a monocentric pattern focused on the PC, and rank-order assumptions produce a polycentric form incorporating IFs as regional centres (Figure 18.8(b) compared to Figure 18.1(b)). As a corollary, the IF share of total population shifts from an initial 11.4 per cent to a terminal 18.3 per cent under rank order assumptions, compared with 7.5 per cent to 6.6 per cent under primate-city assumptions. The difference can be traced to diversion of migration to IFs from the rank-one city, and to the proximity of IFs to smaller entities (largely SFs) which enhances the population drain of the SFs. Nevertheless, the smallest places increase their population share under rank-order size assumptions (Figure 18.10 compared with Figure 18.2). In summary, the postdevelopment urban system emerging from an initial rank-order landscape retains its rank-order distribution, includes intermediate places as important nodes, and thus indicates a spatially integrated economy.

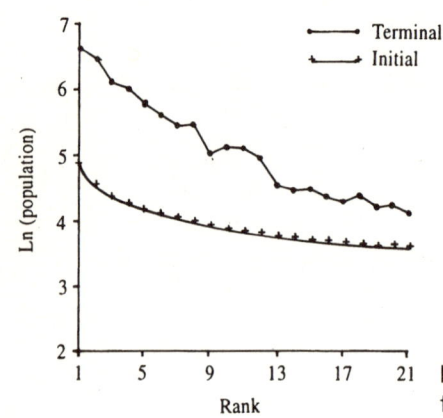

Fig. 18.10 Graphs of city size distribution for the rank-order case

Effects of enhanced diseconomies of urban size

A second landscape variation employs primate-city assumptions with an initial population of 1 300 000 (five times the others) to induce size diseconomies earlier in the development cycle (Figure 18.11(a)). A critical population for the consequent redirection of employment growth and migrants is 2 000 000, which for the largest place occurs at time period 73 in the first primate city simulation, at time period 66 in the rank-order simulation, both having lower total populations, and at time period 43 in the current simulation.[12] This perturbation relates to public policy which (directly or indirectly) 'taxes' urban agglomerations or provides incentives for economic decentralisation.[13]

Under enhanced diseconomies of scale, temporal trends of net migration are similar to the earlier primate city simulation (Figure 18.12 compared with Figure 18.3), but demographic events occur markedly sooner; that is, PC net migration peaks earlier, arrival at and drop off from the peak is more rapid, and other places are affected accordingly. ICs benefit especially, for the following reasons. If knowledge of employment opportunities were correct, migrants

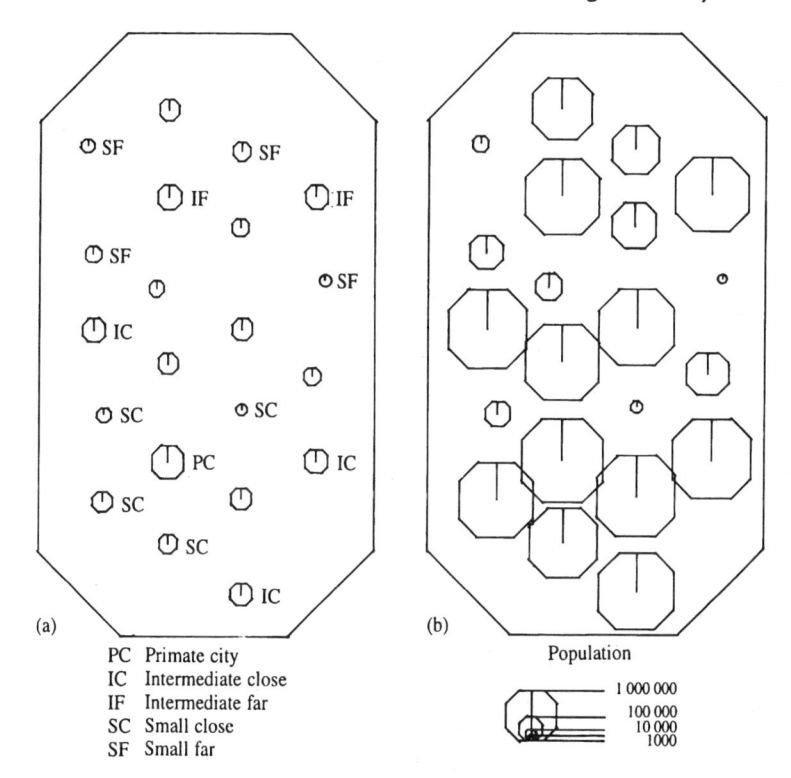

PC Primate city
IC Intermediate close
IF Intermediate far
SC Small close
SF Small far

Fig. 18.11 Population distribution for the case using enhanced diseconomies of scale; (a) initial population distribution, (b) terminal population distribution

would gravitate to rapidly growing ICs and IFs. However, persistent informal communication patterns direct them to the PC, and accordingly, when knowledge improves, the proximity of ICs renders them more attractive. Thus, ICs benefit both by economic attributes and by location.

More generally, enhancing diseconomies of scale greatly magnifies the stage-migration and polarisation-reversal effects observed earlier, evidenced by considerable shifts in population distribution (Figures 18.11 and 18.13). Although the PC remains largest, by approximately 500 000 people, it is rivalled by several ICs; over half of all places exceed a population of 1.5 million, and an 'economic shadow' (Ray, 1965) or 'megalopolis' effect occurs in the PC region. Alternatively, size range among smaller places is greatly extended, similar to the first primate city simulation. Thus, size diseconomies spread growth over more of the urban system, but smaller entities do not benefit proportionately; that is, graphs of city-size distribution (Figure 18.13) shift from 'primate city' to 'convex' form, using Sheppard's (1982) terminology. Interestingly, then, among the conditions examined, only size diseconomies lead to significant change between initial and final population distributions.

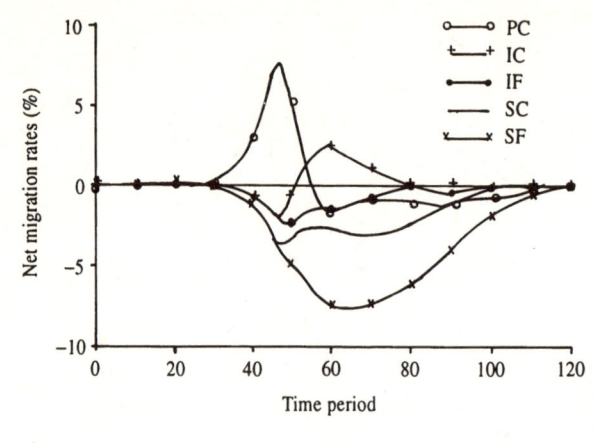

Fig. 18.12 Net migration rates for the case using enhanced diseconomies of scale (*see text* for an explanation of the abbreviations)

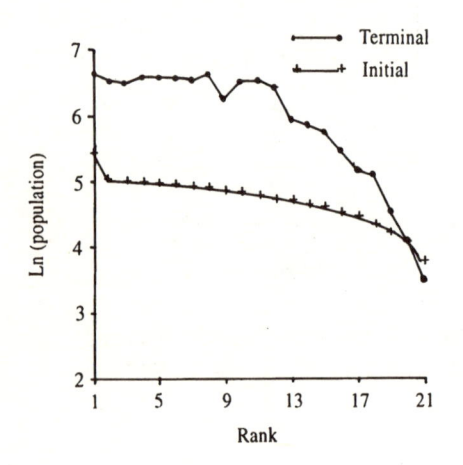

Fig. 18.13 Graphs of city size distribution for the case using enhanced diseconomies of scale

Discussion

To turn to the broad implications of our simulation, one area of concern is the development paradigm of migration (Brown and Sanders, 1981). Its expression over a full development cycle can be seen by modelling, a luxury not available through real-world data. In this regard, migration patterns initially show negligible differentiation by place size and location, a good deal in the middle phases of the development cycle, and relatively little at its end, which agrees with Zelinsky's (1971) hypotheses of mobility transition. Similar patterns characterise shifts in the roles of resource push and informal communications, which peak in significance at about 50 per cent and 60 per cent, respectively, through the development cycle and then diminish. By contrast, development-generated employment pull and formal communications monotonically increase in importance throughout. Within these broad trends, the roles of resource push and informal communications tend to vary inversely with place size and directly with distance from the major city, whereas the opposite holds for development-generated employment and formal communications. However, the major city diverges from these patterns, largely because of urban-size economies and diseconomies. Per-capita employment, for example, is noticeably greater in the primate city than elsewhere early in the development cycle, then reverses to be less, and ends nearly identical with levels in intermediate-size places. These findings agree with suppositions of the development paradigm of migration, except for outcomes related to urban size economies and diseconomies, a factor the paradigm did not consider.

A second area of concern is the link between migration, development, and urbanisation. Although admittedly serendipitous, our simulation falls within the genre of neo-classical regional development models (Richardson, 1979), which are rarely subject to detailed analyses concerning long-run implications for landscape evolution. One exception is Pedersen (1975, pp. 90–109), who addresses the transition from traditional to modern society in terms of likely urban system characteristics and related interactions. To a remarkable degree, his expectations are borne out here.[14] Of additional significance is our finding of polarisation reversal, predicted by growth pole and core–periphery conceptualisations (Richardson, 1980) and observed for settings such as the São Paulo region of Brazil (Townroe and Keen, 1984). Note also the temporal trend in regional disparities, which begins at a low level, peaks somewhat past the midpoint of the simulated development cycle, and then returns to a minimum, as prevailing conceptualisations predict.

Departures from conventional expectations were also found. One is the occurrence of low-level equilibrium traps at small remote places, where continual resource-drain thwarts the agglomeration economies necessary for sustained growth, and only outmigration maintains equilibrium with the consequent resource push. This extends the concept to a space-economy frame of reference, hitherto not addressed (Casetti, 1981a, 1982), and to places as well as to nations.

More generally, distribution of growth and other space-economy attributes are apparently affected by the early character of the urban system. Thus, in the initially rank order landscape, intermediate-size places close to the major city grew proportionally to the system overall; remote intermediates improved their relative position and became regional focal points; and major city dominance increased. The latter also occurred under an initially primate city landscape, but residual growth fell towards intermediate (and some small) places in proximity to the primate city. In the initially primate city landscape with enhanced diseconomies of scale, however, intermediate places benefited considerably, coming to rival the major city in size and creating a megalopolis around it.

A related issue is stability of the city size distribution (Berry and Kasarda, 1977, Chapter 19). Empirical evidence indicates that contemporary and early distributions are often similar; the US urban system, for example, has always been rank order. But theoretical statements suggest transition from primate city to rank order as a corollary of space-economy integration related to development. In a similar vein, Papageorgiou (1980) fits exponential curves to world major city growth, and attributes their inflection point, where growth rapidly accelerates, to disequilibrium (of a catastrophe-theory type) between urban and rural sectors of the surrounding economy. Alternatively, Sheppard (1982) examines convex, S-shaped, rank order, and primate city distributions, concluding that links between their form and development process are groundless. Contributing to the debate, our findings indicate that Papageorgiou's high rates of urban growth are a temporary aberration, and that only urban size diseconomies are likely to alter significantly the size distribution and spatial characteristics of urban systems. Further, small centres fared considerably worse than others in all scenarios, although less so under rank order, an aspect not usually considered but meriting further attention.

Last, the high degree of interrelatedness among space-economy elements should be noted. Models of either migration or development implicitly link those phenomena (Brown and Jones, 1984); explicit linking occurs in Brown and Sanders's (1981) development paradigm of migration, Zelinsky's (1971) hypothesis of the mobility transition, and simultaneous-equations formulations (Gober-Meyers, 1978a, 1978b; Greenwood, 1978; Salvatore, 1981). Other modelling (Ledent, 1980, 1982a; Morrill, 1965; Rogers and Williamson, 1982) adds an urbanisation component. Nevertheless, these phenomena are more customarily treated separately, so that the scope of the simulation, particularly its pertinence to urbanisation, was not anticipated. Accordingly, the broadly focused systems perspective of the model is apparently itself a significant contribution, which should be elaborated in future research.

Acknowledgements

This paper is part of a larger study on interrelationships between migration and development in Third World settings being carried out under Grant SES-8024565 of the National Science Foundation. That support is appreciated. Comments on earlier versions were provided by W. R. Smith of the Ohio State University Department of Geography.

Notes

1 Current work on this theme and reviews of prior research include Brown and Jones (1984), Brown and Kodras (1984), Brown and Lawson (1984a, 1984b, 1985), Brown and Sanders (1981), Conway (1983), Conway and Shrestha (1981), Peek and Standing (1979, 1982), Thapa and Conway (1983), and Urzua (1981).
2 These and other factors commonly found in migration analyses are discussed in the following section of the paper.
3 This relates, of course, to the long-standing 'pattern versus process' issue. For a discussion, *see* Chorley (1962), Amedeo and Golledge (1975), and Harvey (1969).
4 Previous research underlying simulation-component design is reviewed in numerous materials: for example, Brown and Sanders (1981) on migration; United Nations (1973) or Todaro (1981) on population dynamics in a development context; Richardson (1979) or Brown (1981, chapter 8) on regional development processes; and Todaro (1981), Yotopoulos and Nugent (1976) or Mabogunje (1980) on development processes in general. Incidentally, our use of gravity-model-based components and assumption of hierarchical and neighbourhood-

effect patterns of diffusion takes them as surrogates for functional relationships between places. To illustrate, in the real Third World, development generally diffuses from more to less important places, in direct response to accessibility via modes of transport or communications; in our model, however, these functional relationships are represented by population size and distance.

5 Evidence of the prevalence of neo-classical studies in topics addressed here are *Economic Development and Cultural Change* 30 (3) (1982), a special issue titled 'Third World migration and urbanization: a symposium', and *International Regional Science Review* 7 (2) (1982), a special issue titled 'Urbanization and development'.

6 Development-generated employment opportunities stand in contrast to 'traditional' ones and encompass activities in the formal and informal sectors as well as rural nonfarm, small-scale, and Western-style large-scale enterprise. Use of this terminology is to circumvent innuendos associated with more common nomenclature, for example, modern versus traditional sector employment. Further, development-generated employment is treated as a unit in that little is known about the spatial distribution of other than formal activities, and covariance is as reasonable an assumption as any (Sanders, 1980).

7 Here we use the dot notation '.' to represent summation of M_{ijt} over j, that is, $M_{i.t} = \sum_j M_{ijt}$.

8 The decision to use 120 iterations derives from observing convergence of simulated results towards a stable pattern of population distribution and migration.

9 To provide clear breaks in categories, places with 16 000 population or above, except for the primate city, were classified as intermediate; places with 11 000 or less as small; and the seven places which either fall between these in size or are ambiguous locationally were omitted from reported results.

10 Ledent (1982b) reports similar findings for 1950 to 2000, employing both deductive and inductive analyses of actual and projected national population trends.

11 Central-place lattices assume an inverse relationship between size and spacing of places. Accordingly, the second and third largest places, both IFs, are farther from the rank one place than the fourth, fifth, and sixth, which are ICs; and under rank-order assumptions, 'close' and 'far' also indicate urban size.

12 The parameter b_{it}^{EDG} in equation (7) is a quadratic function of the population of place i such that rate of employment growth maximizes at a population of 1 000 000, declines monotonically between populations of 1 000 000 and 2 000 000, then turns negative.

13 Venezuela and Korea, for example, have encouraged industrial decentralisation since the mid-1960s. The myriad of new town, new capital, or frontier development policies serve a similar end. For further discussion, see Renaud (1981) or Richardson (1977).

14 Also relevant to this point are Morrill (1965) and Alonso (1980).

References

Alonso, W. 1980: Five bell shapes in development. *Papers of the Regional Science Association* 45, 5–16.

Amedeo, D. and Golledge, R. G. 1975: *An introduction to scientific reasoning in geography*. New York: John Wiley.

Beaver, S. E. 1975: *Demographic transition theory reinterpreted*. Lexington, Mass.: Lexington Books.

Berry, B. J. L. 1972: Hierarchical diffusion: the basis of developmental filtering and spread in a system of growth centres. In Hansen, N. M. (ed.), *Growth centres in regional economic development*. New York: Free Press, 108–38.

Berry, B. J. L. and Kasarda, J. D. 1977: *Contemporary urban ecology*. New York: Macmillan.

Berry, E. H. 1982: *Migration, fertility, and development: A conceptual note and proposal for research on Ecuador*. Studies on the interrelationships between migration and development in Third World settings, DP-3, Department of Geography, Ohio State University, Columbus, Ohio.

Berry, E. H. 1984: *Regional structure effects on migration and fertility in Ecuador*. Studies on the interrelationships between migration and development in Third World settings, DP-15, Department of Geography, Ohio State University, Columbus, Ohio.

Berry, E. H. 1985: Migration, fertility, and social mobility. *International Migration Review* 18 (forthcoming).

Brown, L. A. 1981: *Innovation diffusion: A new perspective*. New York: Methuen.

Brown, L. A. and Jones, J. P. III. 1984: *Development effects on migration in Third World settings: Conventional modeling with spatially varying parameters via the expansion method*. Studies on the interrelationships between migration and development in Third World settings, DP-13, Department of Geography, Ohio State University, Columbus, Ohio.

Brown, L. A. and Kodras, J. E. 1984: *Migration, human resource transfers, and development contexts: A logit analysis of Venezuelan data*. Studies on the interrelationships between migration and development in Third World settings, DP-21, Department of Geography, Ohio State University, Columbus, Ohio.

Brown, L. A. and Lawson, V. A. 1984a: *Migration, planned growth, and human resource variations within Third World urban systems: A Venezuelan study*. Studies on the interrelationships between migration and development in Third World settings, DP-20, Department of Geography, Ohio State University, Columbus, Ohio.

Brown, L. A. and Lawson, V. A. 1984b: *Rural destined migration in Third World settings: A neglected phenomenon?* Studies on the interrelationships between migration and development in Third World settings, DP-14, Department of Geography, Ohio State University, Columbus, Ohio.

Brown, L. A. and Lawson, V. A. 1985: Migration in Third World settings, uneven development, and conventional modelling: a case study of Costa Rica. *Annals of the Association of American Geographers* 75(1), 29–49.

Brown, L. A. and Moore, E. G. 1970: The intra-urban migration process: a perspective. *Geografiska Annaler*, Series B 52: 1-13; also in *Yearbook of the society for general systems research* 15, 109–22 (1970); abridged version in Bourne, L. S. (ed.) 1971: *Internal structure of the city: Readings on space and environment*: New York: Oxford University Press, 200–9.

Brown, L. A. and Sanders, R. L. 1981: Toward a development paradigm of migration: with particular reference to Third World settings. In De Jong, G. F. and Gardner, R. W. (eds), *Migration decision making: Multidisciplinary approaches to micro-level studies in developed and developing countries*. New York: Pergamon Press, 149–85; also in **Terra 5**, as Hacia un paradigma do desarrollo de la migración con especial referencia a los paises del Tercer Mundo; abbreviated version in *Geographic research on Latin America: Benchmark*, 8 as On the interrelationship between development and migration processes. CLAG Publications; Muncie, Ind., 357–73.

Caldwell, J. C. 1981: The mechanisms of demographic change in historical perspective. *Population Studies* 35, 5–27.

Caldwell, J. C. 1982: *Theory of fertility decline*. New York: Academic Press.
Casetti, E. 1981a: Technological progress, exploitation and spatial economic growth: a catastrophe model. In Griffith, D. A. and Mackinnon, R. D. (eds), *Dynamic spatial models*. Alphen aan den Rijn, The Netherlands: Sijthoff & Noordhoff, 215–27.
Casetti, E. 1981b: The spatial diffusion of migrations and development in modern Europe. In Mandal, R. B. (ed.), *Frontiers in migration analysis*. New Delhi: Concept Publishing, 69–82.
Casetti, E. 1982: The onset of modern economic growth: empirical validation of a catastrophe model. *Papers of the Regional Science Association* 50, 9–20.
Chorley, R. J. 1962: *Geomorphology and general systems theory*. Geological survey professional paper 500-B. Washington, DC: US Government Printing Office.
Connell, J., Dasgupta, B., Laishley, R. and Lipton, M. 1976: *Migration from rural areas: Evidence from village studies*. London: Oxford University Press.
Conway, D. 1983: Rural-to-rural migration and rural fertility: a neglected relationship. Working paper, Department of Geography, Indiana University, Bloomington, Ind.
Conway, D. and Shrestha, N. R. 1981: *Causes and consequences of rural-to-rural migration in Nepal*. Ford Foundation Research Report, Department of Geography, Indiana University, Bloomington, Ind.
Fair, T. J. D. 1982: *South Africa: Spatial frameworks for development*. Cape Town: Juta Publishers.
Findley, S. E. 1981: Rural development programmes: planned versus actual migration outcomes. In Demko, G. J. and Fuchs, R. J. (eds), *Population Studies 75: Population distribution policies in development planning*. Department of International Economic and Social Affairs, United Nations, New York, 144–66.
Gobor-Meyers, P. 1978a: Employment motivated migration and economic growth in post-industrial market economies. *Progress in Human Geography* 2, 207–29.
Gobor-Meyers, P. 1978b: Interstate migration and economic growth: a simultaneous equations approach. *Environment and planning A* 10, 1241–52.
Goldstein, S. and Goldstein, A. 1981: The impact of migration on fertility: an 'own children' analysis for Thailand. *Population Studies* 35, 265–84.
Greenwood, M. J. 1975a: Research on internal migration in the United States: a survey. *Journal of Economic Literature* 8, 397–433.
Greenwood, M. J. 1975b: Simultaneity bias in migration models: an examination. *Demography* 12, 519–36.
Greenwood, M. J. 1978: An econometric model of internal migration and regional economic growth in Mexico. *Journal of Regional Science* 18, 17–31.
Gregory, J. W. and Piche, V. 1981: *The demographic process of peripheral capitalism illustrated with African examples*, WP-29, Centre for Developing Area Studies, McGill University, Montreal.
Hackenberg, R. A. and Magalit, H. F. 1983. *Demographic responses to development: Sources of declining fertility in the Philippines*. Boulder, Colo.: Westview Press.
Harvey, D. 1969: *Explanation in geography*. New York: St Martin's Press.
Havens, A. E. and Flinn, W. L. 1975: Green revolution technology and community development: the limits of action programs. *Economic Development and Cultural Change* 23, 469–81.
Hudson, J. C. 1969: Diffusion in a central place system. *Geographical Analysis* 1, 45–58.
Ledent, J. 1980: Comparative dynamics of three demographic models of urbanisation. *International Institute for Applied Systems Analysis Research Reports* 1, 241–79.
Ledent, J. 1982a: Rural–urban migration, urbanisation, and economic development. *Economic Development and Cultural Change* 30, 507–38.

Ledent, J. 1982b: The factors of urban population growth: net inmigration versus natural increase. *International Regional Science Review* 7, 99–125.

Leibenstein, H. 1957: *Economic backwardness and economic growth.* New York: John Wiley.

Levy, M. and Wadycki, W. 1973: The influence of family and friends upon geographic labour mobility: an international comparison. *Review of Economics and Statistics* 55, 198–203.

Lipton, M. 1976: *Why poor people stay poor: Urban bias and world development.* Cambridge, Mass.: Harvard University Press.

Lipton, M. 1980: Migration from rural areas of poor countries: the impact upon rural productivity and income distribution. *World Development* 8, 1–24; also in Sabot, R. H. (ed.) 1982: *Migration and the labor market in developing countries.* Boulder, Colo.: Westview Press, 191–228.

Luce, R. D. 1959: *Individual choice behavior.* New York: John Wiley.

Mabogunje, A. L. 1970: Systems approach to a theory of rural–urban migration. *Geographical Analysis* 2, 1–17.

Mabogunje, A. L. 1980: *The development process: A spatial perspective.* London: Hutchinson.

Morrill, R. L. 1965: *Lund studies in geography 26: Migration and the spread and growth of urban settlement.* Lund: Gleerup.

Nelson, R. R. 1956: A theory of low level equilibrium trap in underdeveloped economies. *American Economic Review* 46, 896–908.

Papageorgiou, G. J. 1980: On sudden urban growth. *Environment and Planning A* 12, 1035–50.

Pedersen, P. O. 1975: *Urban-regional development in South America: A process of diffusion and integration.* The Hague: Mouton.

Peek, P. and Standing, G. 1979: Rural–urban migration and government policies in low income countries. *International Labour Review* 118, 747–62.

Peek, P. and Standing, G. (eds) 1982: *State policies and migration: Studies in Latin America and the Caribbean.* Beckenham, Kent: Croom Helm.

Ray, D. M. 1965: Market potential and economic shadow: a quantitative analysis of industrial location in southern Ontario. RP-101, Department of Geography, University of Chicago.

Rempel, H. 1980: Determinant of rural-to-urban migration in Kenya. *International Institute for Applied Systems Analysis Research Reports* 2, 281–307.

Rempel, H. 1981: *Rural–urban labour migration and urban unemployment in Kenya.* International Institute for Applied Systems Analysis, Laxenburg, Austria.

Renaud, B. 1981: *National urbanisation policy in developing countries.* London: Oxford University Press.

Richardson, H. W. 1977: City size and national spatial strategies in developing countries. WP 252. Washington, DC: The World Bank.

Richardson, H. W. 1979: *Regional economics.* Urbana: University of Illinois Press.

Richardson, H. W. 1980: Polarization reversal in developing countries. *Papers of the Regional Science Association* 45, 67–85 (*see* this volume, Chapter 11).

Riddell, J. B. 1981: Beyond the description of spatial pattern: the process of proletarianisation as a factor in population migration in West Africa. *Progress in Human Geography* 5, 370–92.

Roberts, B. 1978: *Cities of peasants: The political economy of urbanisation in the Third World.* London: Edward Arnold; Beverly Hills, Calif.: Sage.

Rogers, A. and Williamson, J. G. 1982: Migration, urbanisation, and Third World development: an overview. *Economic Development and Cultural Change* 30, 463–82.

Saint, W. S. and Goldsmith, W. W. 1980: Cropping systems, structural change, and rural–urban migration in Brazil. *World Development* 8, 259–72.
Salvatore, D. 1981: *Internal migration and economic development: A theoretical and empirical study*. Washington, DC: University Press of America.
Sanders, R. L. 1980: The spatial differentiation of informal activities across the urban hierarchy and implications for theory. Ph.D. dissertation, Department of Geography, Ohio State University, Columbus, Ohio.
Santos, M. 1979: *The shared space: The two circuits of the urban economy in underdeveloped countries*. New York: Methuen.
Sheppard, E. 1982: City size distributions and spatial economic change. *International Regional Science Review* 7, 127–51.
Slater, D. 1978: Towards a political economy of urbanisation in peripheral capitalist societies: problems of theory and method with illustrations from Latin America. *International Journal of Urban and Regional Research* 2, 26–52.
Stetzer, F. C. 1983: *ISLAND: A Fortran simulation model of development aspects of migration in Third World settings*. Studies on the interrelationships between migration and development in Third World settings, DP-10, Department of Geography, Ohio State University, Columbus, Ohio.
Swindell, K. 1979: Labour migration in underdeveloped countries: the case of sub-Saharan Africa. *Progress in Human Geography* 3, 239–59.
Taylor, J. E. 1980: Peripheral capitalism and rural-to-urban migration: a study of population movements in Costa Rica. *Latin American Perspectives* 26, 75–90.
Teitelbaum, M. S. 1975: Relevance of demographic transition theory for developing countries. *Science* 188, 420–5.
Thapa, P. and Conway, D. 1983: Internal migration in contemporary Nepal: a set of models which internalise development policies. *Papers of the Regional Science Association* 52, 27–42.
Todaro, M. P. 1976: *Internal migration in developing countries: A review of theory, evidence, methodology, and research priorities*. Geneva: International Labour Office.
Todaro, M. P. 1981: *Economic development in the Third World*, 2nd edition. New York and Harlow, Essex: Longman.
Todaro, M. P. and Stillkind, J. 1981: *City bias and rural neglect: The dilemma of urban development*. New York: Population Council.
Townroe, P. M. and Keen, D. 1984: Polarisation reversal in the state of São Paulo, Brazil. *Regional Studies* 18, 45–54 (*see* this volume, Chapter 13).
Tsui, A. O. and Bogue, D. J. 1978: Declining world fertility trends: causes, implications. *Population Bulletin*, 33-4. Washington DC: Population Reference Bureau.
United Nations. 1973: The determinants and consequences of population trends. *Population Studies 50*. Department of Economic and Social Affairs, United Nations, New York.
Urzua, R. 1981: Population redistribution mechanisms as related to various forms of development. In Demko, G. J. and Fuchs, R. J. (eds), *Population Studies 75. Population distribution policies in development planning*. Department of Economic and Social Affairs, United Nations, New York, 53–69.
Walle, E. Van de and Knodel, J. 1980: Europe's fertility transition: new evidence and lessons for today's developing world. *Population Bulletin*. Washington, DC: Population Reference Bureau, 34–6.
Wrigley, E. A. and Schofield, R. S. 1981: *The population history of England 1541–1871: A reconstruction*. Cambridge, Mass.: Harvard University Press.

Wrigley, E. A. and Schofield, R. S. 1983: English population history from family reconstitution summary results 1600–1799. *Population Studies* 37, 157–84.

Yotopoulos, P. A. and Nugent, J. B. 1976: *Economics of development: Empirical investigations.* New York: Harper & Row.

Zelinsky, W. 1971: The hypothesis of the mobility transition. *Geographical Review* 61, 219–49.

19 H. S. Geyer and T. Kontuly,
'A Theoretical Foundation for the Concept of Differential Urbanization'

From: *International Regional Science Review* 17 (2), 157–77 (1993)

Introduction

Most multi-country population studies on core–periphery migration patterns concentrate on either the developed or the less developed economic environment (Cheshire; Hay, 1986; Cochrane and Vining, 1988; Hall and Hay, 1980; Vining, 1986; Vining and Pallone, 1982). It is rare to find studies that explicitly attempt to transcend the gap between population migration patterns in developed countries and those in less developed countries. Richardson's analysis (1977) and synthesis (1980) of the concept of polarization reversal are exceptions. Although he focused on less developed countries, Richardson went a long way in bridging the gap between deconcentration tendencies in less developed and developed countries.

Drawing from Richardson's observations on the differences between polarization reversal and counterurbanization, as well as from other observations on counterurbanization in developed countries (Berry, 1976; Vining and Kontuly, 1978; Vining and Strauss, 1977), Geyer (1989, 1990) introduced the concept of differential urbanization and offered two propositions. First, he suggested that the urbanization–counterurbanization migration framework results in polarization reversal being introduced as an intermediate phase of urban development between urbanization and counterurbanization. Second, he suggested that disaggregated population migration analysis can either show early signs of deconcentration while concentration forces are still dominant in a country, or exhibit signs of continuing concentration after deconcentration sets in as the predominant migration pattern.

This paper presents a hypothetical model that provides a theoretical foundation for these two propositions. It postulates that groups of large, intermediate-sized, and small cities go through successive periods of fast and slow growth, in a continuum of development that spans the evolution of urban systems in less developed and developed countries. This sequence of fast and slow growth periods illustrates the process of differential urbanization. Also, various stages of differential urbanization are reflected in distinct aspects of dominant or recessive concentration and deconcentration tendencies. The model considers literature on migration in both developed and less developed countries because although migration forces may be generated differently in developed and less developed countries, the spatial effect of those forces on urban development is fundamentally the same.

In addition, questions regarding the beginning (Vining and Strauss, 1977; Gordon, 1979) and end (Cochrane and Vining, 1988; Champion, 1989a, 1989b) of counterurbanization need to be resolved. During the first half of the 1970s, a turnaround was detected in the migration patterns of the United States (Beale, 1975; Berry, 1976), and this discovery motivated researchers to search for the phenomenon in other developed countries, such as Australia, Denmark, France, Germany, Italy, Japan, and the United Kingdom (Champion, 1987; Courgeau, 1986; Court, 1989; Dematteis, 1986; Dematteis and Petsimeris, 1989; Fielding, 1986; Hamnett and Randolph, 1983; Hugo and Smailes, 1985; Hugo, 1988; Kontuly *et al.*, 1986; Kontuly and Vogelsang, 1988; Kontuly, 1991; Ogden, 1985; Robert and Randolph, 1983; Tsuya and Kuroda, 1989). Opinions differed about whether counterurbanization started during this turnaround or not. Some scholars viewed counterurbanization in the United States as an 'unprecedented break with past trends' (Vining and Strauss, 1977, p. 751). Another school of thought contended that as 'the wave of development spreads outward and spills over SMSA lines, a "reversal" is perceived though none may have occurred' (Gordon, 1979, p. 285). After the turnaround of the 1970s was generally accepted (Berry, 1978; Fielding, 1982; Hall, 1987; Morrison and Wheeler, 1976; Vining and Kontuly, 1978; Vining and Pallone, 1982), new evidence from the 1980s suggested a reversal of counterurbanization in certain countries (Cochrane and Vining, 1988; Richter, 1985; Rogerson and Plane, 1985).

Currently, there are several opinions on the status of the counterurbanization process. In his assessment of whether counterurbanization is a temporary phenomenon or a long-term trend, Champion (1989b, p. 241) concludes that the weight of evidence does not favor the idea that counterurbanization is a 'temporary blip in an ongoing process of urbanization.' Cochrane and Vining (1988), on the other hand, contend that the counterurbanization phenomenon, when measured in terms of core to periphery exchanges, is temporary in nature. The turnaround of the 1970s and its reversal during the 1980s motivated Berry (1988) and Mera (1988) to suggest the emergence of migration cycles in developed countries.

Theoretical underpinnings of differential urbanization

Spatial characterization of the model

Drawing on central place theory (Christaller, 1966; Lösch, 1954) and the roles of market forces (Isard, 1972), locational attributes (Richardson, 1973; Ullman,

1958), innovation diffusion (Berry, 1972; Hägerstand, 1965; Pred, 1977; Stephens and Holly, 1980), development axes in the evolution of urban systems (Geyer, 1987; Stewart, 1958; Berry, 1972; Sheppard, 1982), and agglomeration economies (Hirschman, 1958; Myrdal, 1957; Pred, 1966), five propositions can be stated with regard to the development of urban systems. First, many national urban systems initially go through a primate city phase, in which a large proportion of economic development and large numbers of migrants are attracted to one or a few primary centers (Richardson, 1973). Second, as the national urban system expands and matures, new urban centers are added to the lower ranks while many of those that already exist develop and move up through the ranks. In this process, economic development gets dispersed, while the urban system becomes more spatially integrated (Friedmann, 1966; Richardson, 1973). Third, such expanding national urban systems develop various strata of territorially organized subsystems, from the macro-level through the regional and subregional levels to the local or micro-levels (Friedmann, 1972; Bourne, 1975). Fourth, the sequence of tendencies observed in the development of urban systems, first toward concentration and then toward dispersion or deconcentration, is not limited to systems at the national level, but can also manifest itself at each of the lower levels of territorially organized subsystems because the same spatial forces operate at both the national and subnational levels. Fifth, in a growing urban environment, the odds normally favor the development of secondary centers closer to primary centers (van den Berg et al., 1982; Gordon, 1979; Richardson, 1977, 1980; Richter, 1985), unless an outlying center is located in an area with exceptional locational attributes. Using these five propositions as a premise, an impressionistic spatial characterization of differential urbanization was hypothesized, with net migration flows used to identify the different phases of the model (Figure 19.1).[1]

Initially, there is the phase of urbanization during which an increasing proportion of the economic activity and population of a country concentrates in a limited number of rapidly growing centers. In smaller countries there are normally only one or two primate centers, while in larger countries there may be several. During this early phase of urbanization, the urban system expands relatively quickly, with new centers added to the lowest ranks. In establishing new urban centers, favorable localities closer to larger centers stand a better chance of attracting urban development earlier than similar localities farther away. In the model, this phase of urbanization is called the primate city phase.

This initial phase can be subdivided into three stages. First, during the early primate city stage a primate city establishes some degree of overall spatial dominance within an urban system, attracting a relatively large percentage of the net interregional movement (Figure 19.1(a)). Second, in the intermediate primate city stage the primate city is still largely monocentric and growing rapidly with suburbanization as a prominent phenomenon (Figure 19.1(b)). Suburban nodes, the nuclei of the future multi-nodal primate city, may begin to emerge at this time. During this stage the primate city expands at an increasing rate. The rest of the urban system also starts benefiting from net rural-to-urban migration, although it responds rather slowly in the beginning. Because of locational attributes, certain intermediate-sized cities (in the class 2 cate-

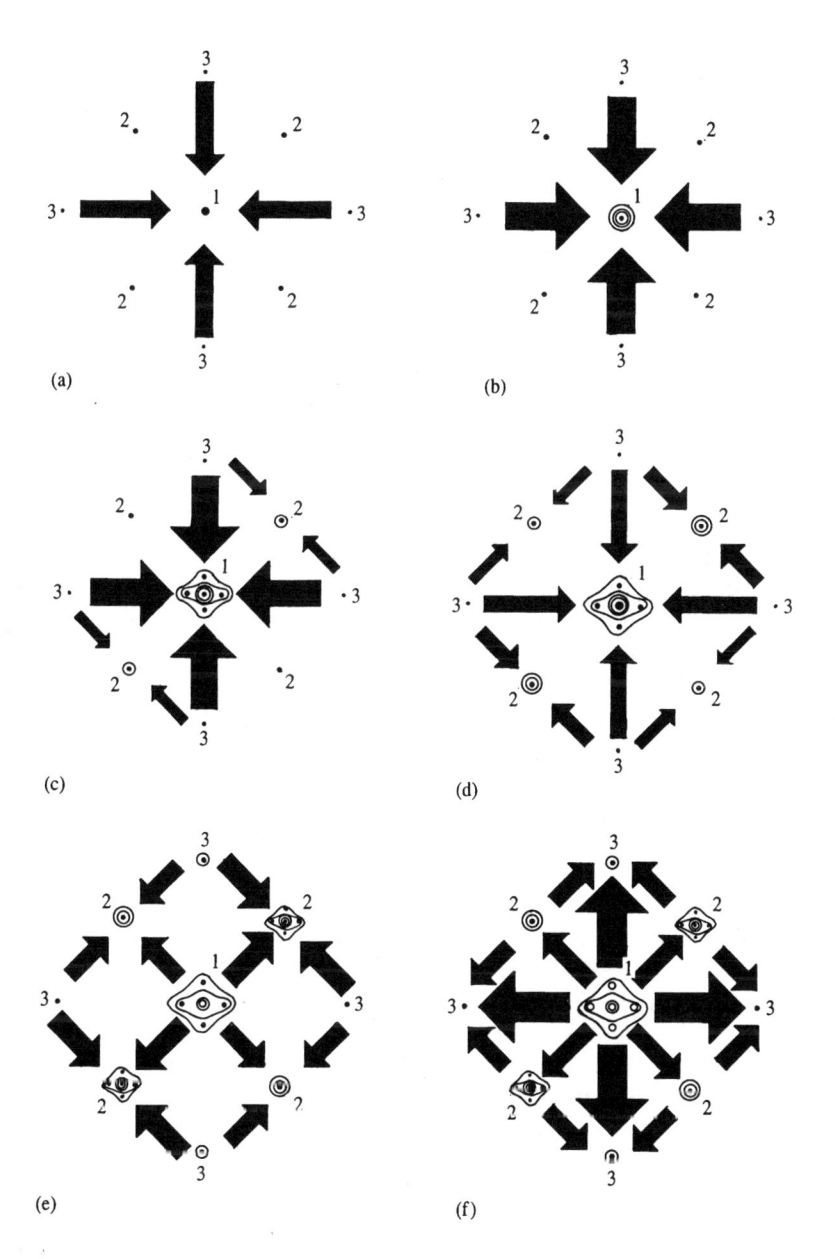

Fig. 19.1 A graphic model of the phases of differential urbanization: mainstream movements; (a) early primate city stage, (b) intermediate primate city stage, (c) advanced primate city stage, (d) early intermediate city stage, (e) advancd intermediate city stage, (f) small city stage

gory) develop relatively faster than others. Finally, the urban system enters the advanced primate city stage when the primate city becomes so large that, owing to agglomeration diseconomies, a monocentric urban structure can no longer prevail. By means of intraregional decentralization within what could now be called the primate region, the primate city develops a multi-centered metropolitan or megalopolitan character, and it totally dominates the rest of the urban system economically and spatially (Figure 19.1(c)). The urban system may expand rapidly at a national level during this phase, with certain existing urban centers entering higher ranks and new centers added to the lower ranks (Friedmann, 1972; Frey and Speare, 1988). The likelihood of one or more intermediate-sized cities entering the primate city category is not excluded. In such an event, the entrance of the second primary center divides the national urban system into two subsystems. When more than one primate city exists, it is highly unlikely that all would be in the same stage of development. While they may all be in the primate city phase, one may be in the early stage, another in the intermediate stage, and another in the advanced stage.

At some point in the development history of most countries, the primate cities start to mature, their growth rates begin to slow down, and the process of spatial deconcentration starts. The aging of the primate city is often accompanied by some growth in several centers close to the primate city, especially intermediate-sized cities. At the national level a turnaround of this nature is known as polarization reversal (Richardson, 1977, 1980) and has been observed in countries such as the United States, France, India, South Africa, South Korea, and Venezuela (Brown and Lawson, 1989; Crook and Dyson, 1982; Fielding, 1989; Frey and Speare, 1988; Geyer, 1990; Kim, 1986; Lee, 1989; Linn, 1978; Richardson, 1977, 1980; Townroe and Keen, 1984).

Although most of the less developed countries today are still confronted with excessive population growth, large-scale primate urbanization, and a continuing urban bias in the allocation of development aid (Brown, 1976; ILO, 1976; Lipton, 1976), primate city net in-migration will not continue at the same high rate indefinitely. Several cities in a number of advanced developing countries and in some less developed countries started to show declining growth rates in the 1970s (Vining, 1986). In many of the advanced developing countries this decline was associated with the growth of certain intermediate-sized cities, especially those in regions adjacent to the primate cities.[2]

Subsequently, various indicators of polarization reversal were specified. These included purely quantitative indicators, such as population (Crook and Dyson, 1982; Renaud, 1981; Townroe and Keen, 1984), and qualitative measures, such as gross geographical product (Linn, 1978), gender, age, educational levels, and occupational status (Brown and Lawson, 1989), income groups (Geyer, 1989, 1990), and a combination of regional growth factors (Lee, 1989).

Richardson's (1980, p. 68) definition of polarization reversal as 'concentrated dispersion' can be extended by assuming that this dispersion, or deconcentration, first occurs in medium-sized cities and then in small cities. In the model, the second distinct phase of urban development is called the intermediate city phase (Figures 19.1(d), 19.1(e)) and the last phase is the small city phase (Figure 19.1(f)). Although these two phases are defined as separate substages in the model, they can be thought of as a continuum, or movement toward deconcentration.

Sometimes the bulk of the initial secondary urban growth takes place in intermediate-sized cities fairly close to the primate cities, giving rise to questions about whether the growth should be regarded as the beginning of deconcentration or seen as continuing urban sprawl (Gordon, 1979; Koch, 1980; Lee, 1989; Vining, 1986; Vining and Kontuly, 1978; Vining and Strauss, 1977). The matter can be resolved by observing two norms. First, the intermediate-sized cities should be independent, that is, they should not be dormitory towns for the primate city, and a majority of their population should be employed within their respective employment catchment areas. Second, the urbanized area of the intermediate-sized city should not be contiguous to the urbanized area of the primate city, because it then becomes physically part of the latter.

After mid-size cities near primate cities have begun to grow, intermediate-sized cities in more distant regions may also start growing faster. Except in cases of exceptional locational attributes, larger intermediate centers seem to have a better chance of developing than smaller ones (Friedmann, 1972).

Two stages are identified within the intermediate city phase (Richardson, 1980). The early intermediate city stage (Figure 19.l(d)) is characterized by an uneven growth of a limited set of intermediate-sized cities which are close but not contiguous to the primate metropolitan region. The primate city is still gaining population in absolute terms, although it is starting to lose in relative terms to the intermediate-sized cities. The suburban centers within the primate metropolitan region are now growing faster than the central city. During the advanced intermediate city stage (Figure 19.l(e)), the intraregional decentralization or suburbanization which characterized the development of the primate city during the advanced primate city stage (Figure 19.1(c)) is repeated in the faster-growing intermediate-sized cities, but on a smaller scale. Also, in contrast to the early intermediate city stage, all centers within the primate metropolitan region begin losing population in absolute terms, with the central city losing more than the suburban centers.

Finally, the urban system enters the small city phase, also known as counterurbanization (Figure 19.1f). During this phase, deconcentration takes place from the primate and intermediate-sized cities toward small urban centers. Initially, those small centers with exceptional locational attributes closer to the former two groups of cities may develop first, but later, similar centers in the periphery may also begin to attract local migrants. Eventually this group of small centers may grow at a faster rate than both the primate cities and the intermediate-size cities.

At the end of the small city stage, the urban system reaches a saturation point. Because of technological and structural limitations, the rural or farming population cannot be reduced much further, and rural-to-urban migration ceases to be a major contributing factor to urbanization. Urban population sizes are now largely supplemented by immigration and the natural increase of the urban population itself. As natural population increase might be very low or even negative at this stage, and immigration is normally kept at a relatively low level, urban growth in general may be slow.

Figure 19.1 characterizes differential urbanization by depicting only the mainstream or dominant net migration flows because the urbanization, polarization reversal, and counterurbanization literature deal only with these large or dominant migration streams. In a later section the model will be expanded by including substream movements as well, which greatly increases the predictive capabilities of the model.

Temporal characterization of the model

Differential urbanization can also be characterized in terms of a temporal sequence of primate city, intermediate-sized city, and small city growth. Figure 19.2 shows this hypothesized temporal sequence as the relationship between net migration rates and time, with time measured on an inverse log scale.[3] It is postulated that both the variation in the net migration rates of each

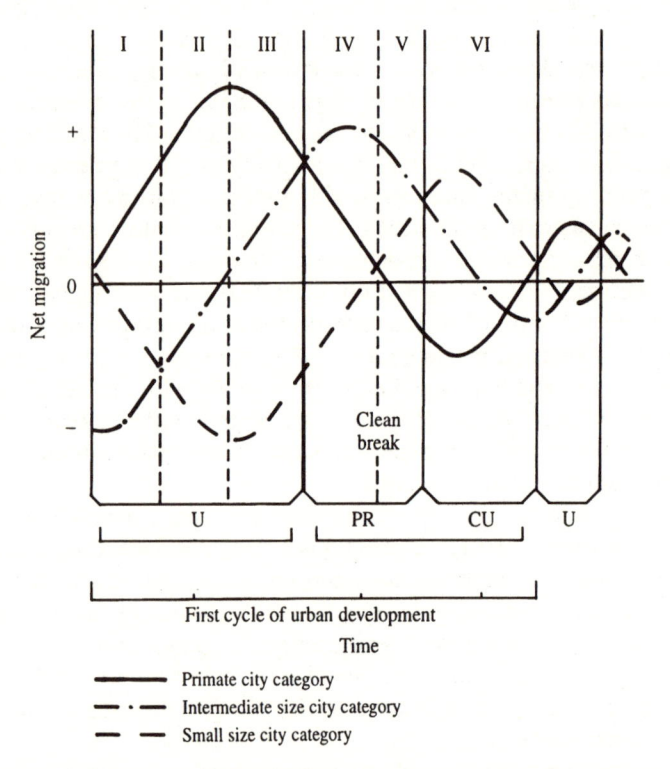

Fig. 19.2 Generalized stages of differential urbanization. I, Early primate city stage (EPC); II, intermediate primate city stage (IPC); III, advanced primate city stage (APC); IV, early intermediate city stage (EIC); V, advanced intermediate city stage (AIC); VI, small city stage (SSC). U = urbanization; PR = polarization reversal; CU = counterurbanization

successive development phase and the duration of each successive growth phase diminish over time. Variation in net migration rates should diminish because the country's population as a whole will grow more slowly as it approaches higher levels of overall development and a high level of urbanization. Continuous improvement and integration of the transportation and communication networks enhance the functional and spatial integration of the urban system and consequently reduce the time required for each successive stage of urban development. As indicated in Figure 19.2, the conclusion of the small city stage ends the first cycle of urban development and signifies the beginning of a new one in which a second sequence of major metropolitan, intermediate-sized city, and small city growth occurs. By the end of the small city stage in the first cycle, the urban system is well developed and spatially integrated to such an extent that the extreme core–periphery differences of the past are greatly reduced or eradicated. Berry's (1990) expectations for the American urban space around the year 2000 suggest that vast differences in the availability of technology and information between the small and large city will no longer exist. Also, natural population growth could be even lower during the second cycle than previously, in which case changes will mostly be caused by variations in internal and foreign migration. Future improvements in transportation and communication technology will be expected to reduce the amplitude and duration of successive growth phases even further.

The set of major metropolitan areas experiencing the highest rates of net in-migration during the urbanization phase of the second cycle may or may not be the same set of large urban areas from the first cycle. Large urban centers able to retain their dominant position in the national and international urban hierarchy, along with a limited group of rapidly growing intermediate-sized urban areas from the first cycle, will constitute this new set of major metropolitan centers.

Testing the temporal characterization of the model

The hypothesized temporal sequence of primate city, intermediate-sized city, and small city growth was tentatively verified using migration and population data for France, South Korea, and India (Figures 19.3, 19.4, and 19.5, respectively). The city size classes for each country were defined to reflect the variability in distinctions between large, medium, and small cities for different spatial and temporal settings. Size classes for France were set such that cities with populations less than 10 000 were regarded as small, those with populations greater than 100 000 were regarded as large, and those in between were regarded as intermediate-sized (Figure 19.3). Lee's (1989) categories of smaller cities, regional metropolitan centers, and metropolitan centers were used as a basis for size differentiation in South Korea (Figure 19.4), and for India, cities greater than 400 000 were regarded as large, less than 50 000 as small, and those in between as intermediate (Crook and Dyson, 1982; Figure 19.5).[4]

During four 7-year periods between 1954 and 1982, France moved from the advanced primate city stage through the early intermediate city and advanced intermediate city stages to the small city stage (Figure 19.3). South Korea

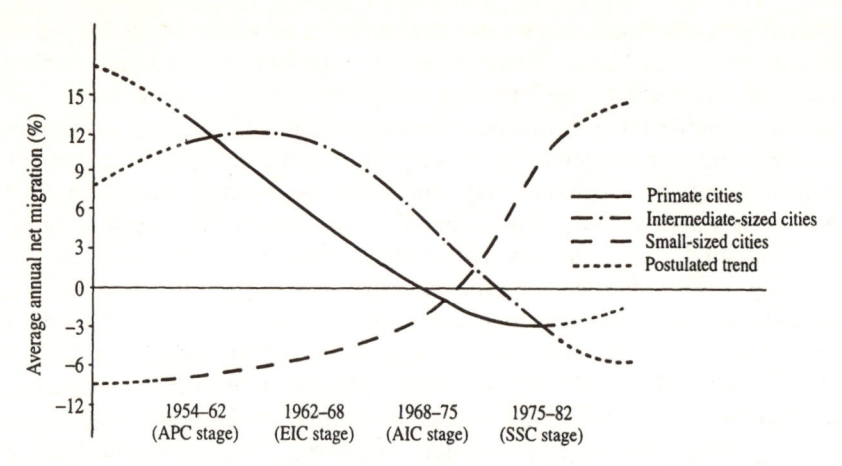

Fig. 19.3 Average net migration to different size groups of cities: France 1954–82. (Source: Fielding, 1989)

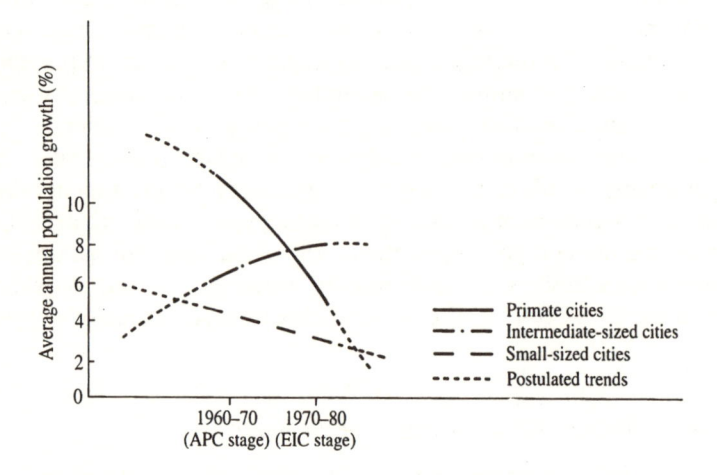

Fig. 19.4 Average population change in different size groups of cities: Republic of Korea 1960–80. (Source: Lee, 1989)

moved from an advanced primate city stage in the 1960s to the early intermediate city stage in the 1970s (Figure 19.4), while India moved through the same stages over approximately the same period (Figure 19.5).

Although Figure 19.3 reflects average annual net migration and Figures 19.4 and 19.5 illustrate average annual population growth rates, the implications of the trends remain the same. The average annual natural increase of the urban population in the different size categories of urban areas over a 10-year period is expected to differ only marginally for each individual country. Therefore, subtracting these model average natural growth rates from the aggregate population growth rates in Figures 19.4 and 19.5 would only change the positions of the curves relative to the zero axes. It would not influence the positions of the curves relative to each other in any significant way.

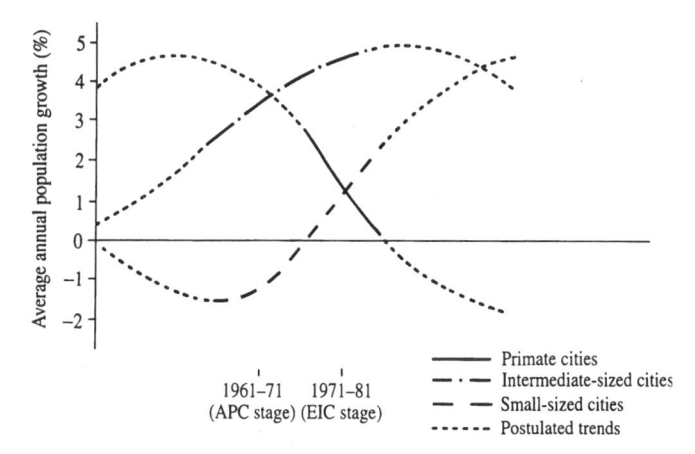

Fig. 19.5 Average population change in different size groups of cities: India 1961–81. (Source: Crook and Dyson, 1982)

Countries in different stages of urban systems development will lie at different positions on the time axis in Figure 19.2. Different migration rates and volumes affect the amplitudes of countries' growth curves differently. Even under reasonably similar circumstances, countries could experience different rates of net in-migration to their large or intermediate-sized urban areas. These facts are illustrated by the slopes of the curves and their positions relative to the zero axes in Figures 19.4 and 19.5. Although the urban system in a particular country may generally follow the growth paths of the different size categories of cities shown in Figure 19.2, the rate at which the size classes move through the various stages may differ greatly from country to country. Also, the rate at which a particular country moves through the various stages is likely to vary for different intervals of time, and the positions of the growth curves of the three groups of cities relative to each other and to the zero axis may differ country by country. The fundamental spatial implications of the differential urbanization still hold, however, despite these individual differences.

Urban concentration and deconcentration

Many factors related to sectoral and spatial economic equilibrium or disequilibrium influence population migration. There is an extensive body of economic development literature that intersects with migration theory. This literature is based on concepts such as economic innovation and information dissemination (Friedmann, 1972; Hägerstrand, 1952; Lazuén, 1972; Perroux, 1950, 1955), the development center or growth pole (Boudeville, 1967; Perroux, 1955; Todd, 1974), urban economic polarization versus dispersion (Friedmann, 1959, 1966; Hirschman, 1958; Myrdal, 1957; Richardson, 1973), the interregional versus intraregional orientation of industries (Alexander, 1954; Leven, 1966; North, 1955; Tiebout, 1956), and economic conversion versus diversion (Adelman and Morris, 1973; Kuznets, 1955; Williamson, 1965). In the chain of cause and effect all of these basic concepts have a direct or indirect impact on migration patterns. They influence the availability of job opportunities, people's income expectations, their educational levels, and their age and sex distributions. These factors, in turn, affect people's living environment, their quality of life, and ultimately their migration patterns. Evidence of these relationships is plentiful (Brown and Lawson, 1989; Frey and Speare, 1988; Gatzweiler, 1975; Ibrahim, 1982; Koch, 1980; Kontuly and Vogelsang, 1988) and is reflected by mainstream demographic and migration theory (Brown and Sanders, 1981; Brown and Stetzer, 1984; Gibbs, 1963; Hugo, 1981; Todaro, 1982; Zelinsky, 1971).

The six stages of differential urbanization referred to in Figure 19.1 can be expanded considerably by further generalizing the relationship between migration theory and the concepts of productionism and environmentalism (Berry, 1978; Hart, 1983; Vining and Pallone, 1982). The expanded model, shown in Figure 19.6 with both mainstream and substream migration flows, postulates that more affluent, better-educated people generally tend to deconcentrate in search of better living conditions (environmentalism), while the less affluent tend to concentrate in search of better prospects of a livelihood (productionism). If migration patterns reflect productionism and environmentalism, a much greater degree of continuity in population migration patterns exists than is generally recognized (Geyer, 1989, 1990).

If a small proportion of a country's population is wealthy and a large proportion is less well-to-do, then a large majority would tend to concentrate, while a minority would tend to deconcentrate.[5] (See the mainstream versus substream movements in Figures 19.6(b), 19.6(c), and 19.6(d).) As the nation becomes more developed and the proportion of the affluent people increases, the mainstream and substream migration tendencies would be expected to reverse (Figures 19.6(e) and 22.6(f)). Thus, disaggregating migration by income, age, or educational levels has the potential to show early signs of deconcentration while concentration forces are still dominant, and could also indicate signs of the continuation of concentration as a substream after deconcentration sets in as the dominant pattern.

One would therefore expect mainstream (net) migration patterns in less developed countries to be consistent with concentration tendencies while still exhibiting undercurrents in the opposite direction. In the most developed countries these tendencies would be reversed. Mainstream migration trends would be toward deconcentration with an undercurrent movement toward concentration.[6] Examples of such mainstream and substream movements are numerous. In West Germany the older, more affluent people tended to deconcentrate during the 1970s and 1980s, while the younger, less well-to-do tended to migrate toward larger centers (Kontuly and Vogelsang, 1989). Similar migration patterns also occurred in Britain (Fielding, 1991a, 1991b) and France (Koch, 1980). In the United States and South Africa there is a tendency for the less affluent, African American and African components of the population to increase in certain larger urban centers, while the white component, as a share of the total population, tends to increase in

Migration Cycles 301

intermediate-sized cities close to the large centers (Berry, 1976; Frey, 1991; Frey and Speare, 1988; Geyer, 1990). The only difference between the tendencies in these two countries is that the Africans form a majority in South Africa, while in the United States the African Americans represent a minority and, therefore, a substream movement.

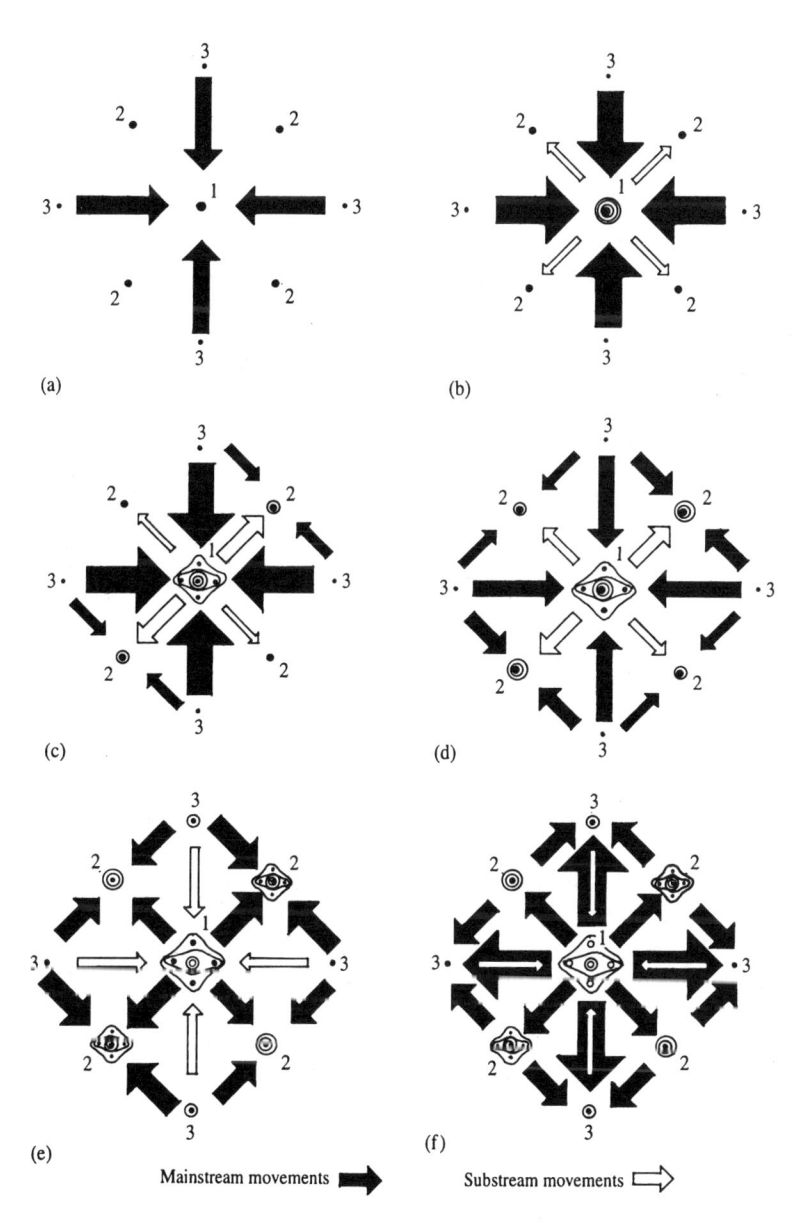

Fig. 19.6 A graphic model of the phases of differential urbanization: mainstream and substream movements; (a) early primate city stage, (b) intermediate primate city stage, (c) advanced primate city stage, (d) early intermediate city stage, (e) advanced intermediate city stage, (f) small city stage

Conclusions

A main task of this article was to integrate the literature on the relationship between interregional migration and urban population growth across the entire development spectrum. This linkage was accomplished by using the notion of differential urbanization. In addition, precise definitions are suggested for a 'clean break' and for the temporal boundaries between urbanization and polarization reversal and between polarization reversal and counterurbanization. A clean break occurs when the net migration rate for small urban areas exceeds that of the primate cities (Figure 19.2). At this point, the small urban areas exhibit a high rate of net migration, while the intermediate-sized urban areas show positive net migration. This point represents a clean break because the large metropolitan areas are no longer the engines of growth and development, and the impetus for growth shifts to the intermediate-sized and small urban areas. The point at which the net migration rate for intermediate-sized urban areas exceeds the rate for intermediate-sized urban areas represents the end of the urbanization phase and the beginning of polarization reversal, while the point at which the net migration rate for small urban areas exceeds the rate for intermediate-sized urban areas represents the end of polarization reversal and the beginning of counterurbanization (Figure 19.2). Counterurbanization ends when the net migration rate for the group of large urban areas begins exceeding the rate for small urban areas, and the net migration rate for intermediate-sized areas reaches a minimum.

The counterurbanization stage, which does not last indefinitely, represents the final phase in the first cycle of urban development. According to the differential urbanization model, counterurbanization will be followed by a concentration or urbanization stage in which net migration once again benefits large metropolitan areas at the expense of mid-sized and small urban areas (Figure 19.2). During the second cycle of urban development, national level changes will be more subtle, and regional-level change will be more significant. At this advanced level of urban development, the differential urbanization model can also be used to characterize the degree of urban development within regions or subregions in a particular developed country. The characterization of the different phases of differential urbanization parallels Berry's (1988) and Mera's (1988) contention that migration cycles are emerging in developed countries. This notion is extended by showing that in the evolution through the primate city, intermediate-sized city, and small city phases, each phase requires less time to complete than the previous phase.

The accuracy of the temporal characterization of the differential urbanization model was tentatively tested using three countries that span the development spectrum. Extensive empirical testing of the spatial characterization and historical accuracy of the model for different countries is still needed. Questions of how to measure small, medium-sized, and large cities over time for one country, and across countries for a single point in time, should also be addressed. The usefulness of the mainstream versus substream dichotomy in the spatial characterization of the model can be verified using migration data disaggregated by age, income, occupation, or educational levels.

Notes

1 The migration patterns depicted in Figure 19.1 do not negate rural-to-rural migration (Brown and Lawson, 1985) or the international role of urban centers, but because the role of internal net rural-to-urban, urban-to-urban and urban-to-rural migration in the development of different city size categories is emphasized, the former issues fall outside the scope of this paper.
2 Various criteria have been used in the past to define intermediate-sized cities. These include the sizes and qualities of their populations, their economic development potential, and their regional functions (Hansen, 1971; Rondinelli, 1983). In this study, the term 'intermediate-sized cities' will imply centers that are lower in rank than the primate cities in an urban system, but which act as prominent regional centers at a lower level of regional disaggregation (Bos and Geyer, 1993).
3 The following brief review of the pre-industrial and industrial urbanization periods provides background information necessary for the differential urbanization model. Six distinct periods of growth and decline are distinguishable in the history of urbanization: (1) the period from 4000 BC to 600 BC when urbanization progressed slowly, (2) the period from 600 BC to AD 400 when the true city in terms of modern standards emerged for the first time, (3) the period of urban decadence, which lasted until the end of the first millennium, (4) the period from 1000 to 1800 when rapid and widespread urbanization was witnessed, and (5) the period of the world cities which began with the industrial revolution (Davis, 1955; Dickinson, 1951; Mumford, 1961). During the latter period of the 'urban region' (Wells, 1902), 'megalopolis' (Gottmann, 1957), 'ecumenopolis' (Doxiadis, 1970), and 'continent city' (Stewart, 1947; Thompson, 1965), urbanization was described as a process in which '[t]he macro-location of industry and population tends towards an ever-increasing concentration in a limited number of areas [and] their micro-location ... towards an increasing diffusion, or "sprawl"' (Clark, 1967, p. 280). During the first half of the 1970s, however, a 'turnaround' was detected in the migration patterns of certain developed countries, accounting for the sixth period in the history of urbanization.
4 In the French example, the size range of intermediate-sized cities was chosen by the authors of this article. In the other two cases, the intermediate city size categories were obtained from the relevant sources. Not all the growth rates of the different categories of cities were quoted in detail in the Indian example. Some of the growth rates given in Figure 19.5 therefore reflect approximations rather than exact rates, but the general tendencies of growth and decline of the different city size categories in India still seem to bear out the hypothesis.
5 Only interregional migration patterns are considered in this model. Intraregional decentralization, or suburbanization, is not.
6 There are bound to be minor departures from these tendencies, but on the whole, the general patterns described seem to hold. Both the general pattern and minor variations are shown through extensive longitudinal studies in Britain and the United States (Fielding, 1991a, 1991b; Fuguitt et al., 1989; Frey and Speare, 1988).

References

Adelman, I. and Morris, C. T. 1973: *Economic growth and social equity in developing countries.* Palo Alto, Calif.: Stanford University Press.

Alexander, J. W. 1954: The basic–nonbasic concept of urban economic functions. *Economic Geography* 30, 246–61.

Beale, C. L. 1975: *The revival of population growth in nonmetropolitan America*. Economic Research Service Publication 605. Washington, DC: US Department of Agriculture.

Berry, B. J. L. 1972: Hierarchical diffusion: the basis of developmental filtering and spread in a system of growth centers. In Hansen, N. M. (ed.), *Growth centers in regional economic development*. New York: Free Press.

Berry, B. J. L. 1976: The counterurbanization process: urban America since 1970. In Berry, B. J. L. (ed.), *Urban Affairs Annual Review*, vol. 11. Beverly Hills, Calif.: Sage Publications (*see* this volume, Chapter 1).

Berry, B. J. L. 1978: The counterurbanization process: how general? In Hansen, N. M. (ed.), *Human settlement systems: International perspectives on structure, change, and public policy*. Cambridge, Mass.: Ballinger.

Berry, B. J. L. 1988: Migration reversals in perspective: the long-wave evidence. *International Regional Science Review* 11, 245–51 (*see* this volume, Chapter 17).

Berry, B. J. L. 1990: Urban systems by the third millennium: a second look. *Journal of Geography* 89, 98–100.

Bos, D. J. and Geyer, H. S. 1993: International perspectives on the definition of intermediate-sized cities: South African applications. *South African Geographer* 20, 46–61

Boudeville, J. 1967: *Problems of regional economic planning*. Edinburgh: Edinburgh University Press.

Bourne, L. 1975: *Urban systems: Strategies for regulation – A comparison of policies in Britain, Sweden, Australia, and Canada*. London: Oxford University Press.

Brown, L. A. and Lawson, V. A. 1985: Rural destined migration in Third World settings: a neglected phenomenon? *Regional Studies* 19, 415–32.

Brown L. A. and Lawson, V. A. 1989: Polarisation reversal, migration related shifts in human resource profiles and spatial growth policies: a Venezuelan study. *International Regional Science Review* 12, 165–88 (*see* this volume, Chapter 15).

Brown, L. A. and Sanders, R. L. 1981: Towards a development paradigm of migration, with particular reference to Third World settings. In De Jong, G. F. and Gardner, R. W. (eds), *Migration decision making*. New York: Pergamon Press.

Brown, L. A. and Stetzer, F. C. 1984: Development aspects of migration in Third World settings: a simulation, with implications for urbanisation. *Environment and Planning A* 16, 1583–603 (*see* this volume, Chapter 18).

Brown, L. R. 1976: The urban prospect: reexamining the basic assumptions. *Population and Development Review* 2, 267–77.

Champion, A. G. 1987: Recent changes in the pace of population deconcentration in Britain. *Geoforum* 18, 379–407.

Champion, A. G. 1989a: Counterurbanization: the conceptual and methodological challenge. In Champion, A. G. (ed.), *Counterurbanization: The changing pace and nature of population deconcentration*. London: Edward Arnold.

Champion, A. G. 1989b: Conclusion: temporary anomaly, long-term trend, or transitional phase? In Champion, A. G. (ed.), *Counterurbanization: The changing pace and nature of population deconcentration*. London: Edward Arnold.

Cheshire, P. and Hay, D. 1986: The development of the European urban system. In Ewers, H. J., Goddard, J. B. and Matzerath, H. (eds), *The future of the metropolis: Berlin, London, Paris, New York, economic aspects*. Berlin: Walter de Gruyter.

Christaller, W. 1966: Central places in southern Germany, trans. Baskin, C. W. Englewood Cliffs, NJ: Prentice-Hall, M9–29.

Clark, C. 1967: *Population growth and land use*. New York: St Martin's Press.

Cochrane, S. G. and Vining, D. R. Jr. 1988: Recent trends in migration between core and peripheral regions in developed and advanced developing countries. *International Regional Science Review* 11, 215–43 (*see* this volume, Chapter 7).
Courgeau, D. 1986: Vers un ralentissement de la 'déconcentration urbaine' en France? *Population et Société* 41, 1–4.
Court, Y. 1989: Denmark: towards a more deconcentrated settlement pattern. In Champion, A. G. (ed.), *Counterurbanization: The changing pace and nature of population deconcentration*. London: Edward Arnold.
Crook, N. and Dyson, T. 1982: Urbanization in India: results of the 1981 Census. *Population and Development Review* 8, 145–55.
Davis, K. 1955: The origin and growth of urbanization in the world. *American Journal of Sociology* 60, 429–37.
Dematteis, G. 1986: Urbanization and counterurbanization in Italy. *Ekistics* 53 (316/317), 26–33.
Dematteis, G. and Petsimeris, P. 1989: Italy: counterurbanization as a transitional phase in settlement reorganization. In Champion, A. G. (ed.), *Counterurbanization: The changing pace and nature of population deconcentration*. London: Edward Arnold.
Dickinson, R. E. 1951: *The West European city: A geographical interpretation.* London: Routledge & Kegan Paul.
Doxiadis, C. A. 1970: Cities of the future. In Bronwell, A. B. (ed.), *Science and technology in the world of the future*. New York: John Wiley & Sons.
Fielding, A. J. 1982: Counterurbanization in Western Europe. In Diamond, D. R. and McLaughlin, J. B. (eds), *Progress in planning*, vol. 17. Oxford: Pergamon Press.
Fielding, A. J. 1986: Counterurbanization in Western Europe. In Findlay, A. and White, P. (eds), *West European population change*. London: Croom Helm.
Fielding, A. J. 1989: Migration and urbanisation in Western Europe since 1950. *Geographical Journal* 155, 60–9 (*see* this volume, Chapter 9).
Fielding, A. J. 1991a: *Migration to and from south east England: Analyses of new data from the National Health Service Central Register and the longitudinal study.* London: Department of the Environment, Draft Working Paper 1.
Fielding, A. J. 1991b: Social and geographical mobility in the non-metropolitan south of England. Brighton: University of Sussex, unpublished manuscript.
Frey, W. H. 1991: Are two Americas emerging? *Population Today* 19 (10), 6–8.
Frey, W. H. and Speare, A. 1988: *Regional and metropolitan growth and decline in the United States.* New York: Russell Sage Foundation.
Friedmann, J. 1959: Regional planning: a problem in spatial integration. *Papers and Proceedings of the Regional Science Association* 5, 167–87.
Friedmann, J. 1966: *Regional development policy: A case study of Venezuela.* Cambridge, Mass.: MIT Press.
Friedmann, J. 1972: A general theory of polarized development. In Hansen, N. M. (ed.), *Growth centers in regional economic development*. New York: Free Press.
Fuguitt, G. V., Brown, D. L. and Beale, C. L. 1989: *Rural and small town America.* New York: Russell Sage Foundation.
Gatzweiler, H. P. 1975: Zur Selektivität interregionaler Wanderungen. *Forschungen zur Raumentwicklung*, Vol. 1. Bonn–Bad Godesberg. Bundesforschungsanstalt für Landeskunde und Raumordnung.
Geyer, H. S. 1987: The development axis as a development instrument in the Southern African Development Area. *Development Southern Africa* 4, 271–301.
Geyer, H. S. 1989: Differential urbanization in South Africa and its consequences for spatial development policy. *African Urban Quarterly* 4, 276–91.

Geyer, H. S. 1990: Implications of differential urbanisation on deconcentration in the Pretoria–Witwatersrand–Vaal triangle metropolitan area, South Africa. *Geoforum* 21, 385–96 (*see* this volume, Chapter 16).

Gibbs, J. 1963: The evolution of population concentration. *Economic Geography* 39, 119–29.

Gordon, P. 1979: Deconcentration without a 'clean break'. *Environment and Planning A* 11, 281–90 (*see* this volume, Chapter 4).

Gottmann, J. 1957: Megalopolis, or the urbanization of the northeastern seaboard. *Economic Geography* 33, 189–200.

Hägerstrand, T. 1952: The propagation of innovation waves: land studies in geography. *Human Geography* 4, 3–19.

Hägerstrand, T. 1965: Aspects of the spatial structure of social communication and the diffusion of information. *Papers of the Regional Science Association* 16, 27–42.

Hall, P. 1987: Metropolitan settlement strategies. In Rodwin, L. (ed.), *Shelter, settlement, and development*. Boston: Allen & Unwin.

Hall, P. and Hay, D. 1980: *Growth centres in the European urban system*. London: Heinemann Educational.

Hamnett, C. and Randolph, W. 1983: The changing population of England and Wales, 1961–81: clean break or consistent progression? *Built Environment* 8, 272–80.

Hansen, N. M. 1971: *Intermediate-sized cities as growth centers: Applications for Kentucky, the Piedmont Crescent, the Ozarks, and Texas*. New York: Praeger Publishers.

Hart, T. 1983: Transport and economic development: the historical dimension. In Button, K. J. and Gillingwater, D. (eds), *Transport location and spatial policy*. Aldershot: Gower.

Hirschman, A. O. 1958: *The strategy of economic development*. New Haven, Conn.: Yale University Press.

Hugo, G. J. 1981: Village-community ties, village norms, and ethnic and social networks: a review of evidence from the Third World. In De Jong, G. F. and Gardner, R. W. (eds), *Migration decision making*. New York: Pergamon Press.

Hugo, G. J. 1988: Counterurbanization in Australia. *Geographical Perspectives* 61, 43–68.

Hugo, G. J. and Smailes, P. J. 1985: Urban–rural migration in Australia: a process view of the turnaround. *Journal of Rural Studies* 1, 11–30.

Ibrahim, S. E. 1982: *A critical review: Internal migration in Egypt*. Cairo: Planning and Family Planning Board, The Supreme Council for Population and Family Planning, Research Monographs, Series No. 5.

ILO. 1976: *Employment, growth, and basic needs: A one world problem*. Geneva: International Labour Office.

Isard, W. 1972: *Location and space-economy*. Cambridge, Mass.: MIT Press.

Kim, I. 1986: The contemporary development of Korean settlement systems. In Bourne, L. S., Cori, B. and Dziewonski, K. (eds), *Progress in settlement systems geography*. Milan: Franco Angeli.

Koch, R. 1980: 'Counterurbanisation' auch in Westeuropa? *Informationen zur Raumentwicklung* 2, 59–69 (*see* this volume, Chapter 8).

Kontuly, T. 1991: The deconcentration theoretical perspective as an explanation for recent changes in the West German migration system. *Geoforum* 22, 299–317.

Kontuly, T. and Vogelsang, R. 1988: Explanations for the intensification of counterurbanization in the Federal Republic of Germany. *Professional Geographer* 40, 42–54.

Kontuly, T. and Vogelsang, R. 1989. Federal Republic of Germany: the intensification of the migration turnaround. In Champion, A. G. (ed.), *Counterurbanization: The changing pace and nature of population deconcentration*. London: Edward Arnold.

Kontuly, T., Wiard, S. and Vogelsang, R. 1986: Counterurbanization in the Federal Republic of Germany. *Professional Geographer* 38, 170–81.
Kutznets, S. 1955: Economic growth and income inequality. *American Economic Review* 45, 1–28.
Lazuén, J. R. 1972: On growth poles. In Hansen, N. M. (ed.), *Growth centers in regional economic development*. New York: Free Press.
Lee, H. 1989: Growth determinants in the core–periphery of Korea. *International Regional Science Review* 12, 165–88 (*see* this volume, Chapter 14).
Leven, C. L. 1966: The economic base and regional growth. In *Research and education for regional and area development*. Ames, Iowa: Iowa State University Press, Center for Agricultural and Economic Development.
Linn, J. F. 1978: Urbanization trends, polarization reversal, and spatial policy in Colombia. Paper no. 12. Münster: Institut für Siedlungs- und Wolnungswesen der Westfalischen Wilhelms-Universität Münster.
Lipton, M. 1976: *Why poor people stay poor: Urban bias and world development.* London: Temple-Smith.
Lösch, A. 1954: *The economics of location*. New Haven, Conn.: Yale University Press.
Mera, K. 1988: The emergence of migration cycles. *International Regional Science Review* 11, 269–75.
Morrison, P. A. and Wheeler, J. A. 1976: *Rural renaissance in America? The revival of population growth in remote areas.* Washington, DC: Population Reference Bureau.
Mumford, L. 1961: *The city in history: Its origins, its transformations, and its prospects.* New York: Harcourt, Brace & World.
Myrdal, G. 1957: *Economic theory and under-developed regions.* London: Duckworth.
North, D. C. 1955: Location theory and regional economic growth. *Journal of Political Economy* 63, 243–58.
Ogden, P. E. 1985: Counterurbanization in France: the results of the 1982 population census. *Geography* 70, 24–35.
Perroux, F. 1950: The domination effect and modern economic theory. *Social Research* 17, 188–206.
Perroux, F. 1955: Note sur la notion de 'pole de croissance.' *Économie Appliqué* 7, 307–20.
Pred, A. R. 1966: The American mercantile city: 1800–1840. In Pred, A. R. (ed.), *The spatial dynamics of US urban-industrial growth.* Cambridge, Mass.: MIT Press.
Pred, A. 1977: *City-systems in advanced economies: Past growth, present processes, and future development options.* New York: John Wiley.
Renaud, B. 1981: *National urbanization policy in developing countries.* New York: Oxford University Press.
Richardson, H. W. 1973: *Regional growth theory.* New York: John Wiley.
Richardson, H. W. 1977. City size and national spatial strategies in developing countries. Staff Working Paper no. 252.Washington, DC: World Bank.
Richardson, H. W. 1980. Polarization reversal in developing countries. *Papers of the Regional Science Association* 45, 67–85 (*see* this volume, Chapter 11).
Richter, K. 1985: Nonmetropolitan growth in the late 1970s: the end of the turnaround? *Demography* 22, 245–63 (*see* this volume, Chapter 5).
Robert, S. and Randolph, W. 1983: Beyond decentralization: the evolution of population distribution in England and Wales, 1961–81. *Geoforum* 14, 75–102.
Rogerson, P. and Plane, D. 1985: Monitoring migration trends. *American Demographics* 7 (2), 27–29, 47.

Rondinelli, D. A. 1983: *Secondary cities in developing countries*. London: Sage.

Sheppard, E. 1982: City size distributions and spatial economic change. *International Regional Science Review* 7, 127–51.

Stephens, J. D. and Holly, B. P. 1980: The changing patterns of industrial corporate control in the metropolitan United States. In Brunn, S. D. and Wheeler, J. O. (eds), *The American metropolitan system: Present and future*. London: Edward Arnold.

Stewart, C. T. Jr. 1958: The size and spacing of cities. *Geographical Review* 48, 222–45.

Stewart, J. Q. 1947: Empirical mathematical rules concerning the distribution and equilibrium of population. *Geographical Review* 37, 461–85.

Thompson, W. R. 1965: *A preface to urban economics*. Baltimore: Johns Hopkins University Press.

Tiebout, C. M. 1956: The urban economic base reconsidered. *Land Economics* 32, 95–9.

Todaro, M. P. 1982: *Economics for a developing world*. Harlow: Longman.

Todd, D. 1974: An appraisal of the development pole concept in regional analysis. *Environment and Planning A* 6, 291–306.

Townroe, P. M. and Keen, D. 1984: Polarisation reversal in the state of São Paulo, Brazil. *Regional Studies* 18, 45–54 (*see* this volume, Chapter 13).

Tsuya N. O. and Kuroda, T. 1989: Japan: The slowing of urbanization and metropolitan concentration. In Champion, A. G. (ed.), *Counterurbanization: The changing pace and nature of population deconcentration*. London: Edward Arnold.

Ullman, E. L. 1958: Regional development and the geography of concentration. *Papers and Proceedings of the Regional Science Association* 4, 129–98.

van den Berg, L., Drewett, R., Klaassen, L. H., Rossi, A. and Vijverberg, C. H. T. 1982: *Urban Europe: A study of growth and decline*. Oxford: Pergamon Press.

Vining, D. R. Jr. 1986: Population redistribution towards core areas of less developed countries, 1950–1980. *International Regional Science Review* 10, 1–45 (*see* this volume, Chapter 12).

Vining, D. R. Jr and Kontuly, T. 1978: Population dispersal from major metropolitan regions: an international comparison. *International Regional Science Review* 3, 49–73 (*see* this volume, Chapter 6).

Vining, D. R. Jr and Pallone, R. 1982: Migration between core and peripheral regions: a description and tentative explanation of the patterns in 22 countries. *Geoforum* 13, 339–410.

Vining, D. R. Jr and Strauss, A. 1977: A demonstration that the current deconcentration of population in the United States is a clean break with the past. *Environment and Planning A* 9, 751–8 (*see* this volume, Chapter 3).

Wells, H. G. 1902: *Anticipations of the reaction of mechanical and scientific progress upon human life and thought*. New York: Harper.

Williamson, J. G. 1965: Regional inequality and the process of national development: a description of the patterns. *Economic Development and Cultural Change* 13, 3–84.

Zelinsky, W. 1971: The hypothesis of the mobility transition. *Geographical Review* 16, 219–49.

20 H. S. Geyer,
'Expanding the Theoretical Foundation of the Concept of Differential Urbanisation'

From: *Tijdschrift voor Economische en Sociale Geografie* 87(1), 44–59 (1996)

Introduction

The wave of research that was sparked off by the migration turnaround of the 1970s in the developed world and then by indications of a turnaround reversal during the 1980s brought forth a set of fresh and seemingly contradictory issues which previously either did not exist, or slipped by without attracting much interest. While issues such as the wave theory versus the clean break were widely debated (Vining and Strauss, 1977; Gordon, 1979; Champion, 1989a, 1989b; Sant and Simons, 1993), both remained controversial and deserve further attention. Other matters deserving further scrutiny are an integration of the literature on interregional and interurban patterns of migration, urban versus rural mainstream and sub-stream migration differentials at the intra- and interregional levels, short-term perspectives versus long-term explanations of migration processes, and an integration of the migration literature on developed and less developed countries.

These were the topics of research in a recent consolidation of the literature on similarities and differences between spatial development patterns in the developed and less developed world (Geyer and Kontuly, 1993). This paper, which is a somewhat revised version of a paper presented at an international conference of the Institut Fédératif de Recherche sur les Économies et les Sociétés Industrielles, held at Lille, France, in March 1994, expands the theoretical foundation of the concept of differential urbanisation with respect to four matters: its graphical properties; the explanations for concentration and deconcentration; the expected characteristics of the advanced counterurbanisation phase; and the criteria for the differentiation between mainstream and sub-stream migration flows.

The concept of differential urbanisation

Generally, the intervals between censuses appear to be an impeding factor in the systematic analysis of migration patterns. Among many researchers this has caused a wait-and-see approach, i.e. an empirical testing of short-term spatial trends at consecutive rounds of census surveys rather than a long-term systematic approach to detect the underlying processes. The issue concerning the beginning and what seems to resemble the end of the turnaround in the developed world are central to this problem. Synthesising from past experience, various attempts have been made to date to address this issue (Champion, 1988, 1992; Mera, 1988; Berry, 1988, 1991; Frey, 1988).

Following on a systematic analysis of long-term trends from US migration data, Vining and Strauss (1977) predicted that developed countries would advance through a more or less fixed spatial development sequence: intraurban diffusion at first, followed by interregional deconcentration from urban to rural regions in a clean break with the previous urbanisation phase, and ending with deconcentration within rural regions.

Having a narrower focus, i.e. on the identifiable sequences in the growth pattern of individual 'urban agglomerations' and distinguishing morphologically between the 'core' and the 'ring'[1] of such urban agglomerations, the Klaassen group (Klaassen *et al.*, 1981; Klaassen and Scimemi, 1981; van den Berg *et al.*, 1981, 1982) identified various phases through which an urban settlement could go successively:

- a phase of 'urbanisation' when certain urban settlements grow at the cost of their 'surrounding countryside';
- a phase of 'suburbanisation', when the ring grows at the cost of the core;
- a phase of 'disurbanisation', when the population loss of the core exceeds the population gain of the ring, resulting in the agglomeration losing population overall; and
- a phase of 'reurbanisation', when either the rate of population loss of the core tapers off, or the core starts regaining population, although the ring might still be loosing population.

Looking at Third World economies globally, Richardson (1980) sees the spatial dimensions of industrial development almost identically to the Vining group's description of migration processes in the First World. According to him, urban-industrial development normally begins in one or two core regions (Friedmann, 1960) in a country. The process of cumulative causation (Myrdal, 1957; Hirschman, 1958) – initially caused by, among other factors, locational constants (Richardson, 1973) – results in the evolving of a core–peripheral spatial economic arrangement. Subsequently, the country enters into a more advanced phase of development when, owing to cost factors related to overcrowding in the primate cities, a monocentric urban structure becomes inefficient and the core develops a multi-centred structure.

The polarisation reversal phase is entered into when increasing agglomeration diseconomies create favourable conditions for deconcentration. These conditions accelerate the industrial decentralisation process and induce population migration to urban destinations outside the core region. This turning point is described by Richardson as the beginning of polarisation reversal.[2] Eventually, deconcentration forces gain enough momentum to cause primate core regions to lose population in absolute terms.

From this discussion it is clear that the concept of polarisation reversal, as developed by Richardson (1977, 1980), essentially deals with forces of industrial agglomeration and dispersal in Third World development. In his discussions of the concept of polarisation reversal, population redistribution patterns are treated as a consequence of industrial development patterns within economic space. As in the case of the paper in which the differential urbanisation model was introduced (Geyer and Kontuly, 1993), this paper presents the concept of polarisation reversal in a slightly different mould. Not only are its spatial implications extended, but also its relevance for migration in both the First and Third Worlds is demonstrated.

Highlighting only the migration component of his description of the urban development process, population accumulates in the primate city at first. This is followed, first, by decentralisation within the core[3] resulting in a multi-nodal structure, then by interregional deconcentration towards a limited number of nodes within the periphery, and later on by decentralisation within these peripheral regions. The advanced stages of this final phase are accompanied by an absolute decrease in the population of the core.[4]

Returning to the Vining–Strauss sequence referred to above, their latter phase of deconcentration within the rural regions never really materialised visibly. Instead, Cochrane and Vining (1988) discovered that, since the 1980s, counterurbanisation unexpectedly either slowed down or came to a halt in certain developed countries. In order to expand the differential urbanisation model, a specific attempt will be made in this paper to analyse the factors influencing migration trends towards the advanced phases of urban development.

In their attempt to explain the puzzling divergence of migration trends in developed countries in the early 1980s, scholars covered all possible options. Assessing the apparent short-lived deconcentration trend of the counterurbanisation era, Cochrane and Vining (1988) regard the rural renaissance of the 1970s as a relatively short-term phenomenon of economic readjustment due to fundamental restructuring. At the other extreme, Champion argues that the reversal of the turnaround could be treated as a 'short-term downward flexure' in the rate of long-term population deconcentration, 'a temporary phenomenon resulting from particular conditions favouring conversion' (Champion, 1988, p. 257). Fielding (1989) covers the middle ground. According to him, few signs point to a simple turnaround from counterurbanisation to urbanisation in Western Europe during the early 1980s; rather, the re-emergence of 'broader regional patterns of growth and decline' is imminent (Fielding, 1989, p. 67).

Berry (1988, 1991) proposes the long-wave theory as a possible explanation for the recent slowing down or termination of counterurbanisation in certain countries. Consecutive migration cycles of concentration and deconcentration are associated with cycles of economic growth and decline. In the USA, 'Each wave of economic growth has produced a wave of urbanward migration' (Berry, 1988, p. 249).

Integrating the relevant evidence on interregional, interurban, and urban–rural migration in the First and Third Worlds, a differential urbanisation model spanning the development spectrum was introduced (Geyer, 1989, 1990; Geyer and Kontuly, 1993). The model expands on the Vining–Strauss sequence of migration phases. During the urbanisation phase rural-to-urban migration dominates. In most countries this is accompanied initially by concentration within the primate and other major cities, and then followed by diffusion or urban sprawl towards the metropolitan fringes. The advanced phase of urbanisation could be accompanied by some deconcentration (sub-stream migration) from the major metropolitan areas, especially towards adjacent intermediate-sized cities. Then, deconcentration from urban to rural regions

312 Differential Urbanization

gains momentum until it becomes the dominant migration pattern. Initially, certain cities adjacent to the major urban regions gain more migrants than others, especially intermediate-sized cities, but subsequently migration to intermediate and small-sized cities in areas further afield increases as well. In the literature, the former process is described as polarisation reversal (Richardson,

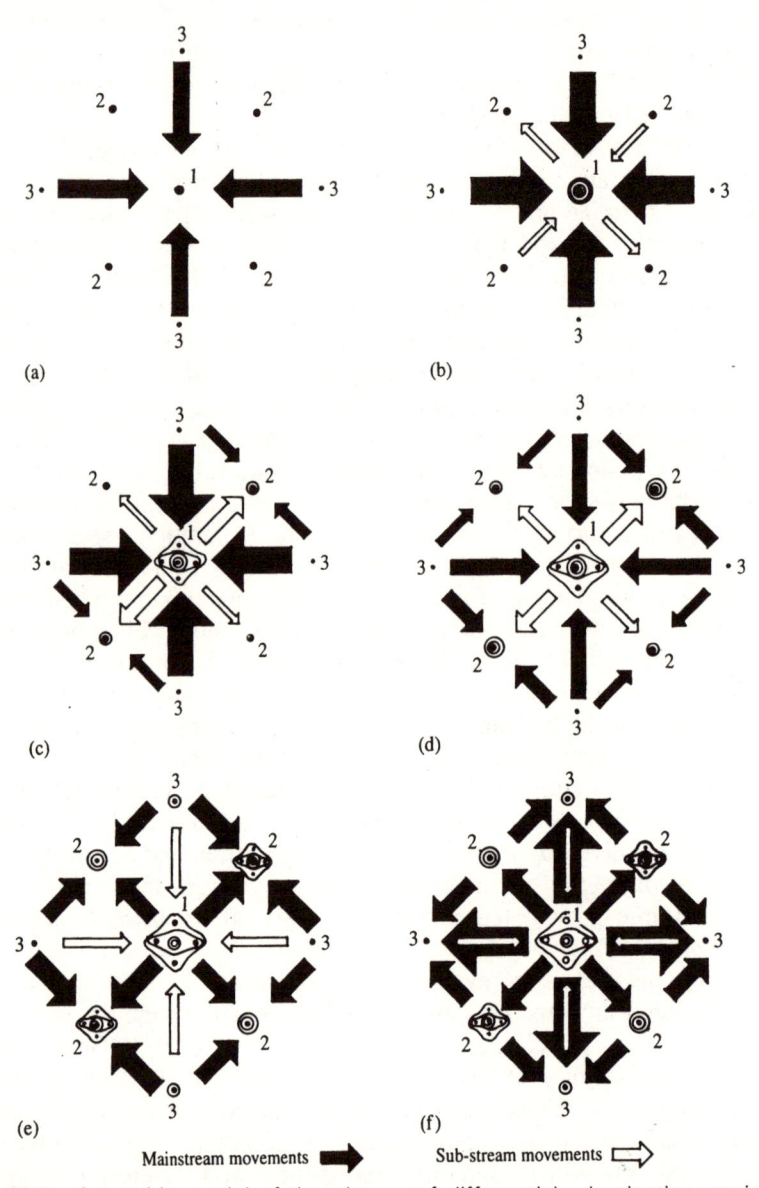

Fig. 20.1 A graphic model of the phases of differential urbanisation: mainstream and sub-stream migration flows; (a) early primate city stage, (b) intermediate primate city stage, (c) advanced primate city stage, (d) early intermediate city stage, (e) advanced intermediate city stage, (f) small city stage (Source: Geyer and Kontuly, 1993)[5]

1977, 1980) and the latter as counterurbanisation (Vining and Strauss, 1977). Throughout the phase of deconcentration, some degree of sub-stream migration from the periphery to the major urban areas may occur. Finally, elements of the first phase of the sequence re-emerge, but now within a functionally and economically integrated spatial system.

This sequence of migration phases is diagrammatically and graphically shown in Figures 20.1 and 20.2. Figure 20.1 indicates the expected directions of mainstream and sub-stream migration flows between the different city size categories from the urbanisation to the counterurbanisation phase. In Figure 20.2 net migration gains and losses are shown for large, intermediate-sized, and small-sized cities on an inverted log scale, compacting time towards the left of the graph.[6] Potentially, Figure 20.2 enables one to identify the positions of different First and Third World countries relative to one another on the same graph at different points in time as they advance through consecutive cycles of urban development. Preliminary tests using French, South Korean, and Indian data bear out the major propositions of the differential urbanisation model (Geyer and Kontuly, 1993).

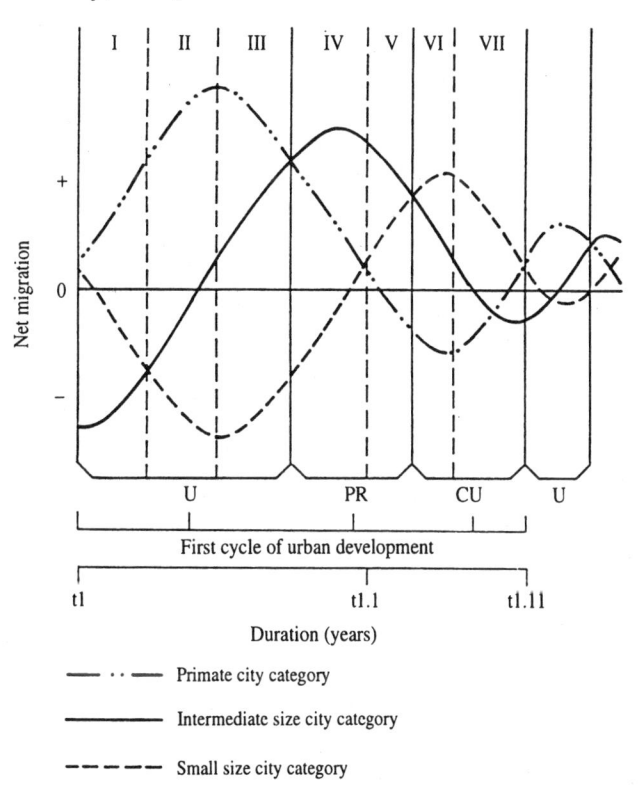

Fig. 20.2 Generalised stages of differential urbanisation. I, Early primate city stage (EPC); II, intermediate primate city stage (IPC); III, advanced primate city stage (APC); IV, early intermediate stage (EIC); V, advanced intermediate city stage (AIC); VI, early small city stage (ESC); VII, advanced small city stage (ASC). U = urbanisation; P = polarisation reversal; CU = counterurbanisation. (Source: Geyer and Kontuly, 1993)

A graphical analysis of differential urbanisation

The multi-faceted nature of differential urbanisation does not make it easy to define the concept. However, differentiating between opposing main- and substream population migration flows that are evident in developed and less developed countries, differential urbanisation can generally be described as a sequence of urban development cycles, each cycle consisting of consecutive phases of urbanisation, polarisation reversal and counterurbanisation. During the phase of urbanisation, mainstream migration supports large city development. This is followed by secondary or regional city development during the phase of polarisation reversal, and finally by small city development during counterurbanisation.

A search of the literature reveals many casual references to several elements of the differential urbanisation model. Pieced together in context, these references serve as tentative proof of the accuracy of the model, especially with regard to the US case. The concurrent occurrence of concentration and decentralisation was observed internationally (Clark, 1967) and in the USA between 1940 and 1950 (Duncan et al., 1961). The absolute decrease in migration towards the large US metropolitan areas in the early 1970s was accompanied by the rapid growth of urban areas in the metropolitan 'shade areas' during the same period (Berry, 1976; Gordon, 1979). In his locational breakdown of non-metropolitan growth in the USA during this time period, Gordon (1979, p. 282) found that non-metropolitan counties most strongly linked to metropolitan centres grew the fastest during the early phase of counterurbanisation. Projecting the wave theory onto the clean break theory, he concluded that, at the time, 'a continued "wave" of urban decentralisation as well as renewed rural growth seemed to be in progress' (Gordon, 1979, p. 281).

Contrasting arguments of the wave theory against the clean break proposition, Vining and Strauss (1977, p. 751) refer to Morrison's finding that 'Nonmetropolitan counties well-removed from the commuting range of SMSAs [Standard Metropolitan Statistical Areas] are growing at a significantly higher rate than these SMSAs themselves, though at a somewhat lower rate than the non-metropolitan counties adjacent to these SMSAs.' Beale (1977, p. 116) sketches the dynamism of the process in the USA as follows: 'The adjacent ... counties have been gaining people at an annual rate about twice as high as they experienced in the 1960s, but the difference in post-1970 annual growth of the adjacent and nonadjacent groups is not great.'

Wherever counterurbanisation occurs, the general trend seems to correspond with what is known as concentrated dispersal. Drawing from evidence internationally, Vining and Strauss (1977) and Vining and Kontuly (1978) came to the conclusion that during the deconcentration process, people tended to concentrate in a limited number of small and intermediate-sized cities in peripheral regions. This conclusion is confirmed by Beale (1977, p. 120), according to whom the USA experienced 'renewed growth or diminished loss of nonmetropolitan population throughout most of the country and increased regional growth of small and medium sized metropolitan areas in the South and the West' during the early 1970s. According to Richter (1985), the growth rate of the smallest SMSAs in the United States, i.e. SMSAs with populations less than 100 000, increased consistently during the 1970s, at almost twice the national growth rate. By the late 1970s most of the growth in the larger metropolitan areas of the USA took place in their fringe counties, but these counties grew at a lower rate than those containing smaller metropolitan areas. Net migration to non-adjacent counties also decreased dramatically. Most migration took place towards adjacent counties (Richter, 1985).

Returning to the differential urbanisation model depicted in Figure 20.2, a clear distinction can be made between the phases of urbanisation and counterurbanisation on the one hand and the phase of polarisation reversal on the other. The former two phases result in concentration and deconcentration respectively. Major cities gain migrants the fastest during the urbanisation phase at the expense of small cities, and vice versa during the counterurbanisation phase. During polarisation reversal intermediate-sized cities grow at the expense of both major and small cities, although major cities may still be gaining population, but at a slower rate now than previously. According to Fielding (1989), there is a positive relationship between the net migration rate and settlement size during urbanisation, and a negative relationship during counterurbanisation. These relationships are indicated in Figure 20.3.

The urbanisation and counterurbanisation curves shown in Figure 20.3 depict only two of the positions in the cycle of urban development outlined in Figure 20.2. A number of intermediate positions in Figure 20.2 are also discernible. The curves in Figure 20.2, representing the relative position of the different city size categories at the transition from one migration stage to the other, are indicated in Figure 20.4. Following Fielding's line of reasoning regarding urbanisation and counterurbanisation, the polarisation reversal phase could be defined as coincident with a symmetrical relationship between net migration rate and settlement size (Figure 20.4(d)). During the polarisation reversal phase intermediate-sized cities grow at a faster rate than large and small-sized cities. The symmetrical relationship can be either parabolic or leptokurtic.

A variety of other relationships, reflecting various stages of polarisation reversal, are also discernible. These include positive or negative skew and bimodal or multimodal curves (Figure 20.5). *Ceteris paribus*, one would

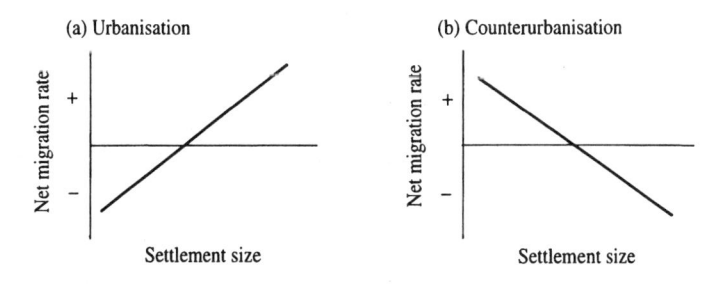

Fig. 20.3 A graphical representation of urbanisation and counterurbanisation. (Source: Fielding, 1989)

316 Differential Urbanization

expect countries to move first into the urbanisation phase (graphically depicted in Figure 20.3(a)), then into a negative skew position, depicting an early phase of polarisation reversal (Figure 20.5(a)), thereafter into a symmetrical position depicting full-scale polarisation reversal (Figure 20.4(d)), then into a positive skew position depicting the transition from polarisation

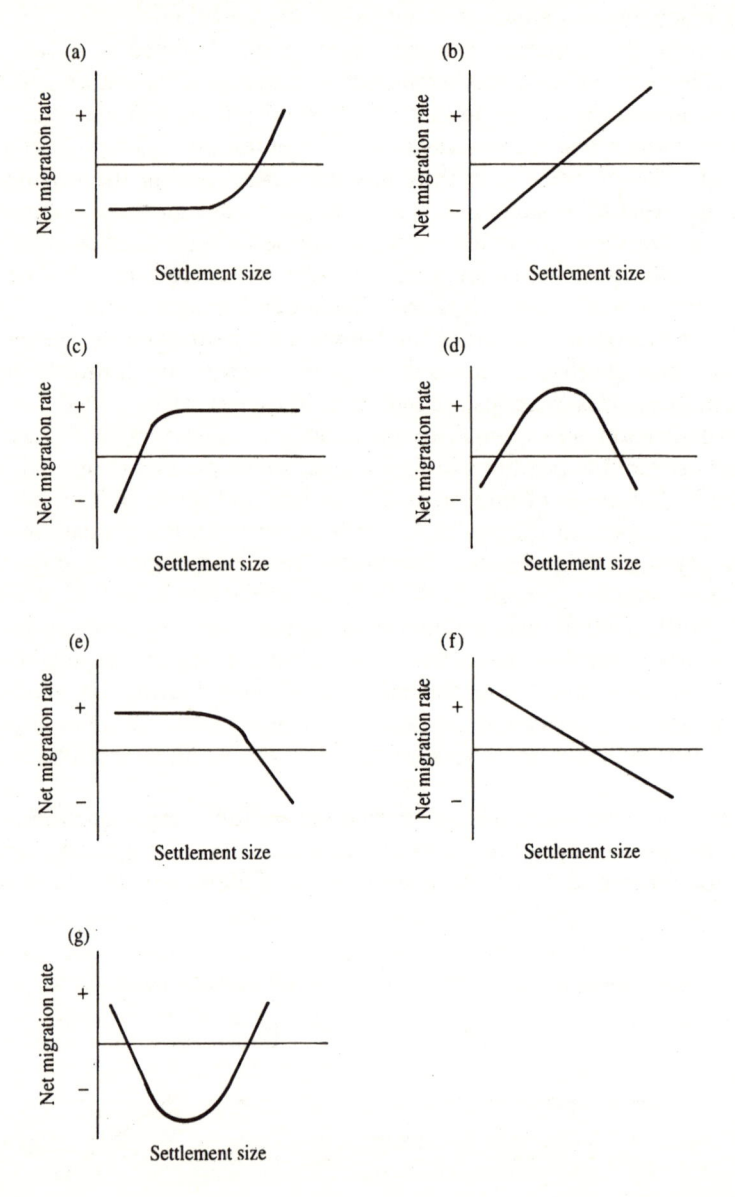

Fig. 20.4 The changing relationships between net migration rate and settlement size during a cycle of urban development; (a) EPC/IPC transition, (b) IPC/APC transition, (c) APC/EIC transition, (d) EIC/AIC transation, (e) AIC/ESC transition, (f) ESC/ASC transition, (g) ASC/EPC transition

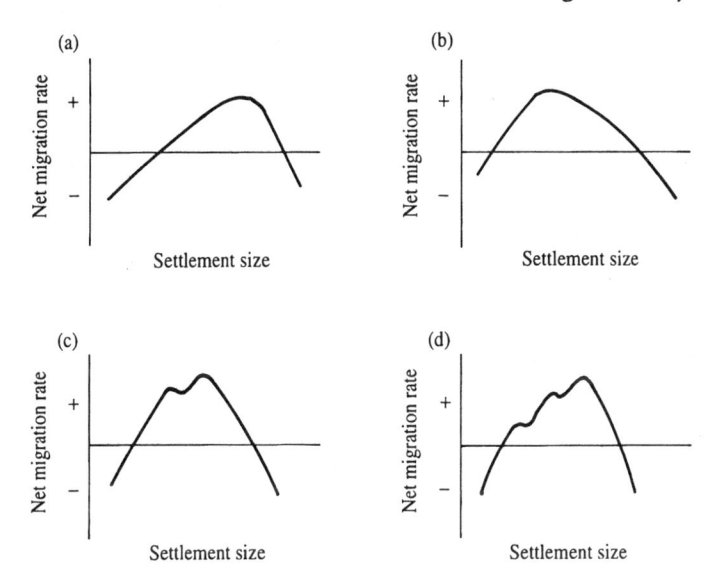

Fig. 20.5 Alternative relationships between net migration rate and settlement size; (a) negative skew, (b) positive skew, (c) bimodal, (d) multi-modal

reversal to counterurbanisation (Figure 20.5(b)), and finally into the counterurbanisation phase indicated in Figure 20.3(b).

Sequentially, this description of how an urban system evolves over time under ideal circumstances could be visualised in two ways: first, in terms of city size categories, and second, in terms of its spatial or geographical dimensions. Continuity in the sequence of city size development could be visualised if the pace of the historical development of an urban system could be stepped up to portray a visual motion similar to a motion picture sped up. The increased pace in the development of different city size categories portrayed in sequence by Figures 20.3(a), 20.5(a), 20.4(d), 20.5(b), and 20.3(b) should then resemble a wave of urban development over time. Initially, major cities grow the fastest, then intermediate-sized cities, first the larger ones, then the smaller ones, until the urban system enters the counterurbanisation phase when the growth rate of smallest cities outstrips that of the larger cities in the system. Spatially, this motion would resemble the ripple effect of an object thrown into a pool of water. First, cities closer to the major areas will start growing, then the ones further afield.

The ideal circumstances described above do not occur in reality, however. Owing to the uneven distribution of natural endowments and the effects policies of state intervention have on spatial development, intermediate-sized cities and specific smaller cities could start developing simultaneously, resulting in bimodal or multi-modal patterns of urban development. This is illustrated by the net migration gains and losses of different size settlements in France between 1954 and 1982 (Figure 20.6).

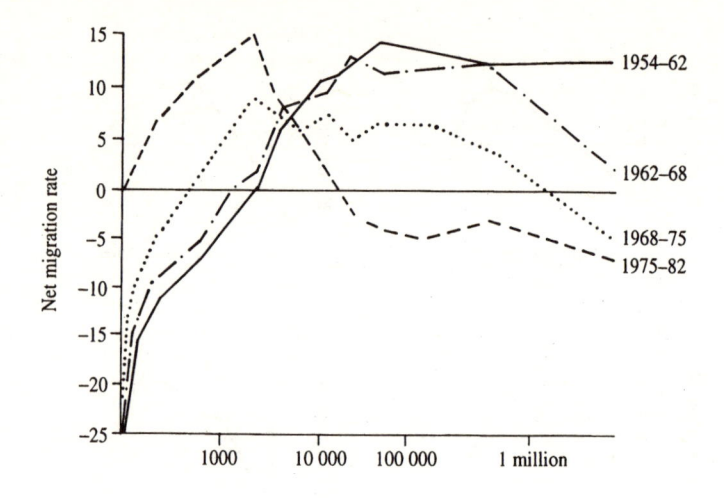

Fig. 20.6 Annual net migration rates by settlement size in France, 1954–82

According to Fielding (1989, p. 61), the 1954–62 curve in Figure 20.6 was 'strongly positive', which in terms of his definition implied urbanisation. However, if one compares the 1954–62 curve in Figure 20.6 with Figure 20.4(d), it is clear that France had already entered the polarisation reversal phase at this time. During the 1962–68 period, France moved deeper into the polarisation reversal phase. During 1962–68, 1968–75, and 1975–82 the curves were bimodal, multi-modal, and positively skewed respectively.

In the following section the concepts of productionism and environmentalism will be discussed.

Forces dictating migration turnarounds

Many attempts have been made to identify the factors that caused the counterurbanisation of the 1970s in developed countries. Explanations given include:

- Economic cyclical circumstances, such as boom and bust periods;
- economic structural changes, such as production changes and changing labour market sizes, including changing spatial divisions of labour;
- spatial economic forces, such as urban agglomeration economies and diseconomies, and deglomeration economies;
- explicit and implicit state policy, such as industrial decentralisation initiatives, rural resource development, and spatial closure or protectionism;
- socio-cultural factors, such as residential preferences, state welfare payments, changing socio-demographic compositions, dispersed educational and social services;
- technological innovations resulting in reduced distance friction; and
- environmental factors such as urban decay, pollution and favourable or adverse climatic conditions.

Some of these factors, including agglomeration diseconomies, could be the cause of deconcentration. Others, such as dispersed educational and social services provided by the State, for instance, are often policy measures in reaction to prevailing population redistribution trends. Consequently, they could be regarded as the result of deconcentration in the beginning, although they could also serve as a cause later on. Factors associated with different phases of the mobility transition continuum could be both cause and effect. Certain economic cyclical factors could be secondary causes of either concentration or deconcentration. Some factors, such as spatial closure and protectionism, could be regarded as instrumental, others, such as residential preferences, as circumstantial, but most are symptomatic. Only a few are fundamental in nature. Two of the fundamental factors underlying the processes of concentration and deconcentration are productionism and environmentalism (Berry, 1978; Hart, 1983). Generically, these explanations given for deconcentration can be classified as fundamental, chronological, cyclical, or structural (Table 20.1).

Table 20.1 A generic classification of explanations given for counterurbanisation in developed countries during the 1970s

Explanations	Explanations		Orientation	
	Cause	Effect	Con	Decon
1. *Chronological*				
Early mobility transition	×	×	×	×
Advanced mobility transition	×	×	×	×
2. *Cyclical*				
Economic cycle	×	×	×	×
3. *Structural*				
Agglomeration economies	×	×	×	×
Deglomeration economies	×	×	×	×
Economic restructuring	×	×	×	×
Changing labour markets	×	×	×	×
Intervention	×	×	×	×
4. *Fundamental*				
Productionism	×		×	
Environmentalism	×			×

Productionism refers to the phase in people's lives when improved job opportunities, education, and income are more important than actual living conditions. After having had the opportunity to reap the benefits of productionism, people normally enter the environmentalism phase. This phase is entered into when the need to improve one's actual living/environmental conditions becomes as important as earning a living. During the environmentalism phase a person would even trade income for pleasant living conditions and in such cases productionism becomes of secondary importance. However, productionism and environmentalism go hand in hand, the former enabling the person to achieve the latter. It could, therefore, be argued that productionism and environmentalism are both driven by the same force; that is, the need to improve one's living conditions. But in productionism, improved living conditions can normally not be realised in the short run, while in environmentalism, improved living and environmental conditions are an immediate need. Paradoxically, people are often willing to endure the most appalling living conditions during the period of productionism to achieve the long-term goal of a better living standard later on.

Productionism and environmentalism hold the potential of explaining many of the root causes of national and regional migration trends in developed and less developed countries, but as in the case of all deterministic indicators,[7] they are not watertight, and should, therefore, be treated with due caution. In spite of exceptions to the rule, there is strong evidence of a general tendency among younger, less educated, less wealthy people in most societies to concentrate spatially, while the opposite groups are prone to deconcentrate. Generally, these tendencies are due to the apparent availability of employment opportunities in the larger urban areas for the former group and the pleasant living conditions of the rural environment for the latter. However, indications are that entrepreneurs are increasingly concentrating on environmentally friendly rural areas for new investments (Klaassen, 1987), which in time could affect the direction of these flows visibly.

Elements of the general trend in productionism- and environmentalism-driven migration have been observed all over the world. Although main and sub-stream migration patterns are not manifested clearly everywhere, indications were found that older people in parts of Western Europe and the USA tended to migrate to non-metropolitan regions, and mostly to those areas closer to the larger urban regions, while younger people and the less developed component[8] of the population migrated towards larger urban areas during the 1970s (Frey and Speare, 1988; Koch, 1980; Lichter *et al.*, 1979). Although local conditions do affect migration patterns significantly, signs of similar trends were observed by Champion (1993a, 1993b), Rees (1993), Robinson (1993), and Fielding (1989) in Britain and by Kontuly and Vogelsang in West Germany (1989) over the past two decades.

Arguing the historical importance of environmentalism in American culture, Berry (1976) and Beale (1977) showed that, while counterurbanisation was in full force in the USA during the early 1970s, a significant proportion of the African Americans and other minorities continued to urbanise. This tendency continued throughout the late 1970s (Richter, 1985), a pattern resembling Figure 20.1(f). Where counterurbanisation flows were observed among African Americans, such flows were found to be inter alia related to increasing employment opportunities resulting from affirmative action – i.e. productionism-related factors (Krieg, 1993). The importance of environmentalism in the explanation of deconcentration in the American example was confirmed by Gordon (1979, p. 288):

> A preference for small-town life has long been used to explain suburbanisation. The data seem to suggest that this trend is as strong as ever and is taking place at ever greater distances from central cities, especially if these central cities are large.

In the Third World there is a general tendency amongst the rural poor to urbanise, despite an oversupply of unskilled and semi-skilled labour in many of the major urban areas. According to Lipton (1990) this tendency is aggravated by the private and public sector's traditional 'urban bias' in investment decisions. Supporting this view, Todaro ascribes the dominating rural–urban migration trend in the Third World to urban–rural differences in expected rather than actual earnings (Todaro, 1982). The productionism principle clearly underlies this migration trend.

As in the case of the other explanations listed above, environmentalism may not always result in deconcentration and productionism not always in concentration. Exceptions to the rule are those cases where people finding themselves in the environmentalism phase remain in the large urban areas during counterurbanisation or return to these areas after a period in the peripheral areas. Whether they remained or returned to these major metropolitan areas it seems logical to expect their residential choice to be determined by environmental factors similar to the factors which prompted the counterurban migration decisions in the first place. More often than not suburbs with the required environmental qualities are found towards the fringes of large metropolitan areas resulting in decentralisation rather then centralisation (Klaasen *et al.*, 1981). Therefore, although these exceptions proved productionism and environmentalism not to be watertight indicators of migration in a specific direction, the former remains a firm indicator of either interregional concentration or intraurban centralisation and the latter an indicator of either intraurban decentralisation or interregional deconcentration.

Differential urbanisation and mobility transition

Differential urbanisation cannot be fully understood without a differentiation between main- and sub-stream migration. In the differential urbanisation model (Figure 20.1) it is clear that large urban areas (mostly primate cities in the Third World) gain rapidly by the urban-bound main migration streams during the primate city stage of the urbanisation phase at the expense of intermediate and smaller sized cities (Figure 20.1(a)–(c)). During the small city stage, i.e. the counterurbanisation phase (Figure 20.1(f)) which ends the first cycle of urban development (Figure 20.2), the migration towards the larger urban areas turns into an undercurrent.

In the introduction of the differential urbanisation model (Geyer and Kontuly, 1993), little is said about what could be expected after the counterurbanisation phase. Following on the observation made earlier that the final migration phase as envisaged by Vining and Strauss (1977) never really materialised on a national scale, an attempt will be made in this section to expand on this issue.

In most of the highly developed societies in Western Europe and North America, the prominent differences between the urban and rural areas of the past have largely been eradicated. In fact, in large parts of these regions, especially in Europe, rural and urban land uses are intermingled to such an extent that the question could well be asked whether the traditional meaning of the term 'rural' is still relevant in such areas today. Because of the social, economic, administrative, technological and physical integration of these societies, indications are that environmentalism-driven migration is being influenced more significantly by what were regarded as less significant push and pull factors of earlier phases of development. For instance, economic boom and bust periods seem to have a more significant effect on the migration of the people in highly developed societies than on migrants in less developed Third World countries.

Linking the concepts of differential urbanisation and transitional mobility (Zelinsky, 1971), a number of general statements can be made regarding urban development and population migration in the First and Third World. In a chain of cause and effect a society's spatial mobility is predetermined by its levels of social, educational, occupational, and technological mobility. There is, therefore, a direct relationship between a society's level of development and its spatial mobility. As a result, first, of the eradication of excessive core–peripheral differences at the national level, and second, the resulting reduction in the gap between the lower- and higher-income groups at advanced stages of development, the contrast between productionism- and environmentalism-driven migration streams can be expected to be less marked towards the beginning of the second urban development cycle than earlier on in the first cycle (Figure 20.2).

The fact that an urban system enters an advanced phase of development at the national level does not mean that individual regions within the national system could not still be in one or more of the earlier phases of development indicated in Figure 20.1. This implies that although a country may be in an advanced phase of development overall, prominent differences in main- and sub-stream migration could still be visible at the sub-national level because certain regional systems of cities could still be in an earlier phase of development than others. It is, therefore, highly unlikely that a country would enter the advanced phase of deconcentration in the manner suggested by Vining and Strauss (1977). Rather, one would expect relatively large areas of the national urban space to be integrated to such an extent socially, economically, technologically, and administratively during the advanced phase that major core–peripheral differences will only be noticeable in certain less developed regions. While less marked differences in productionism- and environmentalism-driven migration could be expected in the more advanced regions, prominent differences may continue to occur in the lagging regions.

These conclusions tie in well with the gradually diminishing duration and amplitudes of the consecutive migration waves of the second cycle of the differential urbanisation model (Figure 20.2). As a country moves into the second urban cycle it is also likely that: (1) the complexity of the factors playing a role in migration will increase, (2) urban- and rural-oriented migration differentials resulting from regional and national differences will become increasingly intermingled, and (3) prominent urban- and rural-oriented migration differentials are likely to be visible in regions which still have a rural character only.

The observations that were made in this section on the relationship between regionalisation and differential urbanisation trends justify a critical evaluation of the regional approach to migration studies.

Regionalisation and differential urbanisation

The study of population redistribution patterns in a regional context is an approach widely used and as widely criticised (Champion, 1988; Sant and Simons, 1993). The concealment of potentially important sub-stream redistribution patterns, the obfuscation of underlying intra- and interregional urban–rural migration gains and losses, and the risk of an over- or under-bounding of core and peripheral regions are three of the most important limitations of the core–peripheral approach to migration analysis as a measure of differential urbanisation.

The differentiation between mainstream and sub-stream migration could play an important role in unravelling the fundamental forces shaping urban systems in the long run, both regionally and supraregionally. In the previous section it was stated that although core–peripheral differences in a country are likely to decrease at the national level by the end of the first urban cycle (Figure 20.2), significant core–peripheral contrasts may still be visible at the regional level. If this is true, the approach that will be needed to pick up the different migration patterns within and between regions at different phases of development will differ fundamentally from the Vining group's approach to migration studies.

The Vining group mainly differentiated between core and peripheral regions at the national level. The group also concentrated on mainstream migration only. For one to differentiate between main- and sub-stream migration on the one hand and between productionism- and environmentalism-driven trends on the other hand, migration will have to be studied in terms of flows between urban settlements within and between urban systems. The core–peripheral approach seems to be too crude a geographical measure to allow one to pick up the subtle spatial intricacies of the concept of differential urbanisation.

The urban system approach to migration studies necessitates a spatial and hierarchical classification of urban settlements. Theoretically, this implies a myriad of possible hierarchical (city size) and spatial (locational) combinations, which further implies that intermediate-sized cities could differ significantly from one another in terms of their population sizes. These size differences could occur from region to region inside a particular country as well as from one country to another. The defining of intermediate-sized cities in terms of their regional function rather than size alone eliminates the old problem of which city size categories are to be regarded as the appropriate ones (Bos and Geyer, 1993). Returning to the bimodal and multi modal curves indicated in Figure 20.5, it speaks for itself that regional centres of different sizes could start developing at the same time, resulting in peaks at different positions in the curve during the polarisation reversal phase.

Therefore, to draw a distinction between intra- and interregional main- and sub-stream migration, as well as determine whether they are productionism- or environmentalism-driven, migration should be disaggregated geographically, thematically, quantitatively, and qualitatively. Promising indicators which potentially comply with all three requirements and allow a differentiation between both sets of criteria are migration data by age groups, income categories, cultural associations, race groups, educational levels, employment groups, core–peripheral regions, urban–non-urban locations, and urban scale.

Depending on how the qualitative and quantitative aspects of the first six indicators are combined with the last three geographical indicators, various aspects of the differential urbanisation concept could be reflected by the same data sets. Total migration, broken down into different city size categories over time, for example, could be used to identify particular stages of urban development within a country. When data on age, race, income, employment, and educational categories are disaggregated to reflect the younger, less developed, less educated, less wealthy, unemployed sections of the population in contrast with the older, wealthier, higher-qualified sections of the population, the data sets could be used to differentiate between environmentalism- and productionism-driven migration streams in a country. This does not mean, of course, that younger people are necessarily less educated or less developed than older, wealthier people; it simply illustrates the importance of particular combinations of migration categories to reflect environmentalism and productionism.

Conclusions

Recently a model was presented outlining a theoretical foundation for the concept of differential urbanisation. Two propositions were suggested. First, that mainstream migration flows tend to favour groups of large, intermediate, and small-sized cities in succession to form a continuum of urban development which spans the development spectrum, from less developed to developed economies. Second, that six successive stages of population concentration and deconcentration occur as urban systems develop over time, every stage reflecting a different combination of opposing main and minor migration streams of differential urbanisation.

Four aspects relevant to the concept of differential urbanisation were examined in this paper with a view to expanding the theoretical underpinnings of the concept. In a graphical analysis of the model, it was suggested that at least seven fundamentally different relationships between net migration rates and settlement size, as well as at least four more variants of the former, are distinguishable. As yet, only two of these phases of urban development have been explored graphically in the literature, urbanisation and counterurbanisation. While urbanisation and counterurbanisation are coincident with a significantly positive or negative relationship between migration rates and settlement size respectively, an intermediate phase of polarisation reversal, coincident with a symmetrical relationship between the two variables, is also suggested. French data support this finding. Although the positive and negative relationships have been associated with migration in the developed world, the relevance of the symmetrical relationships in developed countries has largely been overlooked thus far. The linking of these three basic relationships effectively ties together urban development processes in developed and less developed countries.

The probable effect of mobility transition on societies entering the advanced/post counterurbanisation phase was also discussed. The high levels of territorial, social, economic, and administrative integration typical of highly developed societies will lead to less prominent mainstream and sub-stream differentials at the national level, but not necessarily at the regional level.

A generic discussion of the most important explanations given for the turn-around indicates that they can be associated with more than only the cause of deconcentration. Without exception, they explain either the cause or the effect, or both the cause and effect of concentration and deconcentration. None of these factors singularly explains counterurbanisation adequately. To explain the counterurbanisation process in full, a combination of factors is necessary (Kontuly and Bierens, 1990). The concepts of productionism and environmentalism seem to hold promise for explaining the forces underlying the processes of concentration and deconcentration respectively.

Acknowledgement

Financial assistance by the Centre for Science Development for this research is hereby acknowledged. Opinions expressed and conclusions drawn in this study are those of the author and should not necessarily be attributed to the Centre for Science Development.

Notes

1. The 'core' is defined as the central town, while the surrounding area in which 15 per cent or more of the people work in the core is regarded as the 'ring'.
2. He defines the process of polarisation reversal as the turning point when spatial polarisation trends in the national economy give way to a process of spatial dispersion out of the core region into other regions in the system.
3. For an elaboration on current views on the spatial dynamics inside the metropolitan area, *see* Bourne (1992).
4. It must be stressed that although the migration element has been highlighted in this sequence, demographic space cannot really be separated from economic space. The two are indivisible elements of our human environment. This indivisibility of economic and demographic space is clearly described in Weber's classic explanation of how an initially uninhabited area should develop over time (Isard, 1972). First, the agricultural sector establishes, exploiting a particular range of natural endowments in the area. This layer creates the basic economic framework – i.e. service centres with their corresponding tributary areas – on which the secondary sector, tertiary sector, and quaternary sector as well as what he terms the 'central organization' and 'central dependency' or population layers subsequently develop. In the past the exploitation of resources other than only the output of the agricultural sector often leads to the development of large industrial complexes (Perroux, 1955) and, the subsequent development of primate cities. More often than not, the latter process sets the urbanisation phase in motion.
5. Note the difference in the direction of the sub-stream flows in figures 19.6 and 20.1.
6. As in the case of the urban development sequences described by Berry, Richardson and the Vining group, the Klaassen model and the differential urbanisation model shown in Figure 20.2 have much in common. The Klaassen model describes the urban development cycle of the individual city against the backdrop of the remaining urban environment, while the differential urbanization model attempts to

encompass the development cycles of entire urban systems across the development spectrum. The Klaassen model reveals the morphological changes taking place inside the city – whether it is a small, medium, or major city – as it moves through different phases of development.
7 Although the term 'deterministic' is used here, this is by no means an attempt to revive the concept of geographical determinism (Fisher, 1970; James, 1967; Turnock, 1970). Also, the use of the term 'environmentalism' in this paper must not be confused with the original meaning of the term in a deterministic sense. Here, the term is given a 'modern' meaning, referring to people's purposeful behaviour, not their domination by the natural environment. In planning his/her life a person will intentionally seek to realise his/her personal aims and objectives, which, in terms of the concept of environmentalism, refers to a range of issues dealing with the improvement of a person's quality of life.
8 The less developed component refers to the less educated and poorer section of the population. In the United States a large percentage of the African and Hispanic Americans forms part of this sector.

References

Beale, C. L. 1977: The recent shift of United States population to nonmetropolitan areas, 1970–75. *International Regional Science Review* 2, 113–22 (*see* this volume, Chapter 2).
Berry, B. J. L. 1976: The counterurbanisation process: urban America since 1970. *Urban Affairs Annual Review* 11, 17–30 (*see* this volume, Chapter 1).
Berry, B. J. L. 1978: The counterurbanisation process: how general? In Hansen, N. M. (ed.), *Human settlement systems: International perspectives on structure, change and public policy*. Cambridge, Mass.: Ballinger, 25–47.
Berry, B. J. L. 1988: Migration reversals in perspective: the long-wave evidence. *International Regional Science Review* 11, 245–51 (*see* this volume, Chapter 17).
Berry, B. J. L. 1991: *Long-wave rhythms in economic development and political behaviour*. Baltimore, Md.: Johns Hopkins University Press.
Bos, D. J. and Geyer, H. S. 1993: International perspectives on the definition of intermediate sized cities: South African applications. *South African Geographer* 20, 46–61.
Bourne, L. S. 1992: Population turnaround in the Canadian inner city: contextual factors and social consequences. *Canadian Journal of Urban Research* 1, 66–89.
Champion, A. G. 1988: The reversal of the migration turnaround: resumption of traditional trends? *International Regional Science Review* 11, 253–60.
Champion, A. G. 1989a: Conclusion: temporary anomaly, long-term trend or transitional phase? In Champion, A. G. (ed.), *Counterurbanisation: The changing pace and nature of population deconcentration*. London: Edward Arnold, 230–44.
Champion, A. G. 1989b: Counterurbanisation: the conceptual and methodological challenge. In Champion, A. G. (ed.), *Counterurbanisation: The changing pace and nature of population deconcentration*. London: Edward Arnold, 19–33.
Champion, A. G. 1992: Urban and regional demographic trends in the developed world. *Urban Studies* 29, 461–82.
Champion, A. G. 1993a: A decade of regional and local population change. *Town and Country Planning* 62 (3), 43–5.
Champion, A. G. 1993b: People in Britain – taking stock. *Town and Country Planning* 62 (3), 42.
Clark, C. 1967: *Population growth and land use*. New York: St Martin's Press.

Cochrane, S. G. and Vining, D. R. Jr 1988: Recent trends in migration between core and peripheral regions in developed and advanced developing countries. *International Regional Science Review* 11, 215–243 (*see* this volume, Chapter 7).
Duncan, O., Cuzzort, R. and Duncan, B. 1961: *Statistical geography*. Glencoe, Ill.: Free Press.
Fielding, A. J. 1989: Migration and urbanisation in Western Europe since 1950. *Geographical Journal* 155, 60–9 (*see* this volume, Chapter 9).
Fisher, C. A. 1970: Whither regional geography? *Geography* 55, 373–89.
Frey, W. H. 1988: The re-emergence of core region growth: a return to the metropolis? *International Regional Science Review* 11, 261–7.
Frey, W. H. and Speare, A. 1988: *Regional and metropolitan growth and decline in the United States*. New York: Russell Sage Foundation.
Friedmann, J. 1960: *Regional development policy: A case study of Venezuela*. Cambridge, Mass.: MIT Press.
Geyer, H. S. 1989: Differential urbanisation in South Africa and its consequences for spatial development policy. *African Urban Quarterly* 5, 276–92.
Geyer, H. S. 1990: Implications of differential urbanisation on deconcentration in the Pretoria–Witwatersrand–Vaal triangle metropolitan area, South Africa. *Geoforum* 21, 385–96 (*see* this volume, Chapter 16).
Geyer, H. S. and Kontuly, T. 1993: A theoretical foundation for the concept of differential urbanization. *International Regional Science Review* 15, 157–77 (*see* this volume, Chapter 19).
Gordon, P. 1979: Deconcentration without a 'clean break'. *Environment and Planning A* 11, 281–90 (*see* this volume, Chapter 4).
Hart, T. 1983: Transport and economic development: the historical dimension. In Button, K. J. and Gillingwater, D. (eds), *Transport location and spatial policy*. Aldershot: Gower, 12–22.
Hirschman, A. O. 1958: *The strategy of economic development*. New Haven, Conn.: Yale University Press.
Isard, W. 1972: *Location in space-economy*. Cambridge, Mass.: MIT Press.
James, P. E. 1967: On the origin and persistence of error in geography. *Annals of the Association of American Geographers* 57, 1–24.
Klaassen, L. H. 1987: The future of the larger European towns. *Urban Studies* 24, 251–7.
Klaassen, L. H., Bourdrez, J. A. and Volmuller, J. 1981: *Transport and reurbanisation*. Aldershot: Gower.
Klaassen, L. H. and Scimemi, G. 1981: Theoretical issues in urban dynamics. In Klaassen, L. H., Molle, W. T. M. and Paelinck, J. H. P. (eds), *Dynamics of urban development*. New York: St. Martin's Press.
Koch, R. 1980: 'Counterurbanisation' auch in Westeuropa? *Informationen zur Raumentwicklung* 2, 59–69 (*see* this volume, Chapter 8).
Kontuly, T. and Bierens, H. 1990: Testing the recession theory as an explanation for the migration turnaround. *Environment and Planning A* 22, 253–70.
Kontuly, T. and Vogelsang, R. 1989: Federal Republic of Germany: the intensification of the migration turnaround. In Champion, A. G. (ed.), *Counterurbanisation: The changing pace and nature of population deconcentration*. London: Edward Arnold, 141–61.
Krieg, R. G. 1993: Black–white regional migration and the impact of education: a multinomal logit analysis. *Annals of Regional Science* 27, 211–22.
Lichter, D. T., Heaton, T. D. and Fuguitt, G. V. 1979: Trends in the selectivity of migration between metropolitan and nonmetropolitan areas: 1955–1975. *Rural Sociology* 44, 645–66.

Lipton, M. 1990: Why poor people stay poor: urban bias in world development. In Gugler, J. (ed.), *The urbanisation of the Third World.* New York: Oxford University Press, 40–52.

Mera, K. 1988: The emergence of migration cycles. *International Regional Science Review* 11, 269–75.

Myrdal, G. 1957: *Economic theory and underdeveloped regions.* London: Duckworth.

Perroux, F. 1955: Note sur la notion de 'pole de croissance'. *Économie appliqueé* 7, 307–20.

Rees, P. 1993: The changing geography of age. *Town and Country Planning* 62 (3), 46–9.

Richardson, H. W. 1973: *Economic growth theory.* London: Macmillan.

Richardson, H. W. 1977: *City size and national strategies in developing countries.* Staff Working Report. Washington, DC: World Bank.

Richardson, H. W. 1980: Polarization reversal in developing countries. *Papers of the Regional Science Association* 45, 67–85 (*see* this volume, Chapter 11).

Richter, K. 1985: Nonmetropolitan growth in the late 1970s: the end of the turnaround? *Demography* 22, 245–63 (*see* this volume, Chapter 5).

Robinson, V. 1993: Ethnic minorities and the enduring geography of settlement. *Town and Country Planning* 62 (3), 53–7.

Sant, M. and Simons, P. 1993: The conceptual basis of counterurbanisation: critique and development. *Australian Geographical Studies* 31, 113–26.

Todaro, M. P. 1982: *Economics for a developing world: An introduction to principles, problems, and policies.* Harlow: Longman.

Turnock, D. 1970: The region in modern geography. *Geography* 55, 374–83.

Van den Berg, L., Drewett, R., Klaassen, L. H., Rossi, A. and Vijverberg, C. H. T. 1982: *A study of growth and decline.* Oxford: Pergamon Press.

van den Berg, L., Klaassen, L. H., Molle, W. T. M. and Paelinck, J. H. P. 1981: Synthesis and conclusions. In Klaassen, L. H., Molle, W. T. M. and Paelinck, J. H. P. (eds), *Dynamics of urban development.* New York: St Martin's Press, 251–67.

Vining, D. R. Jr and Kontuly, T. 1978: Population dispersal from major metropolitan regions: An international comparison. *International Regional Science Review* 3, 49–73 (*see* this volume, Chapter 6).

Vining, D. R. Jr and Strauss, A. 1977: A demonstration that the current deconcentration of population in the United States is a clean break with the past. *Environment and Planning A* 9, 751–8 (*see* this volume, Chapter 3).

Zelinsky, W. 1971: The hypothesis of the mobility transition. *Geographical Review* 16, 219–49.

CONCLUSION

Introduction

In this book an attempt was made to integrate the largely compartmentalized migration theory which developed in the First and Third Worlds over the past 15 years. In the articles included in this book various aspects relating to the concepts of urbanization, polarization reversal, counterurbanization, and differential urbanization were analysed.

Because of the pioneering nature of many of the works in the first two parts of the book, few of the authors and co-authors of the original articles had the advantage of hindsight. It is therefore understandable that, individually, none of the articles on counterurbanization and polarization reversal was intended to explain the potential overlap in migration theory applicable to the First and Third Worlds. This was the main objective of this book. Articles were combined in a specific order to create new perspectives on the theoretical linkages between the concepts of urbanization, polarization reversal, counterurbanization, and differential urbanization.

Matters relating to all four of these concepts, intentionally and unintentionally referred to by the authors, will be discussed under the following themes:

- defining urbanization, polarization reversal, counterurbanization, and differential urbanization;
- why counterurbanization occurs;
- criteria for the measurement of counterurbanization, polarization reversal, and differential urbanization;
- the core–periphery approach to the measurement of counterurbanization;
- the link between the First and Third Worlds;
- will polarization reversal and counterurbanization continue?
- differential urbanization;
- a migration cycle approach to urban development.
- the growth of groups of cities;
- mainstream versus undercurrent migration;
- an urban development paradigm;
- productionism versus environmentalism;
- conclusion.

Defining urbanization, polarization reversal, counterurbanization, and differential urbanization

Various attempts have been made to define urbanization, polarization reversal, counterurbanization, and differential urbanization. Fielding defines urbanization as coincident with a significantly positive, and counterurbanization as coincident with a significantly negative, relationship between net migration rate and settlement size. Polarization reversal is defined by Geyer as coincident with a symmetrical relationship between net migration rate and settlement size. Differentiating between mainstream and substream population migration flows, differential urbanization can generally be described as a sequence of urban development cycles, with each cycle consisting of consecutive phases of urbanization, polarization reversal, and counterurbanization. During urbanization, mainstream migration generally supports large city development. This is followed by secondary or regional city development during the phase of polarization reversal, and finally, by small city development during counterurbanization.

Why counterurbanization occurs

Finding adequate explanations for counterurbanization is an important issue in the papers included in this book. Three approaches to the problem have emerged over the years. Certain authors have taken an *ad hoc* approach and attribute counterurbanization to specific factors; others have a more fundamental approach and look at groups of factors, while some look at counterurbanization from a cyclical perspective.

According to Champion, specific explanations that were generally given for counterurbanization were widespread economic recessionary conditions experienced in the world during the 1970s, decentralization policy, rural resource development, urban diseconomies, state welfare payments, reduced friction of distance, changing residential preferences, changing socio-demographic composition, and economic structural changes. He names three groups of factors that were responsible for counterurbanization: improved civil and social infrastructure, economic factors associated with highly qualified labour, and demographic changes linked to socio-economic services. The first group supports centrifugal forces, the second centripetal, and the third supports both forces.

Vining sees population deconcentration as an outcome of regional restructuring.

Aggregate economic conditions seem to be a contributing factor to counterurbanization, but they do not seem to be the sole factor or even the most important factor explaining counterurbanization (*see also* Kontuly and Bierens, 1990). In the cases of Norway and Italy, for instance, Vining and Kontuly found no correlation between economic booms and busts and the rate of deconcentration. Generally, the Vining group sees 'system-wide' changes in the economies of countries as an important mechanism for triggering deconcentration. Richter tested a number of factors explaining counterurbanization in the USA during 1970–80. Factors vary considerably over time. Areas characterized by a high degree of agricultural and military employment, as well as areas with a high Black population component, lost population throughout the decade, while areas with mild temperatures and above average recreation facilities attracted people. Remote areas grew faster during the first half of the period of her study, while adjacent counties kept on attracting more migrants during the second half of the period.

Fielding's analysis of counterurbanization in France and his general analogy of the general state of counterurbanization in the developed world at the time give additional insight into the explanation of the phenomenon. He suggests several specific explanations given for the onset of counterurbanization. These include a decreasing exodus from agriculture, improved communications, government incentives for non-metropolitan development, and retirement migration. In France, however, areas with high agricultural employment tended to lose population.

According to Fielding the reasons that are generally offered as explanations for counterurbanization may be arranged into four groups. First, there is the urban–rural place preference factor, but this factor, he contends, 'exaggerates the choices open to people' and is 'too voluntaristic'. The second factor is labor markets as affected by spatial differentials in unemployment and wage rates. According to him, wage rates and unemployment in large cities and rural areas have remained the opposite of what neoclassical economic migration theory predicts. He also regards them as 'too simplistic'. Public policy is the third factor, but public investment and migration tendencies do not always correspond spatially. He suggests the changing spatial division of labour resulting from economic deconcentration of multi-nationals as the fourth possible factor. In his view the change in the spatial division of labour is mainly due to monopoly ownership and the disintegration of the production process.

Criteria for the measurement of counterurbanization, polarization reversal, and differential urbanization

Another matter arising in the papers included in this book is the criteria that have been used to determine the occurrence or not of counterurbanization. These criteria include metropolitan versus non-metropolitan growth differentials, population deconcentration at different regional scales, core–periphery net migration differentials, and the relationship between net migration rates and settlement density.

Richardson looked at more complex indicators and suggests a weakening of the positive relationship between income and city size, i.e. Williamson's regional income convergence, as one of the indicators of the onset of polarization reversal. Reductions in the core's locational advantage as reflected by reductions in its share of the national manufacturing output, and an increase in the same indicators in the periphery, are also suggested as an indication of the beginning of deconcentration. Other possible indicators suggested by Richardson are, first, a decline in urban primacy, measured in terms of the El-Shakhs index; second, city size distribution, measured by the Pareto coefficient; third, shifts towards equality in personal income distribution – i.e. the Kuznets law – as reflected by Gini coefficients; fourth, a decline in population concentration, primate city growth, and industrialization, as well as differential growth rates of urban sizes.

Preconditions for deconcentration in Third World countries, given by Lo and Salih, are economy-wide full employment, agglomeration diseconomies in manufacturing in the core, interregional linkages which facilitate spread effects, and organizational sophistication in business which permits branching. Richardson questions these conditions. According to him, some of these indicators are difficult to measure empirically, and a decline in the primacy index, according to him, is not necessarily an indication of polarization reversal, although it may be a prelude to polarization reversal. Potentially, many of these indicators could add valuable insight into the relevancy of the concept of polarization reversal within the First World environment.

The core–periphery approach to the measurement of counterurbanization

Champion criticizes the Vining group's core–periphery approach. According to him, the weakest element of the approach lies in the definition of the core, because it seems intuitive and too wide. Second, the number of core regions per country is inconsistent; for instance, three cores were delimited in Sweden, which has 15 million people, and one in Britain, which has 50 million people. Coupled with this issue is the problem of scale – a comparison of metropolitan areas in Sweden with large regions such as the North East, Mid-West, and West in the USA. Fourth, differing degrees of overbounding complicate comparisons. On the positive side, Champion argues that the Vining group confines its comparisons to general tendencies and mostly limits itself to comparisons within particular countries. It also does not indulge in extensive cross-country comparisons.

On the basis of these criticisms the question could be asked whether counterurbanization could safely be regarded as a complete turnaround in migration trends in the First World, or whether it was merely a continuation of the wave-theory. The answer to this question largely depends on the spatial criteria which were used to measure migration trends. Indications are that the position of the boundary between what were regarded as the core and the periphery in each case study played a decisive role.

The link between the First and Third Worlds

Looking at urban development trends in the Third World environment, Richardson discusses public intervention and the effect it may have on polarization reversal. The latter is defined by him as a period when polarization trends give way to dispersion. In his discussion of the concept, Richardson often talks of 'spread' and 'backwash' effects. However, Hirschman and Myrdal's spread and backwash effects are sectoral, and therefore nonspatial, while polarization reversal is demographic-industrial, and therefore, spatial. Richardson sees industrial location and public urban development policy as two of the important factors affecting the direction of migration streams in advanced developing countries. In contrast, Fielding found that migrants in France preferred agricultural regions to industrial regions during the 1970s. Referring to migration destinations in the French example, he remarks: 'the type of area seems to have been more important than whether or not it was policy supported'. In his work Richardson associates population deconcentration in advanced developing countries with high social returns from dispersion. When social returns from dispersion are in parity with, or higher than, the social returns from concentration, polarization reversal could set in. The same could be said of counterurbanization, only the destinations of migrants in developed countries are different owing to different priorities.

In less developed countries it seems as if industrial development has a marked influence on migration because employment opportunities are still a priority to a majority of potential migrants. In contrast, a good living environment is an important pull factor for the more developed communities of the First World. Migration destinations in developed countries indicate the importance attached to environmental factors while the importance of employment opportunities is stressed in migration in the Third World. This illustrates the fundamental difference between deconcentration in developed countries and in those less developed countries where population deconcentration has occurred.

Richardson points out that although polarization reversal and counterurbanization may seem similar in many respects, polarization reversal occurs under circumstances of slowed metropolitan growth while counterurbanization is associated with an absolute decrease in the growth of core regions. The last contribution on differential urbanization in this book examined the time period when the counterurbanization phase actually started in France. The discussion it was graphically demonstrated that France actually moved into a polarization reversal phase prior to the onset of counterurbanization in the country. This clearly illustrates the relevance of polarization reversal within the First World economic environment, a factor that has not been recognized until now.

Will polarization reversal and counterurbanization continue?

In the 1980s and 1990s, spatial population concentration and deconcentration trends in the developed world have not been straightforward

(Champion, 1994). During the 1980s, Austria, Belgium, Canada, the old Federal Republic of Germany, the Netherlands, Sweden, and the United States showed a return to traditional patterns of regional population concentration, while Finland and Norway never experienced a turnaround (Cochrane and Vining, 1988; Frey, 1993; Kontuly and Schön, 1994; Richter, 1985; Rogerson and Plane, 1985). In contrast, the turnaround continued in Australia, France, Great Britain, and Italy (Cochrane and Vining, 1988; Champion, 1994; Winchester and Ogden, 1989).

The slowdown in counterurbanization patterns in certain developed countries and the end of the turnaround in others during the 1980s raised the question of whether counterurbanization really represented the beginning of a new era. Cochrane and Vining gave an extensive review of migration patterns in developed and advanced developing countries during the early 1980s and conclude that counterurbanization came to an end in certain countries, while in others the tendencies are indicating a new turnaround.

Champion, on the other hand, treats the reversal of the turnaround as a short-term downward fluctuation in the rate of population deconcentration. He regarded the slowdown of counterurbanization during the early 1980s as merely a temporary phenomenon resulting from particular conditions favouring concentration and contends that deconcentration remains the dominant long-term trend.

Richter found that the growth of non-metropolitan counties in the USA slowed down during the late 1970s, but their growth rate was still higher than that of the metropolitan counties. Most of this decline was experienced in peripheral counties where growth rates fell below those of metropolitan counties. Counties adjacent to the metropolitan areas maintained a higher growth rate than the metropolitan areas. This represents an unequivocal return to the urbanization wave tendencies of the pre-counterurbanization era.

Fielding found few signs of a simple turnaround from counterurbanization to urbanization in Western Europe in the early 1980s. Rather, there appears to be a re-emergence of 'broader regional patterns of growth and decline'. The lack of a general return to traditional patterns of regional concentration in the developed world appears to be confirmed by recent research on Sweden and the USA, which indicates a 'new turnaround' or renewed counterurbanization during the first part of the 1990s (Beale and Fuguitt, 1990; Borgegård et al., 1995; Johnson, 1993; Johnson and Beale, 1994; Fuguitt and Beale, 1995). Within the Third World, Vining found an association between economic development and population concentration in non-Western countries, similar to the basic pattern observed in Western nations during the urbanization era. He found very little evidence of deconcentration in advanced Third World countries, because of what he regarded as the increased absorptive capacity of the agricultural sector. Various contributions describe the onset of polarization reversal in more advanced Third World countries. Strong indications of the end of urbanization and the beginning of polarization reversal are found in the works of Townroe and Keen and Lee. The onset of polarization reversal seems to be associated with a decline in the growth of primate cities and an increase in the growth of secondary cities.

Differential urbanization

The Klaassen–Paelink group made a valuable contribution to understanding the life-cycle of individual urban agglomerations within a developed economic environment. Borrowing from this group's work, as well as from several other sources, the differential urbanization model was developed. First, the differential urbanization model expands the urban life-cycle model to encompass the total urban system of a country. Second, it ties together the polarization reversal and counterurbanization concepts within the same paradigm. The model distinguishes between groups of small, medium and large-sized cities without referring to the actual size of cities. This makes it possible to compare the phase of urban development in two countries, which otherwise would not be possible owing to differences in their levels of development, the sizes of their populations, and the magnitude of their urban systems. Third, the differential urbanization model also provides a dynamic framework which allows comparisons over time. Not only can one trace the changes that occurred in the development of small, medium and large-sized cities in a particular urban system over time, but one can also record the evolving urban development processes of two diverse countries on the same graph.

The link between deconcentration tendencies and the level of economic development of individuals is detected in Richardson's question when he asks to what extent polarization reversal lags behind the turning point in Williamson's inverted-U curve of income. It is further illustrated in Vining's study of migration in Third World countries when he concludes that deconcentration could become dominant when per capita GNP levels in these countries have reached a certain cut-off point. At the same time migrants tended to prefer agricultural regions in France to old industrial regions receiving state subsidies during the early 1970s. Fielding says: 'the type of area seems to have been more important than whether or not it was policy supported'. Prima facie, these two groups of observations, one dealing with migration in the Third World and the other with migration in the First World, do not have anything in common. However, they do illustrate the relevancy of the concepts of 'productionism' and 'environmentalism' as explained in the contributions on differential urbanization in this book. Vining ties the deconcentration of people to income and Fielding ties it to areas of high visual quality. In the discussion of the factors which are fundamentally responsible for deconcentration both these issues are linked to the concept of environmentalism.

A migration cycle approach to urban development

Authors regard Berg, Klaassen and Paelink as the urban life-cycle pioneers. Vining and Strauss also expected countries to move through a spatial development sequence. First, it was predicted that decentralization will take place within the urban regions. This phase will have been followed by decentralizaton from urban to rural regions, and finally, decentralization would have occurred within the rural regions. The differential urbanization model suggests the following expansion of the Vining–Strauss sequence of dominant migration phases. First, concentration within the urban regions dominates. Second, decentralization within the urban regions starts. Third, deconcentration from urban to rural regions begins to set in, initially more to urban settlements adjacent to the major urban regions and later also to rural regions further afield. This process is followed by decentralization within the rural regions, and finally the beginning of the second phase of the sequence.

Berry proposes long-wave theory as an answer to the problem of identifying the beginning and end of counterurbanization. He links the concept of migration long waves in the United States with indicators such as GNP growth and decline cycles and contends that each wave of economic growth produces a wave of net urban migration. Gordon also argues along the same lines. One of the underlying issues in his article is the relevance of the migration wave in the USA during the early 1970s, despite the onset of counterurbanization.

Richardson also recognizes cyclical elements in the urban development process. Although Richardson states that Friedmann's core–periphery model is more of a paradigm than a deterministic predictive model he sees a link between polarization reversal and the third to fourth stages of the core–periphery model.

According to Fielding, an economic 'system change' could have been responsible for the disappearance of counterurbanization in Western Europe. At the time he argued that the mass production/marketing economics of the past were being replaced by what he termed 'flexible accumulation'. This includes a number of interrelated processes, notably vertical and lateral industrial disintegration and the computer revolution, both of which played an important role in the successful socio-economic linking of certain regions internationally.

The growth of groups of cities

According to Vining and Strauss, deconcentration to non-metropolitan counties in the USA from 1970 to 1974 was largely characterized by dispersed concentration. Non-metropolitan counties which did show a decline in concentration were mostly more urban in character than the former. Also, Vining and Kontuly refer to dispersed concentration in a limited number of small and medium-sized cities in Western Europe and Japan in the early 1970s. According to Beale, the USA had experienced renewed growth or diminished loss of non-metropolitan population and increased regional growth of small and medium-sized metropolitan areas in the South and the West during the early 1970s. According to Richter, the growth rate of the smallest Standard Metropolitan Statistical Areas (less than 100000) in the United States increased consistently during the 1970s, at almost twice the national growth rate. These are all direct or indirect references to the growth, stagnation or decline of groups of cities during a specific phase of urban development. This is also one of the fundamental arguments put forward in the articles on differential urbanization. During the urbanization phase, the larger the city the more favourable the chances for it to attract migrants. During the polarization reversal phase, the growth of large cities tapers off, while intermediate-sized cities start to expand. Finally, during the counterurbanization phase small cities start attracting population while primate cities start losing population in absolute terms.

Mainstream versus undercurrent migration

In the differential urbanization model a distinction is made between mainstream and undercurrent migration, the former referring to the migration processes dominating redistribution trends in a country at any point in time, and the latter to those migration streams that are important, but less obvious. Such a differentiation is for instance being made by Champion when he distinguishes between family-age adults in Britain, who, during the period covered by his study, tended to deconcentrate, while young adults tended to concentrate. Koch made similar observations in Europe during the 1970s. In his article, Berry observes that while counterurbanization took place in the USA during the early 1970s, certain minority groups such as the Blacks and other minority groups continued to concentrate. This fact was confirmed by Beale and Richter.

An urban development paradigm

In the Geyer and Kontuly article on differential urbanization, five statements regarding the development of urban systems were made. It was contended that as national urban systems develop, many of them go through a primate city stage. This is the phase when urban growth in a country is dominated by a few centres which have more growth potential than most others, while new centres are being added to the system from below. This process results in a hierarchy of urban centres forming territorially organized subsystems at various levels of aggregation. The same spatial forces operating at the national level are also at work at the subregional level, and therefore each subsystem has the potential of displaying the same types of population redistribution patterns the urban system displays at the national level. *Ceteris paribus*, centres closer to large cities have a better chance to develop than those further away. This corresponds with Berry's and Gordon's observations in the USA that urban areas and regions closer to metropolitan areas developed faster during the early 1970s than those further away. This reopens the debate on the importance of wave theory explanations. According to all observers, the United States did experience an absolute decrease in migration to metropolitan areas in the early 1970s. In his locational breakdown of non-metropolitan growth in the USA during the early 1970s, Gordon found that 'growth is greatest in those non-metropolitan counties which are most linked to the metropolitan centres'. Projecting the 'wave theory' onto the 'clean break theory', he concludes that 'a continued "wave" of urban decentralization as well as renewed rural growth seem to be in progress'.

Also, Vining and Strauss deal with views of the overspill/wave proponents and come to the conclusion that non-metropolitan counties well removed from the commuting range of SMSAs were growing at a significantly higher rate than the SMSAs themselves during the early 1970s, 'though at a somewhat lower rate than the nonmetropolitan counties adjacent to these SMSAs'. This view is shared by Beale in the statement 'The adjacent [counties] have been gaining people at an annual rate about twice as high as they experienced in the 1960s, but the difference in post-1970 annual growth of the adjacent and nonadjacent groups . . . is not great.'

The general counterurbanization trend seemed to have been one of concentrated dispersal. Vining and Kontuly say: 'It is our impression that people in moving back to the peripheral regions have tended to concentrate in a limited number of small and medium sized cities there.' Vining and Strauss refer to Duncan's reference to the concomitant occurrence of concentration and diffusion in the USA between 1940 and 1950.

According to Richter, most of the growth in the larger metropolitan areas of the USA took place in their fringe counties during the late 1970s, but these fringe counties grew slower than those of the smaller metropolitan areas. Net migration to nonadjacent counties also decreased dramatically. Most migration took place towards adjacent counties. These observations prove the relevance in the United States of the five propositions made in terms of the concept of differential urbanization.

Productionism versus environmentalism

Berry argues the historical importance of environmentalism in the American culture. In his article, Gordon says the following of migration trends in the USA:

> 'A preference for small-town life has long been used to explain suburbanisation. The data seem to suggest that this trend is as strong as

ever and is taking place at ever greater distances from central cities, especially if these central cities are large.'

Richter also refers to Lichter's observation in 1979 that the counterurban stream generally consists of older people compared to the urban stream.

Conclusion

The differential urbanization concept transcends the void between the migration literature which developed within a First World environment and the literature written for a less developed economic context. It is a model which brings together three issues. First, it creates the vision for a direct longitudinal link between migration and urban development in the First and Third Worlds, two very diverse economic environments. Second, it differentiates between mainstream and undercurrent migration as a potential explanation of fundamentally different redistribution trends which could occur in the same nation at the same time. It shows the potential similarity between the considerations driving mainstream migration in the First World and undercurrent migration in the Third World. Third, it creates a framework for linking the concepts of productionism and environmentalism in both the First and Third Worlds. Albeit in a different mould, productionism has for a long time been regarded as a major factor in the explanation of migration trends in the Third World. It was a key element, for instance, in the formulation of Todaro's migration model. At the same time it also links the concept of productionism with undercurrent migration trends in the First World. When combined, these three elements bring to the fore concepts which are not new *per se in* migration studies, but are thereby put into a new perspective.

References

Beale, C. L. and Fuguitt, G. V. 1990: Decade of pessimistic nonmetro population trends ends on optimistic note. *Rural Development Perspectives* 6 (3), 14–18.

Borgegård, L.-E., Håkansson, J. and Malmberg, G. 1995: Population redistribution in Sweden: Long term trends and contemporary tendencies. *Geografiska Annaler* 77B (1), 31–45.

Champion, A. G. (1994) Population change and migration in Britain since 1981: evidence for continuing deconcentration. *Environment and Planning A* 26: 1501–20.

Cochrane, S. G. and Vining, D. R. Jr. 1988: Recent trends in migration between Core and peripheral regions in developed and advanced developing Countries. *International Regional Science Review* 11, 215–43 (*see* this volume, Chapter 7).

Frey, W. H. 1993: The new urban revival in the United States. *Urban Studies* 30 (4/5), 741–74.

Fuguitt, G. V. and Beale, C. L. 1995: Recent trends in nonmetropolitan migration:

toward a new turnaround? Center for Demography and Ecology Working Paper no. 95-07. University of Wisconsin – Madison. May.

Johnson, K. 1993: Demographic change in nonmetropolitan America, 1980–1990. *Rural Sociology* 58, 347–65.

Johnson, K. and Beale, C. L. 1994: The recent revival of widespread population growth in nonmetropolitan areas of the United States. *Rural Sociology* 59, 655–67.

Kontuly, T. and Bierens, H. 1990: Testing the recession theory as an explanation for the migration turnaround. *Environment and Planning A* 22, 253–70.

Kontuly, T. and Schön, K. P. 1994: Changing western German internal migration systems during the second half of the 1980s.' *Environment and Planning A* 26, 1521–43.

Richter, K. 1985: Nonmetropolitan growth in the late 1970s: the end of the turnaround? *Demography* 22, 245–63 (*see* this volume, Chapter 5).

Rogerson, P. and Plane, D. 1985: Monitoring migration trends. *American Demographics* 7 (2), 27–9, 47.

Winchester, H. P. M. and Ogden, P. E. 1989: France: decentralization and deconcentration in the wake of late urbanization. In Champion, A. G. (ed.), *Counterurbanization: The changing pace and nature of population deconcentration*, London: Edward Arnold.

SUBJECT INDEX

Absorptive capacity 146–7, 334
Ageing 118
Agglomeration *see* major cities and metropolis
Agglomeration
   diseconomies 38, 46, 150, 310, 319
   economies 144, 190, 203, 292, 318
Agriculture 177
   restructuring 127
Apartheid 243, 250

Backwash effects 153, 154
Bantustans 245
Birth rate 112
Black migration 11, 24, 239

Central place theory 291
Clean break theory 29
Colonization xvii, 70
Concentrated dispersion 29, 144, 202, 214
Concentration *see* urbanization
Conurbation *see* megalopolis
Core
   fringe 239
   inner 210
   intermediate 241
   outer 241
   region 123, 133, 143, 153, 192, 207, 225
   periphery xviii, 91, 108, 205, 266, 290, 297, 332
   periphery model 153
Counterurbanization xviii, 7, 37, 112, 123, 133, 155, 188, 238, 290
Cumulative causation 202
Cycles 138

Decentralization 12, 56, 86, 135, 144
Declining birth rate 111
Deconcentration 31, 68, 122, 164, 244

Definition
   counterurbanization 313, 315, 330
   differential urbanization 314, 330
   environmentalism 319
   polarization reversal 134, 188, 195, 294, 312, 316, 330
   productionism 319
   urbanization 311, 315, 330
Determinism 145, 320, 326
Development
   axis 249, 250, 291
   centre *see* growth pole
   corridor *see* development axis
   cycle 276, 290
Differential urbanization 239, 252, 290, 335
Diseconomies of scale 86, 135, 270, 280
Distance friction 135
Disurbanization 310
Division of labour 127, 128, 131
Dominant regions 179
Dualistic economics 239, 266

Economic
   conversion/diversion 300
   cycles 74, 77, 85, 115, 135, 249, 262, 318
   dispersion 144
   growth 176, 262, 266
   recession/depression *see* economic cycles
   restructuring 207, 318
Ecumenopolis xvii
El-shakhs index 150
Elasticity of supply of housing 146
Employment 58, 59, 60, 295
Environmentalism 14, 15, 24, 48, 64, 126, 131, 151, 174, 191, 218, 300, 319, 338
Explaining migration 135, 136, 144, 158, 318, 330
External economies *see* agglomeration economies

Subject Index

Fertility 122
First world/Third world 333
Fordism 127, 131
Foreign migration *see* international migration
Functional urban regions 41–42

Gini-coefficients 150, 156, 223
Green Revolution 177
Growth pole 151, 155, 213, 214, 220, 266, 299
Guest workers 217

Hierarchical
  diffusion 150
  interaction 212
Hinterlands 44, 120
Hoover index 29, 40
Human resources 218

Implicit policy 148
Income convergence 152
Industrial
  decentralization 222
  development xvii, 108, 176, 205, 244
  restructuring 127
  Revolution xvii, 7
Informal sector 251, 265
Infrastructure 118, 119
Innovation diffusion 266, 292, 299
Interaction 208
Intermediate sized cities 27, 112, 128, 144, 148, 155, 169, 189, 198, 202, 220, 233, 239, 273, 292, 295
International migration 73
Intervention 145
Investment 147

Kuznets' law 150

Labour
  market 117, 126, 129, 189, 315
  productivity 218
Large metropolitan areas *see* major metropolitan areas
Linkages 150, 155
Location advantages 143, 205, 291
Lagging regions 191
Long wave 259

Mainstream migration xix, 239, 293, 296, 301, 313, 337
Major cities 115, 117, 123, 143, 144, 150, 189, 212, 239, 273
Market restructuring xviii, 119, 144
Mass production 127

Measuring migration *see* migration
Medium sized cities *see* intermediate sized cities
Megalopolis xvii, 68, 70, 78, 118
Metropolis 112, 179, 265
Metropolitan
  areas xviii, 11, 13, 22
  region 69
  shadow area 283, 292, 314
Migration
  age 25, 48, 56, 72, 115, 118, 122, 129, 136, 223
  measuring 331
  recreation 62, 119
  cycle 336
  income 191
  mainstream/substream 239, 300, 337
  chain 265
  rate 122
Mobility transition 321
Monocentric cities 144
Morality 122
  rate 112
Multi-national companies 127

Natural increase 72
Neo-classical economic theory 126
Non-metropolitan growth 11, 13, 23, 53

Outmigration 118
Overspill 28

Pareto-coefficient 153, 156
Peripheral areas 33, 69, 72, 112, 115, 117, 144, 166, 174, 214
Polarization (*see also* urbanization)
  reversal xx, 143, 188, 203, 220, 232, 239, 259, 275, 290, 310
  stabilization 195
Population (*see also* urbanization)
  concentration 32
  deconcentration 31, 68, 122, 164, 244
  redistribution 123
  spillover 68
Primate city distribution 270, 278
Primate city *see* major cities
Proximity of centres 209, 292
Production restructuring 122, 131
Productionism 119, 300, 319, 338
Productivity 174–5
Public policy 126

Regional
  balance 130
  cities *see* intermediate sized cities
  specialization 187
  system 240, 264

Regionalization 323
Residential preferences (*see also* environmentalism)
Reurbanization 310
Rural
　development 166
　regions 70
　Renaissance 135

Scale economies 143, 144
Scenic beauty 118
Secondary cities *see* intermediate sized cities
Segregation 17, 27
Selective spatial closure 153, 154
Settlement 122, 123
Small cities 150, 152, 214, 297
SMSA 19, 51, 63
Space economy 143
Spatial division of labour 119, 120, 128
Spatial planning 152
Spillover growth 38
Spread effects 38, 150, 153, 211, 232, 266
Suburbanization 14, 310
System of cities *see also* urban system 250, 251

Technological innovation 318
Third World 162
Trickling-down effects *see* spread-effects
Turnaround xvii, xix, 50–1

Unemployment 117
Urban
　agglomerations *see* migration and metropolis
　bias 294
　decay 16
　deconcentration index 193
　development model 265, 291, 296, 337
　evolution 48
　life cycle 136, 313–8
　scale 86
　sprawl 67, 118
　systems 114, 191, 195, 203, 233, 265, 292, 295, 334
Urbanization xvii, 7, 14, 29, 67, 112, 122, 133, 290

Wave theory (*see also* Long wave) 28, 37–9, 45, 134, 291
Welfare payments 135
Williamson's hypothesis 150

INDEX OF PLACE NAMES

Algeria 168
Argentina 163, 165, 176
Australia 291
Austria 43
Bangladesh 169, 176
Belgium 43, 80, 91, 92, 104, 113
Botswana 170
Brazil 167, 175, 176, 188
Britain 43, 76, 96, 102, 105, 118, 128, 134, 300
Canada 68, 91, 97, 106, 108
Chilli 166, 176, 179
Colombia 151, 167, 176,
Cuba 171, 179
Czechoslovakia 91, 102
Denmark 43, 81, 91, 92, 104, 117, 291
Dominican Republic 167
Ecuador 218
Egypt 166, 176, 179
Ethiopia 170
Finland 43, 82, 91, 97, 133
France 43, 75, 91, 94, 112, 123, 239, 252, 291, 297, 300, 317
Germany 33, 43, 78, 91, 92, 102, 112, 115, 116, 117, 123, 239, 252, 291
Greece 165, 175, 176
Guinea-Bassau 170
Haiti 170
Hong Kong 175
Hungary 43, 82, 91
Iceland 91
India 169, 176, 178, 297
Indonesia 176, 178
Ireland 43, 112, 117, 165, 175, 176
Italy 43, 73, 74, 91, 99, 105, 112, 117, 123, 133, 291
Japan 32, 37, 43, 45, 69, 91, 99, 105, 134, 150, 151, 161, 175, 291

Jordan 171
Korea 82, 91, 102, 106, 175, 297
Libya 168
Malaysia 168, 175
Mexico 168, 175, 176
Netherlands 43, 80, 74, 91, 94, 104, 112, 117
New Zealand 81, 91
Norway 43, 73, 91, 97, 105, 134
Pakistan 169, 176, 178
Panama 168
Peru 166, 179
Philippines 163, 168, 178
Poland 43, 82, 91
Portugal 43, 176
Scotland 112
Sierra Leone 170
Singapore 175
South Africa 176, 239, 252, 300
South Korea 164, 169, 176
Spain 43, 82, 91, 99, 105, 123, 134, 165, 176
Sri-Lanka 176, 178, 179
Sweden 43, 70, 91, 99, 134
Switzerland 43
Syria 172
Taiwan 82, 91, 102, 106, 169, 175
Tanzania 170
Thailand 169, 175
Tunisia 168
Turkey 169, 176
Uruguay 165
USA 33, 45, 90, 91, 95, 106, 112, 116, 134, 238, 252, 259, 291, 300
Venezuela 165, 166, 270
Yugoslavia 188
Zambia 170